Lecture Notes in Computer Science 1556

Edited by G. Goos, J. Hartmanis and J. van Leeuwen

Springer

Berlin
Heidelberg
New York
Barcelona
Hong Kong
London
Milan
Paris
Singapore
Tokyo

Stafford Tavares   Henk Meijer (Eds.)

Selected Areas
in Cryptography

5th Annual International Workshop, SAC'98
Kingston, Ontario, Canada, August 17-18, 1998
Proceedings

 Springer

Series Editors

Gerhard Goos, Karlsruhe University, Germany
Juris Hartmanis, Cornell University, NY, USA
Jan van Leeuwen, Utrecht University, The Netherlands

Volume Editors

Stafford Tavares
Henk Meijer
Department of Electrical and Computer Engineering
Queen's University
Kingston, Ontario K7L 3N6, Canada
E-mail: tavares@ee.queensu.ca
        henk@qucis.queensu.ca

Cataloging-in-Publication data applied for

Die Deutsche Bibliothek - CIP-Einheitsaufnahme

Selected areas in cryptography : 5th annual international workshop ; proceedings
/ SAC '98, Kingston, Ontario, Canada, August 17 - 18, 1998. Stafford Tavares ;
Henk Meijer (ed.). - Berlin ; Heidelberg ; New York ; Barcelona ; Hong Kong ;
London ; Milan ; Paris ; Singapore ; Tokyo : Springer, 1999
(Lecture notes in computer science ; Vol. 1556)
ISBN 3-540-65894-7

CR Subject Classification (1998): E.3, G.2.1, D.4.6, K.6.5, F.2.1-2, C.2, J.1

ISSN 0302-9743
ISBN 3-540-65894-7 Springer-Verlag Berlin Heidelberg New York

Typesetting: Camera-ready by author
SPIN: 10693237   06/3142 – 5 4 3 2 1 0   Printed on acid-free paper

Preface

SAC '98 is the fifth in a series of annual workshops on Selected Areas in Cryptography. SAC '94 and SAC '96 were held at Queen's University in Kingston and SAC '95 and SAC '97 were held at Carleton University in Ottawa. The purpose of the workshop is to bring together researchers in cryptography to present new work on areas of current interest. It is our hope that focusing on selected topics will present a good opportunity for in-depth discussion in a relaxed atmosphere. The themes for the SAC '98 workshop were:
- Design and Analysis of Symmetric Key Cryptosystems
- Efficient Implementations of Cryptographic Systems
- Cryptographic Solutions for Internet Security
- Secure Wireless/Mobile Networks

Of the 39 papers submitted to SAC '98, 26 were accepted and two related papers were merged into one. There were also two invited presentations, one by Alfred Menezes entitled "Key Agreement Protocols" and the other by Eli Biham entitled "Initial Observations on SkipJack: Cryptanalysis of SkipJack-3XOR". There were 65 participants at the workshop.

The Program Committee members for SAC '98 were Carlisle Adams, Tom Cusick, Howard Heys, Henk Meijer, Doug Stinson, Stafford Tavares, Serge Vaudenay, and Michael Wiener. We also thank the following persons who acted as reviewers for SAC '98: Zhi-Guo Chen, Mike Just, Liam Keliher, Alfred Menezes, Serge Mister, Phong Nguyen, David Pointcheval, Thomas Pornin, Guillaume Poupard, Yiannis Tsioumi, Amr Youssef, and Robert Zuccherato.

This year, in addition to the Workshop Record distributed at the workshop, the papers presented at SAC '98 are published by Springer-Verlag in the Lecture Notes in Computer Science Series. Copies of the Springer Proceedings are being sent to all registrants.

The organizers of SAC '98 are pleased to thank Entrust Technologies for their financial support and Sheila Hutchison of the Department of Electrical and Computer Engineering at Queen's University for administrative and secretarial help. Yifeng Shao put together the Workshop Record and provided invaluable assistance in the preparation of these Proceedings. We also thank Laurie Ricker who looked after registration.

November 1998   Stafford Tavares and Henk Meijer
                                                         SAC '98 Co-Chairs

Organization

Program Committee

Carlisle Adams	Entrust Technologies
Tom Cusick	SUNY Buffalo
Howard Heys	Memorial University of Newfoundland
Henk Meijer	Queen's University
Doug Stinson	University of Waterloo
Stafford Tavares	Queen's University
Serge Vaudenay	Ecole Normale Supérieure/CNRS
Mike Wiener	Entrust Technologies

Local Organizing Committee

Stafford Tavares (Co-Chair)	Queen's University
Henk Meijer (Co-Chair)	Queen's University

Table of Contents

Cryptographic Systems

Public Key Cryptosystems

Design and Implementation of Secret Key Cryptosystems

Attacks on Secret Key Cryptosystems

Invited Talks

Author Index

Feistel Ciphers with L_2-Decorrelation

Serge Vaudenay

Ecole Normale Supérieure – CNRS
Serge.Vaudenay@ens.fr

Abstract. Recently, we showed how to strengthen block ciphers by decorrelation techniques. In particular, we proposed two practical block ciphers, one based on the $GF(2^n)$-arithmetics, the other based on the $x \bmod p \bmod 2^n$ primitive with a prime $p = 2^n(1 + \delta)$. In this paper we show how to achieve similar decorrelation with a prime $p = 2^n(1 - \delta)$. For this we have to change the choice of the norm in the decorrelation theory and replace the L_∞ norm by the L_2 norm. We propose a new practical block cipher which is provably resistant against differential and linear cryptanalysis.

At the STACS'98 conference, the author of the present paper presented the technique of decorrelation which enables to strengthen block ciphers in order to make them provably resistant against the basic differential and linear cryptanalysis [13].[1] So far, this analysis which is based on Carter and Wegman's paradigm of universal functions [3, 17], has been used with the L_∞-associated matrix norm in order to propose two new practical block cipher families which are provably resistant against those cryptanalysis: COCONUT98 and PEANUT98. This technique has been shown to enable to propose real-life encryption algorithms as shown by the Advanced Encryption Standard submission [5] and related implementation evaluations on smart cards [9]. In this paper we present some earlier results based on the L_2 norm in order to make a new practical block cipher PEANUT97.[2]

1 Basic Definitions

We briefly recall the basic definitions used in the decorrelation theory. Firstly, let us recall the notion of d-wise distribution matrix associated to a random function.

Definition 1 ([13]). *Given a random function F from a given set $\mathcal{A}$ to a given set $\mathcal{B}$ and an integer d, we define the "d-wise distribution matrix" $[F]^d$ of F as a $\mathcal{A}^d \times \mathcal{B}^d$-matrix where the (x, y)-entry of $[F]^d$ corresponding to the multi-points $x = (x_1, \ldots, x_d) \in \mathcal{A}^d$ and $y = (y_1, \ldots, y_d) \in \mathcal{B}^d$ is defined as the probability that we have $F(x_i) = y_i$ for $i = 1, \ldots, d$.*

---

[1] A full paper version [14] is available on the web site [15].

[2] The decorrelation technique with the L_2 norm happens to be somewhat less easy than the L_∞ norm, which is why is has not been published so far.

S. Tavares and H. Meijer (Eds.): SAC'98, LNCS 1556, pp. 1–14, 1999.
© Springer-Verlag Berlin Heidelberg 1999

Secondly, we recall the definition of two matrix norms: the L_∞-associated norm denoted $|||.|||_\infty$, and the L_2-norm denoted $||.||_2$.

Definition 2. *Given a matrix A, we define*

$$||A||_2 = \sqrt{\sum_{x,y} (A_{x,y})^2} \tag{1}$$

$$|||A|||_\infty = \max_x \sum_y |A_{x,y}| \tag{2}$$

where the sums run over all the (x,y)-entries of the matrix A.[3]

Finally, here is the definition of the general d-wise decorrelation distance between two random functions.

Definition 3 ([13]). *Given two random functions F and G from a given set $\mathcal{A}$ to a given set $\mathcal{B}$, an integer d and a matrix norm $||.||$ over the vector space $\mathbf{R}^{\mathcal{A}^d \times \mathcal{B}^d}$, we call $|||[F]^d - [G]^d||$ the "d-wise decorrelation $||.||$-distance" between F and G. In addition, we call "d-wise decorrelation $||.||$-bias" of a random function (resp. permutation) F its d-wise decorrelation $||.||$-distance to a random function (resp. permutation) with a uniform distribution.*[4]

We consider block ciphers on a message-block space $\mathcal{M}$ with a key represented by a random variable K as a random permutation C_K defined by K over $\mathcal{M}$. Since the subscript K is useless in our context we omit it and consider the random variable C as a random permutation with a given distribution. Ideally, we consider the Perfect Cipher C^* for which the distribution of C^* is uniform over the set of the permutations over $\mathcal{M}$. Hence for any multi-point $x = (x_1, \ldots, x_d)$ with pairwise x_is and any multi-point $y = (y_1, \ldots, y_d)$ with pairwise y_is we have

$$[C^*]^d_{x,y} = \Pr[C^*(x_i) = y_i; i = 1, \ldots, d] = \frac{1}{\#\mathcal{M} \ldots (\#\mathcal{M} - d + 1)}.$$

We are interested in the decorrelation bias $||[C]^d - [C^*]^d||$ of a practical cipher C.

We recall that $|||.|||_\infty$ and $||.||_2$ are matrix norms (*i.e.* that the norm of any matrix-product $A \times B$ is at most the product of the norms of A and B) which makes the decorrelation bias a friendly measurement as shown by the following Lemma.

---

[3] The strange $|||.|||_\infty$ notation used in [13] comes from the fact that this norm is associated to the usual $||.||_\infty$ norm over the vectors defined by $||V||_\infty = \max_x |V_x|$ by

$$|||A|||_\infty = \max_{||V||_\infty = 1} ||AV||_\infty.$$

[4] It is thus important to outline that we are considering a function or a permutation.

Lemma 4. *Let $||.||$ be a norm such that $||A \times B|| \le ||A||.||B||$ for any matrix A and B. For any independent random ciphers denoted C_1, C_2, C_3, C_4, C^* (where C^* is perfect), the following properties hold.*

$$||[C_1 \circ C_2]^d - [C^*]^d|| \le ||[C_1]^d - [C^*]^d||.||[C_2]^d - [C^*]^d|| \tag{3}$$

$$||[C_1 \circ C_2]^d - [C_1 \circ C_3]^d|| \le ||[C_1]^d - [C^*]^d||.||[C_2]^d - [C_3]^d|| \tag{4}$$

$$||[C_1 \circ C_2]^d - [C_3 \circ C_4]^d|| \le ||[C_1]^d - [C^*]^d||.||[C_2]^d - [C_4]^d||$$
$$+ ||[C_1]^d - [C_3]^d||.||[C_4]^d - [C^*]^d|| \tag{5}$$

Those properties come from the easy facts $[C_1 \circ C_2]^d = [C_2]^d \times [C_1]^d$ and $[C^*]^d \times [C_1]^d = [C^*]^d$.

Feistel Ciphers are defined over $\mathcal{M} = \mathcal{M}_0^2$ for a given group $\mathcal{M}_0$ (*e.g.* $\mathcal{M}_0 = \mathbb{Z}_2^{\frac{m}{2}}$) by round functions $F_1, \ldots, F_r$ on $\mathcal{M}_0$. We let $C = \Psi(F_1, \ldots, F_r)$ denote the cipher defined by $C(x^l, x^r) = (y^l, y^r)$ where we iteratively compute a sequence (x_i^l, x_i^r) such that

$$x_0^l = x^l \text{ and } x_0^r = x^r$$
$$x_i^l = x_{i-1}^r \text{ and } x_i^r = x_{i-1}^l + F_i(x_{i-1}^r)$$
$$y^l = x_r^r \text{ and } y^r = x_r^l$$

(see Feistel [4]).

To illustrate the problem, we stress out that perfect decorrelation (*i.e.* decorrelation bias of zero) is achievable on a finite field (no matter which norm we take). For instance, a random $(d-1)$-degreed polynomial with a uniform distribution is a perfectly d-wise decorrelated function. A random affine permutation with a uniform distribution is a perfectly pairwise decorrelated permutation. (Perfect decorrelation of higher degree is much more complicated.) Finite field arithmetic is however cumbersome in software for the traditional characteristic two. This is why we studied decorrelation biases.

2   Previous Security Results

Decorrelation enables to quantify the security of imperfectly decorrelated ciphers. Here we consider the security in the Luby–Rackoff model [6]. We consider opponents as Turing machines which have a limited access to an encryption oracle device and whose aim is to distinguish whether the device implements a given practical cipher $C_1 = C$ or a given cipher C_2 which is usually $C_2 = C^*$. When fed with an oracle c, the Turing machine $\mathcal{T}^c$ returns either 0 or 1. If we want to distinguish a random cipher C from C^*, we let p (resp. p^*) denote $\Pr[\mathcal{T}^C = 1]$ (resp. $\Pr[\mathcal{T}^{C^*} = 1]$) where the probability is over the distribution of the random tape of $\mathcal{T}$ and the distribution of the cipher. We say the attack is successful if $|p - p^*|$ is large. On the other hand, we say that the cipher C resists against the attack if we have $|p - p^*| \le \epsilon$ for some small ϵ. This model is quite powerful, because if we prove that a cipher C cannot be distinguished from the Perfect Cipher C^*, then any attempt to decrypt a ciphertext provided by C will also be applicable to the cipher C^* for which we know the security. (For more motivation on this security model, see Luby-Rackoff [6].)

Inspired by Biham and Shamir's attack [2] we call *differential distinguisher* with the (fixed) characteristic (a, b) and complexity n the following algorithm:

Input: a cipher c, a complexity n, a characteristic (a, b)
1. for i from 1 to n do
   (a) pick uniformly a random X and query for $c(X)$ and $c(X \oplus a)$
   (b) if $c(X \oplus a) = c(X) \oplus b$, stop and output 1
2. output 0

Similarly, inspired by Matsui's attack [7] we call *linear distinguisher* with the characteristic (a, b) and complexity n the following algorithm:[5]

Input: a cipher c, a complexity n, a characteristic (a, b), a set A
1. initialize the counter value t to zero
2. for i from 1 to n do
   (a) pick a random X with a uniform distribution and query for $c(X)$
   (b) if $X \cdot a = c(X) \cdot b$, increment the counter t
3. if $t \in A$, output 1, otherwise output 0

Both linear and differential distinguishers are particular cases of iterative distinguisher attacks (see [14]).

Theorem 5 ([14]). *Let C be a cipher on the space $\mathcal{M} = \mathbf{Z}_2^m$, let C^* be the Perfect Cipher, and let $\epsilon = ||\,||[C]^2 - [C^*]^2||\,||_\infty$. For any differential distinguisher between C and the Perfect Cipher C^* with complexity n, the advantage $|p - p^*|$ is at most $\frac{n}{2^m - 1} + n\epsilon$. Similarly, for any linear distinguisher, the advantage is such that*

$$\lim_{n \to +\infty} \frac{|p - p^*|}{n^{\frac{1}{3}}} \leq 9.3 \left(\frac{1}{2^m - 1} + 2\epsilon \right)^{\frac{1}{3}}.$$

This theorem means that C is immune against any differential or linear distinguisher if $||\,||[C]^2 - [C^*]^2||\,||_\infty \approx 2^{-m}$. In this paper, we show we can obtain similar results with the L_2-decorrelation and that we can use them for an efficient real-life cipher.

3   Security by L_2-Decorrelation

It is well known that differential and linear cryptanalysis with characteristic (a, b) respectively depend on the following measurements. If C is a random cipher on $\mathbf{Z}_2^m$ where $\oplus$ denotes the group operation (the bitwise XOR) and $\cdot$ denotes the dot product (the parity of the bitwise *and*), we denote

---

[5] For differential and linear cryptanalysis, we assume that the message space $\mathcal{M}$ is $\mathbf{Z}_2^m$ so that the addition $+$ is the bitwise exclusive *or* and the dot product $\cdot$ is the parity of the bitwise *and*.

$$\text{EDP}^C(a,b) = \underset{C}{E}\left(\underset{X}{\Pr}[C(X \oplus a) = C(X) \oplus b]\right) = 2^{-m} \sum_{\substack{x_1 \oplus x_2 = a \\ y_1 \oplus y_2 = b}} [C]_{x,y}^2$$

$$\text{ELP}^C(a,b) = \underset{C}{E}\left(\left(2\underset{X}{\Pr}[X \cdot a = C(X) \cdot b] - 1\right)^2\right) = 1 - 2^{2-2m} \sum_{\substack{x_1 \cdot a = y_1 \cdot b \\ x_2 \cdot a \neq y_2 \cdot b \\ x_1 \neq x_2, y_1 \neq y_2}} [C]_{x,y}^2$$

where $x = (x_1, x_2)$ and $y = (y_1, y_2)$, and X is uniformly distributed.[6] In [14], Theorem 5 comes from the upper bounds

$$|\text{EDP}^C(a,b) - \text{EDP}^{C^*}(a,b)| \leq |||[C]^2 - [C^*]^2|||_\infty$$
$$|\text{ELP}^C(a,b) - \text{ELP}^{C^*}(a,b)| \leq 2|||[C]^2 - [C^*]^2|||_\infty.$$

The same inequalities hold with the L_2 norm. (These are the consequence of Cauchy-Schwarz Inequality.) We can thus adapt Theorem 5 with the L_2 bounds without any more argument.

Theorem 6. *Theorem 5 remains valid if we replace $|||.|||_\infty$ norm by the $||.||_2$ norm.*

This means that if $\epsilon = ||[C]^2 - [C^*]^2||_2$ is small (*i.e.* if $\epsilon < 2^{-m}$), the complexity of any basic differential or linear cryptanalysis is close to 2^m, thus no more efficient than exhaustive search.

In the following sections we show how to construct a practical cipher with a relatively small $||[C]^2 - [C^*]^2||_2$. For this we first study how to bound the decorrelation L_2-bias of a Feistel Cipher from the decorrelation of its round functions. Then we construct round functions with relatively small decorrelation L_2-bias and a corresponding dedicated cipher.

4 L_2-Decorrelation of Feistel Ciphers

Here we show how to measure the decorrelation L_2-bias of a Feistel cipher from the decorrelation of its round functions. We first study the case of a 2-round Feistel Cipher.

Lemma 7. *Let $\mathcal{M}_0$ be a group and let $\mathcal{M} = \mathcal{M}_0^2$. Let F_1, F_2, F_1^* and F_2^* be four independent random functions on $\mathcal{M}_0$ where F_1^* and F_2^* have a uniform distribution. If we have $||[F_i]^d - [F_i^*]^d||_2 \leq \epsilon$ then we have*

$$||[\Psi(F_1, F_2)]^d - [\Psi(F_1^*, F_2^*)]^d||_2 \leq \epsilon\sqrt{\epsilon^2 + 2P_d}$$

where P_d is the number of partitions of $\{1, \ldots, d\}$.

---

[6] Those notations are inspired from Matsui's [8]. Actually, Matsui defined DP and LP and we use here their expected values over the distribution of the cipher in order to measure the average complexity of the attacks.

Proof. Let $x = (x_1, \ldots, x_d)$ (resp. $y = (y_1, \ldots, y_d)$)be a multi-point with $x_i = (x_i^l, x_i^r)$ (resp. $y_i = (y_i^l, y_i^r)$). We recall that the relation $y_i = \Psi(g_1, g_2)(x_i)$ means that $y_i^r = x_i^l + g_1(x_i^r)$ and $y_i^l = x_i^r + g_2(y_i^r)$. We thus have

$$\Pr[\Psi(G_1, G_2)(x_i) = y_i; i]$$
$$= \Pr[G_1(x_i^r) = y_i^r - x_i^l; i] \Pr[G_2(y_i^r) = y_i^l - x_i^r; i].$$

The 1–1 relation between (x^l, x^r, y^l, y^r) and $(x^r, y^r - x^l, y^r, y^l - x^r)$ is an important point. In the following, we let (t, u, v, w) denote this family. Let us write the previous equation

$$\Pr_{G_1 G_2}[x \mapsto y] = \Pr_{G_1}[t \mapsto u] \Pr_{G_2}[v \mapsto w].$$

Let $\Delta \Pr$ denotes $\Pr_F - \Pr_{F^\bullet}$ with obvious notations. We have

$$\Delta \Pr_{12}[x \mapsto y] = \Delta \Pr_1[t \mapsto u] \Delta \Pr_2[v \mapsto w] + \Pr_{F_1^\bullet}[t \mapsto u] \Delta \Pr_2[v \mapsto w]$$
$$+ \Pr_{F_2^\bullet}[v \mapsto w] \Delta \Pr_1[t \mapsto u].$$

Now we have

$$\|[\Psi(F_1, F_2)]^d - [\Psi(F_1^*, F_2^*)]^d\|_2^2 = \sum_{x,y} \left(\Delta \Pr_{12}[x \mapsto y] \right)^2.$$

We note that

$$\sum_{t,u} \Pr_{F_1^\bullet}[t \mapsto u] \Delta \Pr_1[t \mapsto u] = 0$$

(and a similar property for $\Pr_2$), thus we have

$$\|[\Psi(F_1, F_2)]^d - [\Psi(F_1^*, F_2^*)]^d\|_2^2 = \epsilon_1^2 \epsilon_2^2 + \epsilon_1^2 \sum_{v,w} \left(\Pr_{F_2^\bullet}[v \mapsto w] \right)^2$$
$$+ \epsilon_2^2 \sum_{t,u} \left(\Pr_{F_1^\bullet}[t \mapsto u] \right)^2$$

where $\epsilon_j = \|[F_j]^d - [F_j^*]^d\|_2^2$. Hence

$$\|[\Psi(F_1, F_2)]^d - [\Psi(F_1^*, F_2^*)]^d\|_2^2 \leq \epsilon^4 + 2\epsilon^2 \sum_{t,u} \left(\Pr_{F_1^\bullet}[t \mapsto u] \right)^2.$$

For any partition $\mathcal{P} = \{O_1, \ldots, O_k\}$ of $\{1, \ldots, d\}$ into k parts, let

$$\mathcal{M}_\mathcal{P} = \{t; \forall i, j \ (t_i = t_j \Leftrightarrow \exists k \ i, j \in O_k)\}.$$

We have

$$\sum_{t,u} \left(\Pr_{F_1^\bullet}[t \mapsto u] \right)^2 = \sum_{\substack{\mathcal{P} \text{ into} \\ k \text{ parts}}} \sum_{t \in \mathcal{M}_\mathcal{P}} \sum_u \left(\Pr_{F_1^\bullet}[t \mapsto u] \right)^2.$$

We have M^k u-terms for which the probability is not zero. Namely it is $1/M^k$. The number of t-terms which correspond to this partition is $M(M-1)\ldots(M-k+1)$ thus

$$\sum_{t,u}\left(\Pr_{F_1^\bullet}[t\mapsto u]\right)^2 = \sum_{\substack{\mathcal{P} \text{ into} \\ k \text{ parts}}} \frac{M(M-1)\ldots(M-k+1)}{M^k}$$

which is less than P_d.   □

In order to measure the decorrelation distance between a 2-round Feistel Cipher and the Perfect Cipher, we thus have to study the case of a truly random 2-round Feistel Cipher.

Lemma 8. *Let $\mathcal{M}_0$ be a group and let $\mathcal{M} = \mathcal{M}_0^2$. Let F_1^* and F_2^* be two independent random functions on $\mathcal{M}_0$ with a uniform distribution and let C^* be the Perfect Cipher on $\mathcal{M}$. We have*

$$||[\Psi(F_1^*, F_2^*)]^d - [C^*]^d||_2 \leq \sqrt{P_d(P_d - 1)}$$

where P_d is the number of partitions of $\{1, \ldots, d\}$.

Proof. With obvious notations we have

$$||[\Psi(F_1^*, F_2^*)]^d - [C^*]^d||_2^2 = \sum_{x,y}\left(\Pr_{\Psi(F_1^*, F_2^*)} - \Pr_{C^\bullet}\right)^2 [x\mapsto y].$$

The sums $\sum \Pr_{C^\bullet}^2[x\mapsto y]$ and $\sum \Pr_{\Psi(F_1^\bullet, F_2^\bullet)}\Pr_{C^\bullet}[x\mapsto y]$ are equal to P_d. (We observe it by fixing the partition associated to x and making the sum over all ys.) For the remaining sum, we use same ideas as in the previous proof:

$$\sum_{x,y}\left(\Pr_{\Psi(F_1^\bullet, F_2^\bullet)}[x\mapsto y]\right)^2 = \sum_{t,u,v,w}\left(\Pr_{F_1}[t\mapsto u]\Pr_{F_2}[v\mapsto w]\right)^2$$

which is less than P_d^2.   □

Lemma 8 may look useless because the decorrelation bias of is greater than one (so we cannot consider product cipher and get efficient bounds). We can however use it to study the case of a 4-round Feistel Cipher. From Lemma 7 and Lemma 8 and from Equation (5) we obtain the following Lemma in a straightforward way.

Lemma 9. *Let $\mathcal{M}_0$ be a group and let $\mathcal{M} = \mathcal{M}_0^2$. Let $F_1, \ldots, F_4, F_1^*, \ldots, F_4^*$ be eight independent random functions on $\mathcal{M}_0$ where the F_i^*s have a uniform distribution. If we have $||[F_i]^d - [F_i^*]^d||_2 \leq \epsilon \leq \sqrt{2}$ then we have*

$$||[\Psi(F_1, F_2, F_3, F_4)]^d - [\Psi(F_1^*, F_2^*, F_3^*, F_4^*)]^d||_2 \leq 2\sqrt{2}(P_d)^{\frac{3}{2}}\epsilon$$

where P_d is the number of partitions of $\{1, \ldots, d\}$.

It thus remains to study the decorrelation distance between a truly random 4-round Feistel Cipher and the Perfect Cipher: once we know that

$$||[\Psi(F_1, F_2, F_3, F_4)]^d - [\Psi(F_1^*, F_2^*, F_3^*, F_4^*)]^d||_2 \leq u_d$$

we obtain from Equation (3) that

$$||[\Psi(F_1, \ldots, F_{4r})]^d - [\Psi(F_1^*, \ldots, F_{4r}^*)]^d||_2 \leq \left(2\sqrt{2}(P_d)^{\frac{3}{2}}\epsilon + u_d\right)^r$$

where $\epsilon = \max_i ||[F_i]^d - [F_i^*]^d||_2 \leq \sqrt{2}$. Unfortunately, the problem of obtaining a general result on the d-wise decorrelation of a truly random 4-round Feistel Cipher is still open.[7] In the next section we propose a construction in the $d = 2$ case for which we can evaluate the decorrelation.

5   A Dedicated Construction

In a general finite field $GF(q)$, an obvious way to construct pairwise decorrelated functions (resp. permutations) consists of taking

$$F(x) = a.x + b$$

where (a, b) is a random pair uniformly distributed in $GF(q)^2$ (resp. $GF(q)^* \times GF(q)$). Unfortunately, the traditional message space $\mathbf{Z}_2^m$ requires that we use finite fields of characteristic two. If we aim to implement a cipher in software on a modern microprocessor, it looks cumbersome to implement a poor characteristic-two multiplication since there already is a built-in integer multiplication. For this reason we can think of the

$$F(x) = ((ax + b) \bmod p) \bmod 2^{\frac{m}{2}}$$

imperfectly decorrelated function to be inserted at the input of each round function of a Feistel Cipher, where p is a prime close to $2^{\frac{m}{2}}$.

In [14] the (m, r, d, p)-PEANUT Cipher Family is defined to be the set of all r-round Feistel Ciphers over $\mathbf{Z}_2^m$ in which all round functions can be written

$$F(x) = g\left(\sum_{i=1}^{d} k_i.x^{d-i} \bmod p \bmod 2^{\frac{m}{2}}\right)$$

where $(k_1, \ldots, k_d)$ is an (independent) round key which is uniformly distributed in $\{0, \ldots, 2^{\frac{m}{2}} - 1\}^d$, p is a prime, and g is a permutation. For $p > 2^{\frac{m}{2}}$, F has a friendly d-wise decorrelation $|||.|||_\infty$-bias which is roughly $2d\delta$ when $p = (1+\delta)2^{\frac{m}{2}}$. For $p < 2^{\frac{m}{2}}$, the $|||.|||_\infty$-decorrelation is poor for $d \geq 2$. For instance, in the case $d = 2$, for $x = (0, p)$ we have

$$\sum_{y=(y_1,y_2)} \left| \Pr \begin{bmatrix} g(k_2 \bmod p) = y_1 \\ g(k_2 \bmod p) = y_2 \end{bmatrix} - \Pr \begin{bmatrix} F^*(0) = y_1 \\ F^*(p) = y_2 \end{bmatrix} \right| = 2 - 2^{1-\frac{m}{2}} + \delta.$$

---

[7] This problem has been solved in [13, 14] with the $|||.|||_\infty$ norm. This is why the L_2 norm looks less friendly.

Hence $|||[F]^2 - [F^*]^2|||_\infty \approx 2$. The $p < 2^{\frac{m}{2}}$ case can however be studied with the L_2 norm. In the following, we consider a PEANUT Cipher construction with $d = 2$ and $p < 2^{\frac{m}{2}}$.

Lemma 10. *Let A and B be two independent random variables with a uniform distribution over $\{0, \ldots, 2^{\frac{m}{2}} - 1\}$. We let $F(x) = Ax + B \bmod p$ where $p = (1 - \delta)2^{\frac{m}{2}}$ is a prime for $1/14 \geq \delta \geq 0$. Let F^* be a random function uniformly distributed over the same set. We have*

$$\|[F]^2 - [F^*]^2\|_2 \leq 2\sqrt{2\delta}.$$

Proof. We let $N = 2^{\frac{m}{2}}$. We want to upper bound the sum

$$\sum_{\substack{x=(x_1,x_2) \\ y=(y_1,y_2)}} \left([F]_{x,y}^2 - [F^*]_{x,y}^2\right)^2.$$

Table 1 shows all the possible cases for x and y, the number of times they occur and an upper bound for the probability difference.

For instance if $x_1 = x_2 \not\equiv 0 \pmod p$ and $y_1 = y_2 < p$, we have

$$[F]_{(x_1,x_2),(y_1,y_2)}^2 = \Pr[Ax_1 + B \bmod p = y_1]$$

and $[F^*]_{(x_1,x_2),(y_1,y_2)}^2 = N^{-1}$. We let a (resp. b, c, d) be the number of $(A \bmod p, B \bmod p)$ pairs such that $Ax_1 + B \bmod p = y_1$ and

- $A \bmod p < \delta N$ and $B \bmod p < \delta N$ (resp.
- $A \bmod p < \delta N$ and $B \bmod p \geq \delta N$,
- $A \bmod p \geq \delta N$ and $B \bmod p < \delta N$,
- $A \bmod p \geq \delta N$ and $B \bmod p \geq \delta N$).

We have $a + b = \delta N$, $a + c = \delta N$ and $a + b + c + d = p$. Hence

$$[F]_{(x_1,x_2),(y_1,y_2)}^2 = \frac{4a + 2b + 2c + d}{N^2} = \frac{N + \delta N + a}{N^2}.$$

Since we have $0 \leq a \leq \delta N$, we have

$$(1 + \delta)N^{-1} \leq [F]_{(x_1,x_2),(y_1,y_2)}^2 \leq (1 + 2\delta)N^{-1}.$$

The $x_1 \not\equiv x_2$ case is split into four cases which depend on x_1 and x_2. The last case is $y_1 \geq p$ or $y_2 \geq p$ for which $[F]_{x,y}^2 = 0$. The three other cases correspond to cases on $(A \bmod p, B \bmod p)$ with $y_i = Ax_i + B \bmod p$.

Case 1: $A \bmod p < \delta N$, $B \bmod p < \delta N$. We have $[F]_{x,y}^2 = 4N^{-1}$.
Case 3: $A \bmod p \geq \delta N$, $B \bmod p \geq \delta N$. We have $[F]_{x,y}^2 = N^{-1}$.
Case 2: other values. We have $[F]_{x,y}^2 = 2N^{-1}$.

We can now upper bound the whole sum. We obtain that the decorrelation bias $\|[F]^2 - [F^*]^2\|_2^2$ is less than

$$7\delta + 14\delta^2 - 6\frac{\delta}{N} - 4\delta^3 - 24\frac{\delta^2}{N} + 4\frac{\delta}{N^2} - 8\delta^4 + 16\frac{\delta^3}{N} - 8\frac{\delta^2}{N^2}$$

which is less than 8δ when $\delta \leq 1/14$. □

case x	case y	num. x	num. y	$\|[F]^2_{x,y} - [F^*]^2_{x,y}\| \leq$
$x_1 = x_2 \equiv 0$	$y_1 = y_2 < \delta N$	2	δN	N^{-1}
	$y_1 = y_2 \geq (1-\delta)N$		δN	N^{-1}
	other cases		$N^2 - 2\delta N$	0
$x_1 = x_2 \not\equiv 0$	$y_1 = y_2 \geq (1-\delta)N$	$N-2$	δN	N^{-1}
	$y_1 = y_2 < (1-\delta)N$		$(1-\delta)N$	$2\delta N^{-1}$
	$y_1 \neq y_2$		$N^2 - N$	0
$x_1 \neq x_2, x_1 \equiv x_2 \equiv 0$	$y_1 = y_2 < \delta N$	2	δN	$2N^{-1} - N^{-2}$
	$y_1 = y_2 \geq (1-\delta)N$		δN	$N^{-1} - N^{-2}$
	other $y_1 = y_2$		$(1-2\delta)N$	N^{-2}
	$y_1 \neq y_2$		$N^2 - N$	N^{-2}
$x_1 \neq x_2, x_1 \equiv x_2 \not\equiv 0$	$y_1 = y_2 < (1-\delta)N$	$2\delta N - 2$	$(1-\delta)N$	$(1+2\delta)N^{-1} - N^{-2}$
	other cases		$N^2 - (1-\delta)N$	N^{-2}
$x_1 \not\equiv x_2$	case 1		$\delta^2 N^2$	$3N^{-2}$
	case 2	$N^2 - (1+2\delta)N$	$2(1-2\delta)\delta N^2$	N^{-2}
	case 3		$(1-2\delta)^2 N^2$	0
	y_1 or $y_2 \geq (1-\delta)N$		$(2-\delta)\delta N^2$	N^{-2}

Table 1. Decorrelation of $A.x + B \bmod (1-\delta)N$

Lemma 11. *Let* $\mathcal{M} = \mathbf{Z}_2^m$. *Let* $F_1^*, \ldots, F_4^*$ *be four independent random functions on* $\mathbf{Z}_2^{\frac{m}{2}}$ *with a uniform distribution and let* C^* *be the Perfect Cipher on* $\mathcal{M}$. *We have*

$$\|[\Psi(F_1^*, F_2^*, F_3^*, F_4^*)]^2 - [C^*]^2\|_2 \leq \sqrt{2.2^{-m} + 4.2^{-\frac{3m}{2}}}.$$

Proof. For each input pair $x = (x_1, x_2)$ we have $x_i = (x_i^l, x_i^r)$. Similarly, for each output pair $y = (y_1, y_2)$ we have $y_i = (y_i^l, y_i^r)$. All (x, y) pairs can be split into 10 cases:

1. $y_1^r \neq y_2^r$, $x_1^r \neq x_2^r$
2. $y_1^r \neq y_2^r$, $x_1^r = x_2^r$, $x_1^l \oplus x_2^l \notin \{0, y_1^r \oplus y_2^r\}$
3. $y_1^r \neq y_2^r$, $x_1^r = x_2^r$, $x_1^l \oplus x_2^l = y_1^r \oplus y_2^r$
4. $y_1^r \neq y_2^r$, $x_1 = x_2$
5. $y_1^r = y_2^r$, $y_1^l \neq y_2^l$, $x_1^r \oplus x_2^r \notin \{0, y_1^l \oplus y_2^l\}$
6. $y_1^r = y_2^r$, $y_1^l \neq y_2^l$, $x_1^r \oplus x_2^r = y_1^l \oplus y_2^l$
7. $y_1^r = y_2^r$, $y_1^l \neq y_2^l$, $x_1^r = x_2^r$, $x_1^l \neq x_2^l$
8. $y_1^r = y_2^r$, $y_1^l \neq y_2^l$, $x_1 = x_2$
9. $y_1^r = y_2^r$, $y_1^l = y_2^l$, $x_1 = x_2$
10. $y_1 = y_2$, $x_1 \neq x_2$

Each case requires a dedicated study.

We consider a truly random $2r$-round Feistel cipher for $r \geq 1$ denoted $C = \Psi(F_1^*, \ldots, F_{2r}^*)$. We have

$$[C]_{x,y}^2 = \begin{cases} A_r^1 = \frac{1}{N^2(N^2-1)}\left(1 - \frac{1}{N^{2r}}\right) & \text{if case 1} \\ A_r^2 = \frac{1}{N^2(N^2-1)}\left(1 - \frac{1}{N^{r-1}} - \frac{1}{N^r} + \frac{1}{N^{2r-1}}\right) & \text{if case 2} \\ A_r^3 = \frac{1}{N^2(N^2-1)}\left(1 + \frac{1}{N^{r-2}} - \frac{1}{N^{r-1}} - \frac{2}{N^r} + \frac{1}{N^{2r-1}}\right) & \text{if case 3} \\ 0 & \text{if case 4} \\ A_r^5 = \frac{1}{N^2(N^2-1)}\left(1 - \frac{1}{N^{r-1}} - \frac{1}{N^r} + \frac{1}{N^{2r-1}}\right) & \text{if case 5} \\ A_r^6 = \frac{1}{N^2(N^2-1)}\left(1 + \frac{1}{N^{r-2}} - \frac{1}{N^{r-1}} - \frac{2}{N^r} + \frac{1}{N^{2r-1}}\right) & \text{if case 6} \\ A_r^7 = \frac{1}{N^2(N^2-1)}\left(1 - \frac{1}{N^{2r-2}}\right) & \text{if case 7} \\ 0 & \text{if case 8} \\ N^{-2} & \text{if case 9} \\ 0 & \text{if case 10} \end{cases}$$

where $N = 2^{\frac{m}{2}}$. We prove this by an easy induction. Namely we show that

$$\begin{pmatrix} A_r^1 \\ A_r^2 \\ A_r^3 \end{pmatrix} = \begin{pmatrix} \frac{(N-1)^2}{N^2} + \frac{1}{N} & \frac{N-2}{N^2} & \frac{1}{N^2} \\ 1 - \frac{1}{N} & \frac{1}{N} & 0 \\ 1 - \frac{1}{N} & 0 & \frac{1}{N} \end{pmatrix}^{r-1} \begin{pmatrix} \frac{1}{N^4} \\ 0 \\ \frac{1}{N^3} \end{pmatrix}$$

and

$$\begin{pmatrix} A_r^5 \\ A_r^6 \\ A_r^7 \end{pmatrix} = \begin{pmatrix} \frac{(N-1)(N-2)}{N^2} + \frac{1}{N} & \frac{N-1}{N^2} & \frac{N-1}{N^2} \\ \frac{(N-1)(N-2)}{N^2} & \frac{N-1}{N^2} + \frac{1}{N} & \frac{N-1}{N^2} \\ 1 - \frac{2}{N} & \frac{1}{N} & \frac{1}{N} \end{pmatrix}^{r-1} \begin{pmatrix} 0 \\ \frac{1}{N^3} \\ 0 \end{pmatrix}$$

For instance, if $r = 1$ and $y_1^r \neq y_2^r$, $x_1^r = x_2^r$, $x_1^l \oplus x_2^l = y_1^r \oplus y_2^r$, the probability corresponds to the fact that $F_1^*(x_1^r)$ XORs the good value on both x_1^l and x_2^l (with probability $1/N$) and that both $F_2^*(y_1^r)$ and $F_2^*(y_2^r)$ XOR the good values on x_1^r and x_2^r respectively (with probability $1/N^2$).

To prove the matrix relations, we let x denote the input of C, y denote the output of the first two rounds and z denote the output. We have

$$z = \Psi(F_1^*, F_2^*, F_3^*, \ldots, F_{2r}^*)(x) = \Psi(F_3^*, \ldots, F_{2r}^*)(y)$$
$$(y^r, y^l) = \Psi(F_1^*, F_2^*)(x).$$

For instance, transition from case 2 ($z_1^r \neq z_2^r$, $y_1^r = y_2^r$, $y_1^l \oplus y_2^l \notin \{0, z_1^r \oplus z_2^r\}$) to case 1 ($z_1^r \neq z_2^r$, $x_1^r \neq x_2^r$) corresponds to the $N(N-2)$ possibilities for y_1^l and y_2^l (all but for $y_1^l = y_2^l$ or $y_1^l \oplus y_2^l = z_1^r \oplus z_2^r$), all with probability $1/N^2$ (since $F_1^*(x_1^r)$ and $F_1^*(x_2^r)$ are independent), mixed with the N possibilities for $y_1^r = y_2^r$, all with probability $1/N^2$, which gives $(N-2)/N^2$. This means A_r^1 includes a term $\frac{N-2}{N^2} A_{r-1}^2$ which represents all possible ys coming from case 2.

With this result we can compute the pairwise decorrelation bias of C. We have

$$\|[C]^2 - [C^*]^2\|_2^2 = n_1(\Delta A_r^1)^2 + n_2(\Delta A_r^2)^2 + n_3(\Delta A_r^3)^2$$
$$+ n_5(\Delta A_r^5)^2 + n_6(\Delta A_r^6)^2 + n_7(\Delta A_r^7)^2$$

where $\Delta A_r^i = A_r^i - \frac{1}{N^2(N^2-1)}$ and n_i is the number of (x,y) pairs in case i. We obtain

$$\frac{2N^{4-2r} - 6N^{2-2r} - 4N^{1-2r} + N^{4-4r} + 2N^{3-4r} + N^{2-4r}}{(N-1)^2}.$$

For $r = 2$ (four rounds), this is less than $2N^{-2} + 4N^{-3}$.   □

We can now define the PEANUT97 Cipher construction. It consists of a $(m, 4r, 2, p)$-PEANUT Cipher, i.e. a $4r$-round Feistel Cipher on m-bit message blocks which is characterized by some prime $p \leq 2^{\frac{m}{2}}$. Each round function of the cipher must be with the form

$$F_i(x) = g_i(K_{2i-1}x + K_{2i} \bmod p)$$

where $(K_1, \ldots, K_{8r})$ is uniformly distributed in $\mathbf{Z}_2^{4mr}$ and g_i is a (possibly independently keyed) permutation on the $\frac{m}{2}$-bit strings. The lemmata 9, 10 and 11 proves the following theorem.

Theorem 12. *Let C be a $(m, 4r, 2, p)$-PEANUT97 Cipher such that $p = (1 - \delta)2^{\frac{m}{2}}$ with $0 \leq \delta \leq \frac{1}{14}$. Let C^* be the Perfect Cipher. We have*

$$\|[C]^2 - [C^*]^2\|_2 \leq \left(16\sqrt{2\delta} + \sqrt{2.2^{-m} + 4.2^{-\frac{3m}{2}}}\right)^r.$$

For instance, with $m = 64$ and $p = 2^{32} - 5$, we obtain $\|[C]^2 - [C^*]^2\|_2 \leq 2^{-10r}$. Thus for $r = 7$ we have $\|[C]^2 - [C^*]^2\|_2 \leq 2^{-70}$. Theorem 6 thus shows that $|p - p^*| \leq 0.1$ for any differential distinguisher with complexity $n \leq 2^{60}$ and any linear distinguisher with complexity $n \leq 2^{44}$.

This PEANUT97 construction has been tested on a Pentium in assembly code. A 28-round 64-bit encryption required less than 790 clock cycles, which yields an encryption rate of 23Mbps working at 300MHz. The table below compares it with the PEANUT98 construction, for which the $\|\|.\|\|_\infty$-decorrelation theory enables to decrease the number of rounds (see [13]) and the DFC AES candidate which is a PEANUT98 128-bit block cipher (see [5]). All ciphers have similar security against differential and linear cryptanalysis. We remark that one PEANUT97 is much faster than the other rounds, so PEANUT97 may be faster than PEANUT98 if we can get tighter bounds in order to decrease the number of rounds.

cipher	PEANUT97	PEANUT98	DFC
block length	64	64	128
number of rounds	28	9	8
cycles/encryption	788	396	754
cycles/round	28	44	94
enc. rate at 300MHz	23Mbps	46Mbps	49Mbps
pairwise decorrelation	2^{-70} (L_2)	2^{-76} $(\|\|.\|\|_\infty)$	2^{-112} $(\|\|.\|\|_\infty)$
reference	[12], here	[13,14]	[5,9]

6   Conclusion

We have shown how to use the $ax + b \bmod p$ pairwise decorrelation primitive for $p \leq 2^{\frac{m}{2}}$. It requires that we use the L_2 norm in the decorrelation technique, which leads to more complicated computations than for the $|||.|||_\infty$ norm.

When used at the input of Feistel Ciphers, this primitive enables to protect it against differential and linear cryptanalysis. For 64-bit message block, it however requires at least 28 rounds.

Some extensions of the $|||.|||_\infty$-decorrelation results to the L_2-decorrelation is still open: it is not clear how to state results with higher degrees of decorrelation ($d > 2$) and how to prove the security of decorrelated ciphers against general iterated attacks as in [14].

References

1. E. Biham. A Fast New DES Implementation in Software. In *Fast Software Encryption*, Haifa, Israel, Lectures Notes in Computer Science 1267, pp. 260–272, Springer-Verlag, 1997.
2. E. Biham, A. Shamir. *Differential Cryptanalysis of the Data Encryption Standard*, Springer-Verlag, 1993.
3. L. Carter, M. Wegman. Universal Classes of Hash Functions. *Journal of Computer and System Sciences*, vol. 18, pp. 143–154, 1979.
4. H. Feistel. Cryptography and Computer Privacy. *Scientific American*, vol. 228, pp. 15–23, 1973.
5. H. Gilbert, M. Girault, P. Hoogvorst, F. Noilhan, T. Pornin, G. Poupard, J. Stern, S. Vaudenay. Decorrelated Fast Cipher: an AES Candidate. Advanced Encryption Standard Submissions, US Department of Commerce, 1998.
6. M. Luby, C. Rackoff. How to Construct Pseudorandom Permutations from Pseudorandom Functions. *SIAM Journal on Computing*, vol. 17, pp. 373–386, 1988.
7. M. Matsui. The First Experimental Cryptanalysis of the Data Encryption Standard. In *Advances in Cryptology CRYPTO'94*, Santa Barbara, California, U.S.A., Lectures Notes in Computer Science 839, pp. 1–11, Springer-Verlag, 1994.
8. M. Matsui. New Structure of Block Ciphers with Provable Security Against Differential and Linear Cryptanalysis. In *Fast Software Encryption*, Cambridge, United Kingdom, Lectures Notes in Computer Science 1039, pp. 205–218, Springer-Verlag, 1996.
9. G. Poupard, S. Vaudenay. Decorrelated Fast Cipher: an AES Candidate Well Suited for Low Cost Smart Cards Applications. Submitted to CARDIS'98.
10. C. E. Shannon. Communication Theory of Secrecy Systems. *Bell system technical journal*, vol. 28, pp. 656–715, 1949.
11. A. Shamir. How to Photofinish a Cryptosystem? Presented at the Rump Session of Crypto'97.
12. S. Vaudenay. A Cheap Paradigm for Block Cipher Security Strengthening. Technical Report LIENS-97-3, Ecole Normale Supérieure, 1997.
13. S. Vaudenay. Provable Security for Block Ciphers by Decorrelation. In *STACS 98*, Paris, France, Lectures Notes in Computer Science 1373, pp. 249–275, Springer-Verlag, 1998.

14. S. Vaudenay. Provable Security for Block Ciphers by Decorrelation. (Journal Version.) Submitted.
15. S. Vaudenay. The Decorrelation Technique Home-Page.
    URL:http://www.dmi.ens.fr/~vaudenay/decorrelation.html
16. G. S. Vernam. Cipher Printing Telegraph Systems for Secret Wire and Radio Telegraphic communications. *Journal of the American Institute of Electrical Engineers*, vol. 45, pp. 109–115, 1926.
17. M. N. Wegman, J. L. Carter. New Hash Functions and their Use in Authentication and Set Equality. *Journal of Computer and System Sciences*, vol. 22, pp. 265–279, 1981.

Key-Dependent S-Box Manipulations

Sandy Harris[1] and Carlisle Adams[2]

[1] Kaya Consulting, 6 Beechwood Avenue, Suite 16
Vanier, Ontario,Canada, K1L 8B4
sandy.harris@sympatico.ca
[2] Entrust Technologies, 750 Heron Road
Ottawa, Ontario, Canada, K1V 1A7
cadams@entrust.com

Abstract. This paper discusses a method of enhancing the security of block ciphers which use s-boxes, a group which includes the ciphers DES, CAST-128, and Blowfish. We focus on CAST-128 and consider Blowfish; Biham and Biryukov [2] have made some similar proposals for DES.

The method discussed uses bits of the primary key to directly manipulate the s-boxes in such a way that their contents are changed but their cryptographic properties are preserved. Such a strategy appears to significantly strengthen the cipher against certain attacks, at the expense of a relatively modest one-time computational procedure during the set-up phase. Thus, a stronger cipher with identical encryption / decryption performance characteristics may be constructed with little additional overhead or computational complexity.

1 Introduction

Both carefully-constructed and randomly-generated s-boxes have a place in symmetric cipher design. Typically, a given cipher will use one or the other paradigm in its encryption "engine". This paper suggests that a mixture of the two paradigms may yield beneficial results in some environments. In our examples, we use the four 8×32 s-boxes which the Blowfish and CAST-128 ciphers employ, but variations of this technique could be applied to any cipher using s-boxes, whatever their number and sizes.

We propose using strong s-boxes and applying key-dependent operations to them at the time of key scheduling, before the actual encryption begins. The goal is to get the benefits of strong s-boxes (as in CAST-128) and of key-dependent s-boxes (as in Blowfish) without the drawbacks of either.

The technique can be powerful. If a basic cipher can be broken in a second by exhaustive search over the key space and if key-dependent operations on the s-boxes add 32 bits to the effective key length, then breaking the improved cipher by brute force takes 2^{32} seconds (just over a century). If these operations add 64 effective bits, then it would take 2^{32} centuries to break the improved cipher.

Key-dependent operations on s-boxes can use large numbers of bits. For 8×32 s-boxes, XORing constants into the inputs and outputs can use 40 bits per s-box.

S. Tavares and H. Meijer (Eds.): SAC'98, LNCS 1556, pp. 15–26, 1999.

Permuting the inputs and outputs can use $\log_2(8!) + \log_2(32!) > 130$ bits per s-box. A cipher with four such s-boxes could use 680 bits of additional key with these operations (although it is recognized that this will not necessarily be the increase in the effective key length).

We start with the CAST-128 cipher and propose using between 148 and 256 additional key bits for s-box transformations. The increase in effective key length (although difficult to compute precisely) is likely to be considerably lower than this, but the transformations still appear to be worthwhile in at least some applications. In particular, the cost is moderate and the resulting cipher appears to be more resistant to attacks that rely upon knowledge of the s-box contents.

It is important to note that the technique is inherently efficient in one sense: all the s-box transformations are done at set-up time. Thus, there is no increase in the per-round or per-block encryption time of the strengthened cipher.

2   Considerations

2.1   The Extra Key Bits

The additional bits required for this proposal may come from one of two sources: derived key bits or primary key bits. As an example of the former, CAST-128 [1] expands the 128-bit key to 1024 bits but does not use all of them. The actual encryption uses 37 bits per round (592 in the full 16 rounds) so that 432 bits are generated by the key scheduling algorithm but are unused by the cipher. These currently unused bits may be good candidates for the bits needed for s-box manipulation.

Alternatively, additional primary key bits may be used for the proposal in this paper. This has the advantage of increasing the key space for exhaustive search attacks, at the cost of increased key storage and bandwidth requirements. Note, however, that in some environments the bandwidth required for key transfer or key agreement protocols need not increase. In one common use of symmetric ciphers, session keys are transmitted using a public key method such as RSA [5] or Diffie-Hellman [3]. Public key algorithms use large numbers of bits so that to transmit a 128-bit session key, you may need to encrypt, transmit and decrypt a full public-key block of 1024 bits or more. In such a case, any key up to several hundred bits can be transmitted with no additional cost compared with a 128-bit key.

Using derived key bits has no impact on primary key size, but depends upon a key scheduling algorithm that generates extra (i.e., currently unused) bits. Using primary key bits places no such requirement on the key scheduling algorithm, but has storage and bandwidth implications, and may show some susceptibility to chosen-key-type attacks (since the two pieces of the primary key are "separable" in some sense).

2.2   CAST's Strong S-Boxes

The CAST design procedure uses fixed s-boxes in the construction of each specific CAST cipher. This allows implementers to build strong s-boxes, using bent

Boolean functions for the columns and choosing combinations of columns for high levels of s-box nonlinearity and for other desirable properties. Details can be found in Mister and Adams [4].

For example, the CAST-128 s-boxes appear to be strong but they are fixed and publicly-known. This may allow some theoretical attacks (e.g., linear or differential cryptanalysis) to be mounted against a given CAST cipher which uses these s-boxes, although the computational cost of these attacks can be made to be infeasibly high with a suitable choice in the number of rounds.

2.3   Blowfish's Key-Dependent S-Boxes

Blowfish generates key-dependent s-boxes at cipher set-up time. This means the attacker cannot know the s-boxes, short of breaking the algorithm that generates them.

There are at least two disadvantages, which can to some extent be traded off against each other. One is that generating the s-boxes has a cost. The other is that the generated s-boxes are not optimized and may even be weak.

A Blowfish-like cipher might, with some increase in set-up cost, avoid specific weaknesses in its s-boxes. Schneier discusses checking for identical rows in Blowfish [6, page 339] but considers this unnecessary. In general, it is clearly possible to add checks which avoid weaknesses in randomly-generated s-boxes for Blowfish-like ciphers, but it is not clear whether or when this is worth doing.

On the other hand, generating cryptographically strong s-boxes at run time in a Blowfish-like cipher is impractical, at least in software. Mister and Adams [4] report using 15 to 30 days of Pentium time to generate one 8×32 s-box suitable for CAST-128, after considerable work to produce efficient code. This is several orders of magnitude too slow for a run-time operation, even for one used only at set-up time and not in the actual cipher.

2.4   Resistance to Attack

Schneier [6, p.298] summarizes the usefulness of randomly-generated s-boxes with respect to the most powerful statistical attacks currently known in his introduction to the Biham and Biryukov work on DES with permuted s-boxes [2]:

"Linear and differential cryptanalysis work only if the analyst knows the composition of the s-boxes. If the s-boxes are key-dependent and chosen by a cryptographically strong method, then linear and differential cryptanalysis are much more difficult. Remember, though, that randomly-generated s-boxes have very poor differential and linear characteristics, even if they are secret."

This inherent dilemma leads to the proposal presented in this paper: we suggest s-boxes that are key-dependent but are not randomly generated.

3   The Proposal

Start with carefully-prepared strong s-boxes, such as those described for CAST in Mister and Adams [4] and apply key-dependent operations to them before use. The goal is to introduce additional entropy so that attacks which depend on knowledge of the s-boxes become impractical, without changing the properties which make the s-boxes strong.

We apply the operations before encryption begins and use the modified s-boxes for the actual encryption, so the overhead is exclusively in the set-up phase. There is no increase in the per-block encryption cost.

It can be shown that important properties of strong s-boxes are preserved under carefully-chosen key-dependent operations. Given this, it is possible to prepare strong s-boxes off-line (as in CAST-128) and manipulate them at cipher set-up time to get provably strong key-dependent s-boxes (in contrast with ciphers such as Blowfish).

The question is what operations are suitable; that is, what operations are key-dependent, reasonably efficient, and guaranteed not to destroy the cryptographic properties of a strong s-box.

The first two requirements can be met relatively easily; simultaneously achieving the third is somewhat more difficult. However, several classes of operations may be used.

- Permuting s-box columns
    - this has the effect of permuting output bits.
- Adding affine functions to s-box columns
    - this has the effect of complementing output bits, possibly depending upon the values of other output bits.
- Permuting s-box inputs
    - this has the effect of producing certain s-box row permutations.
- Adding affine functions to s-box inputs
    - this has the effect of producing other s-box row permutations, possibly depending upon the values of other input bits.

In general, then, the Boolean function for an s-box column may be modified from

$$f(\overline{x}) = f(x_1, x_2, x_3, \ldots, x_m) \ ,$$

for binary variables x_i, to

$$f\left(P\left(g_1\left(\overline{x}\right), g_2(\overline{x}), g_3(\overline{x}), \ldots, g_m(\overline{x})\right)\right) \oplus h(\overline{x}) \ ,$$

for some Boolean functions $g_i(\overline{x})$ and $h(\overline{x})$ and a permutation P. The set of columns may then be further permuted. We will consider the above classes of operations in the order presented.

3.1   Permuting S-Box Columns

Permuting s-box columns can be accomplished by permuting each row in the same way (done in one loop through the s-box).

Various important properties are conserved under this operation. In particular, if a column is bent, it will clearly remain so when moved; if a group of columns satisfies the bit independence criterion, it will still do so after being permuted. Finally, since s-box nonlinearity is defined to be the minimum nonlinearity of any function in the set of all non-trivial linear combinations of the columns (see [4], for example), then s-box nonlinearity is also conserved through a column permutation.

In carefully designed s-boxes, rearranging the columns in a key-dependent way does not degrade cryptographic strength. However, such an operation can make it significantly more difficult to align characteristics in a linear cryptanalysis attack and so can increase the security of the cipher by raising the computational complexity of mounting this attack.

3.2   Adding Affine Functions to S-Box Columns

In the extreme case in which the affine functions are simply all-one vectors, the addition can be done by XORing a constant into all rows (done in one loop through the s-box). More generally, other techniques (perhaps involving storage of specific affine vectors) may be necessary to accomplish this addition.

Various important properties are conserved under this operation. Because the nonlinearity of a Boolean function is unchanged by the addition of an affine function, s-box column bentness, s-box bit independence criterion, and s-box nonlinearity are all conserved.

The addition of affine functions, therefore, does nothing to degrade cryptographic security in the s-boxes. However, such an operation, by modifying the contents of the s-boxes in a key-dependent way, can make it significantly more difficult to construct characteristics in a differential cryptanalysis attack (because it cannot be computed in advance when the XOR of two given s-box outputs will produce one value or another). Hence, this operation can increase the security of the cipher by raising the computational complexity of mounting this attack.

3.3   Permuting S-Box Inputs

Permuting the rows of an s-box seems attractive because of its potential for thwarting linear and differential cryptanalysis. However, it is not always possible to permute rows without compromising desirable s-box properties. In particular (e.g., for CAST-128 s-boxes), not all row permutations are permissible if column bentness is to be preserved.

Biham and Biryukov [2] made only one small change to the s-box row order in the DES s-boxes: they used one key bit per s-box, controlling whether or not the first two and the last two rows should be swapped. However, an operation that used more key bits and that provably could not weaken strong s-boxes may be preferable in some environments.

One such operation is to use the subset of row permutations that result from a permutation on the s-box inputs. We will show that these do not damage the desirable s-box properties.

Mister and Adams [4] introduce the notion of *dynamic distance of order j* for a function f

$$f : \{0, 1\}^m \to \{0, 1\}$$

and define it as

$$DD_j(f) = \max_{\overline{d}} \frac{1}{2} \left| 2^{m-1} - \sum_{\overline{x}} (f(\overline{x}) \oplus f(\overline{x} \oplus \overline{d})) \right|$$

where both $\overline{d}$ and $\overline{x}$ are binary vectors of length m, $\overline{d}$ ranges through all values with Hamming weight $1 \le \text{wt}(\overline{d}) \le j$ and $\overline{x}$ ranges through all possible values.

It is shown in [4] that cryptographic properties such as Strict Avalanche Criterion (SAC) and Bit Independence Criterion (BIC), higher-order versions of these (HOSAC and HOBIC), maximum order versions of these (MOSAC and MOBIC), and distances from these (DSAC, DBIC, DHOSAC, DHOBIC, DMOSAC, and DMOBIC) can all be defined in terms of dynamic distance.

In an s-box, all bits are equal. There is no most- or least-significant bit in either the input or the output. Thus, permuting the bits of x in the formula above does not change the value of the summation term for a given d, provided we apply the same permutation to d. Hence it does not change the maximum (the value of the dynamic distance).

Therefore, column properties defined in terms of dynamic distance (DSAC, DHOSAC, and DMOSAC) remain unchanged. In particular, if the columns are bent functions (i.e., DMOSAC = 0) then permuting inputs preserves bentness.

Furthermore, the s-box properties DBIC, DHOBIC, and DMOBIC also remain unchanged because these are defined in terms of dynamic distance of a Boolean function f comprised of the XOR of a subset of s-box columns. (Note that permuting the inputs of each of the column functions with a fixed permutation P is identical to permuting the inputs of the combined function f using P.) By a similar line of reasoning, s-box nonlinearity is also unaffected by a permutation of its input bits.

Permuting inputs, therefore, does nothing to degrade cryptographic security in the s-boxes. However, such an operation, by rearranging the order of the s-box rows in a key-dependent way, can make it significantly more difficult to construct linear or differential characteristics (because specific outputs corresponding to specific inputs cannot be predicted in advance). Hence, this operation can increase the security of the cipher by raising the computational complexity of mounting these attacks.

3.4   Adding Affine Functions to S-Box Inputs

Adding selected affine functions to s-box inputs is another effective way of producing a subset of row permutations that does not reduce the cryptographic security of the s-box.

In the extreme case in which the affine functions are constant values, the addition simply complements some of the s-box input bits. Inverting some input bits is equivalent to XORing a constant binary vector into the input, making the summation in the dynamic distance

$$\sum_{\overline{x}} \left(f(\overline{x} \oplus \overline{c}) \oplus f(\overline{x} \oplus \overline{c} \oplus \overline{d}) \right)$$

Clearly $(\overline{x} \oplus \overline{c})$ goes through the same set of values that $\overline{x}$ goes through, so this does not change the sum and, consequently, does not change the dynamic distance. Therefore, column properties and s-box properties are unchanged.

In the more general case in which the affine functions are not constant values, the addition conditionally complements some of the s-box inputs (depending upon the particular values of some subset of input variables). Consider the following restriction. Choose any k input variables and leave these unchanged. For the remaining m-k input variables, augment each with the same randomly-chosen, but fixed, affine function of the chosen k input variables. For example, in a $4 \times n$ s-box, we may choose input variables x_1 and x_2 to be unchanged and augment x_3 and x_4 with the affine function $g(x_1, x_2) = x2 \oplus 1$ so that the Boolean function $f_i(x_1, x_2, x_3, x_4)$ defining each s-box column i becomes

$$f_i(x_1, x_2, x_3 \oplus g(x_1, x_2), x_4 \oplus g(x_1, x_2)) = f_i\left(x_1, x_2, (x_3 \oplus x_2 \oplus 1), (x_4 \oplus x_2 \oplus 1)\right).$$

With the operation restricted in this way it is not difficult to see that as the chosen k variables go through their values, at each stage the remaining m-k variables go through all their values (either all simultaneously complemented, or all simultaneously not complemented, depending upon the binary value of the affine function). Thus, rewriting the summation in the dynamic distance equation as

$$\sum_{\overline{x}} \left(f(\overline{x}') \oplus f(\overline{x}' \oplus \overline{d}) \right)$$

where $\overline{x}'$ is in accordance with the restriction as specified, we see that $\overline{x}'$ goes through the full set of values that $\overline{x}$ goes through, so the sum is unchanged and the resulting dynamic distance is unchanged.

Adding affine functions (restricted as specified above[1]) to s-box inputs, therefore, does not degrade cryptographic security in the s-boxes. Like permuting inputs, this operation, by rearranging the order of the s-box rows in a key-dependent way, can make it significantly more difficult to construct linear or

---

[1] Note that other restrictions on the type and number of affine functions that may be added to preserve s-box properties may also exist. This area is for further research.

differential characteristics. The security of the cipher may therefore be increased by raising the computational complexity of mounting these attacks.

3.5   Other Possible Operations

Other key-dependent operations that preserve s-box properties are also possible. For example, it is theoretically possible to construct strong s-boxes with more than 32 columns and select columns for actual use at set-up time, but this would likely be prohibitively expensive in practice since Mister and Adams [4] report that s-box generation time doubles for each additional column.

Another possibility is to order the s-boxes in a key-dependent way. This is not particularly useful with only four s-boxes since only 4! orders are possible, adding less than five bits of entropy to the key space. However, with the eight s-boxes in CAST-128, this operation becomes somewhat more attractive. A CAST-143 might be created in a very straightforward way: $\log_2(8!) = 15$ bits of key puts the eight s-boxes into some key-dependent order (cheaply by adjusting pointers), and then key expansion and encryption proceeds exactly as in CAST-128 except with the s-boxes in the new order. The overhead (set-up time) is quite low and the new cipher uses 15 extra bits of unexpanded key.

3.6   Limitations in Key-Dependent Operations

Ciphers With XOR-Only Round Functions   A cipher which combines s-box outputs with XOR, such as the CAST example in Applied Cryptography [2, page 334]), does not work well with some types of s-box manipulation. For example, permuting the order of the four round function s-boxes is of no benefit in such a cipher, since XOR is commutative.

XORing different constants into the four s-boxes in such a cipher has exactly the same effect as XORing a single constant into each round function output, or into any one s-box.

Furthermore, if the cipher's round function combines its input and the round key with XOR, then XORing a constant into the output of one round is equivalent to XORing that constant into the key of the next round. If the round keys are already effectively random, unrelated, and unknown to the attacker (as they should be), then XORing them with a constant does not improve them.

In terms of the difficulty of an attack, then, the net effect of XORing four constants into the s-boxes is equivalent to XORing a single constant into the output of the last round, for a cipher which uses XOR both to combine s-box outputs and to combine round input with the round key.

Ciphers With Mixed Operations Combining S-Box Outputs   A cipher which uses operations other than XOR to combine s-box outputs, such as Blowfish or CAST-128, will give different round outputs if the order of the s-boxes is changed or if a constant is XORed into each row. This makes these operations more attractive in such ciphers.

Even in such ciphers, however, the precise cryptographic strength of XORing a constant into the rows is unclear. Certainly it is an inexpensive way to mix many key bits (128 if the cipher uses four $m \times 32$ s-boxes) into the encryption, but it is not clear exactly how much this increases the effective key length.

Ciphers With Mixed Operations Combining Key and Input In Blowfish and in some CAST ciphers, the round input is XORed with the round key at the start of a round, then split into four bytes which become inputs to the four s-boxes. XORing an 8-bit constant into each s-box input is equivalent to XORing a 32-bit constant (the concatenation of the 8-bit constants) into each of the round keys.

Suppose an attack exists that discovers the round keys when the s-boxes are known. Then the same attack works against the same cipher with s-boxes that are known but have had their rows permuted in this way. The attack discovers a different set of round keys equivalent to the real ones XORed with a 32-bit constant, but it still breaks the cipher, and with no extra work for the attacker.

However, for ciphers that use other operations to combine the round input and the round key (CAST-128, for example, which uses addition and subtraction modulo 2^{32} for input masking in some of its rounds), such an operation seems to add value.

Options For both inputs and outputs, the addition of affine functions appears stronger than just XORing with a constant, and performing permutations appears to be stronger again (but at much higher computational cost). In a practical cipher, however, there appears to be no disadvantage to using XOR (for both input and output if mixed operations are used everywhere in the round function) because it is inexpensive and offers some protection against the construction of iterated characteristics.

4 Practical Considerations

4.1 Stage One

Since XORing a constant into the s-box rows is the cheapest way to bring many extra key bits into play; we should do that if we're going to use this approach at all. The cipher's round function should use operations other than XOR to mix s-box outputs so that this will be effective.

If we are iterating through the s-box rows for that, it makes sense to permute the columns in the same loop. We suggest simply rotating each row under control of 5 bits of key. A CAST-128 implementation will have code for this, since the same rotation is used in the round function, and rotation is reasonably efficient.

At this point, we have used 37 bits per s-box, 148 bits in all. In many applications, this will be quite sufficient.

Costs of this are minimal: 1024 XOR and rotation operations. This is much less than CAST-128's round key generation overhead, let alone Blowfish's work to generate s-boxes and round keys.

4.2 Stage Two

To go beyond that, you can add affine functions to s-box columns or you can permute s-box rows in a manner equivalent to permuting the input bits.

The choice would depend upon the relative strength of these two methods, along with the relation between their overheads and the resources available in a particular application. In our suggested implementation, using affine functions requires more storage while permuting the rows involves more computation. Neither operation looks prohibitively expensive in general, but either might be problematic in some environments.

For purposes of this paper, we will treat permuting the inputs as the next thing to add, and then go on to look at adding affine functions.

To permute the rows in a manner equivalent to permuting the input bits we add the following mechanism. We use a 256-row array, each row composed of an 8-bit index and a 32-bit output. We can rearrange rows as follows:

- put the (256*32)-bit s-box in the output part of the array;
- fill the index part with the 8-bit values in order from hex 00 to FF;
- operate in some way on the index parts (without affecting the 32-bit s-box rows) so as to give each row a new index;
- sort the 256 rows so that the index parts are again in order (00 to FF), moving the s-box rows along with the indexes so they are in a new order;
- discard the index portion.

This results in a cryptographically identical s-box with rows in the new order. The operations permitted in the third step for changing the 8-bit values are just those which are equivalent to permuting and inverting the s-box inputs. We can XOR a constant into all index rows or we can permute index columns. Neither operation alters the XOR difference between index rows, so cryptographic properties are conserved as shown earlier.

XORing a constant into each index row is of little benefit. This is also true of rotation, which uses only 3 bits per s-box (hardly enough to justify the overhead of sorting the s-boxes).

To operate usefully on the inputs, then, we should do a full permutation on the index columns. In code, this would need a function to permute 8-bit values under control of a 15-bit key. It would use 15 key bits per s-box.

At this point, we are using 52 bits per s-box, 208 bits in all, and are permuting both rows and columns or both inputs and outputs. Again, this would be quite sufficient for many applications.

4.3 Stage Three

We can, however, go further by adding affine functions to the columns.

There are exactly 512 affine Boolean functions of 8 variables. In theory, it would be possible to add a key-selected affine function to each s-box column, using 9 bits of key per column, or 1152 bits for a set of four 8×32 s-boxes, but this seems unacceptably expensive in practice.

Since the inverse of an affine function is also affine, we need only store half of the 512 possible functions to have them all available. Consider a (256*256)-bit Boolean array with affine functions in all columns and no two columns either identical or inverses. From this, create four 256×32 arrays. This can be done using $\log_2 \binom{256}{128}$ key bits, but implementing this would also be expensive. As a more practical alternative, using $\log_2 \binom{16}{8} \approx 13$ bits of key to select eight of sixteen "chunks of 16 columns" from the original array may be a reasonable compromise.

4.4 Putting It All Together

Given four 256×32 arrays of affine functions, we add a few operations inside the loop that runs through each s-box. The inner loop ends up as follows (in pseudo-C with "<<<" for rotation and "^" for XOR):

```
unsigned *s, *a ;   // pointers to s-box & affine array
unsigned char *p;   // pointer into index array
// initialize pointers here...
for( i = 0 ; i < 256 ; i++, s++, a++, p++ )
   {
   *s = (*s <<< k1) ^ (*a <<< k2) ^ k3; // 5+5+32 key bits
   *p = permute8(*p, k4) ;              // 15 key bits
   }
qsort(.....) ;
```

This uses 57 key bits per s-box inside the loop. With the 13 used outside the loop setting up the A-boxes, and another 15 used in re-arranging the original 8 s-boxes, we have 256 key bits in total exclusively used for s-box manipulations.

5 Further Work

Further work in this area can be done both on the theoretical side and on the practical side. For example, a formal characterization of the set of affine functions that can be added to an s-box without reducing its cryptographic strength (beyond the subset specified in this paper) would be of interest. As well, since factorials do not correspond to powers of 2, more precise practical specifications need to be given to convert expressions such as "$\log_2(8!)$" to bit lengths (i.e., it needs to be stated exactly which particular permutation corresponds to each value of a 15-bit key segment).

6 Conclusions

This paper has proposed the concept of using key-dependent s-box manipulations to strengthen specific block ciphers against attacks which depend upon knowledge of the s-box contents (such as linear and differential cryptanalysis and their variations). The manipulations described include a permutation of the output bits, a permutation of the input bits, the addition of affine functions to the s-box columns, and the addition of a restricted set of affine functions to the s-box inputs. It has been shown (using the concept of dynamic distance [4]) that such manipulations do not degrade the cryptographic properties of carefully-constructed s-boxes, and therefore do not degrade the cryptographic strength of the corresponding ciphers with respect to existing analysis. On the contrary, it is possible that cryptographic strength may be substantially increased by such manipulations because the most effective cryptanalytic attacks to date would appear to require a significant exhaustive search phase in addition to their current complexity in order to be mounted against ciphers with such "hidden" s-box contents.

Some implementation considerations for this proposal were also discussed, and options were presented with respect to the level of complexity that might be employed in various environments.

References

1. C. Adams, "Constructing Symmetric Ciphers Using the CAST Design Procedure", Designs, Codes and Cryptography, vol.12, no.3, Nov. 1997, pp.71-104.
2. E. Biham and A. Biryukov, "How to Strengthen DES Using Existing Hardware" Advances in Cryptology - ASIACRYPT 94 Proceedings.
3. W. Diffie and M. Hellman, "New Directions in Cryptography", IEEE Transactions on Information Theory, vol.22, 1976, pp.644-654.
4. S. Mister and C. Adams, "Practical s-box Design", Workshop Record of the Workshop on Selected Areas in Cryptography (SAC '96), Queen's University, Kingston, Ontario, Aug. 1996, pp.61-76.
5. R. Rivest, A. Shamir, and L. Adelman, "A Method for Obtaining Digital Signatures and Public-Key Cryptosystems", Communications of the ACM, v. 21, n. 8, Feb. 1978, page 120.
6. B. Schneier, Applied Cryptography, Second Edition, John Wiley & Sons, 1996.

On the Twofish Key Schedule

Bruce Schneier[1], John Kelsey[1], Doug Whiting[2], David Wagner[3],
Chris Hall[1], and Niels Ferguson[1]

[1] Counterpane Systems, 101 E Minnehaha Parkway
Minneapolis, MN 55419, USA
{schneier,kelsey,hall,niels}@counterpane.com
[2] Hi/fn, Inc., 5973 Avenida Encinas Suite 110, Carlsbad, CA 92008, USA
dwhiting@hifn.com
[3] University of California Berkeley, Soda Hall, Berkeley, CA 94720, USA
daw@cs.berkeley.edu

Abstract. Twofish is a new block cipher with a 128 bit block, and a
key length of 128, 192, or 256 bits, which has been submitted as an AES
candidate. In this paper, we briefly review the structure of Twofish, and
then discuss the key schedule of Twofish, and its resistance to attack. We
close with some open questions on the security of Twofish's key schedule.

1 Introduction

NIST announced the Advanced Encryption Standard (AES) program in 1997
[NIST97a]. NIST solicited comments from the public on the proposed standard,
and eventually issued a call for algorithms to satisfy the standard [NIST97b].
The intention is for NIST to make all submissions public and eventually, through
a process of public review and comment, choose a new encryption standard to
replace DES.

Twofish is our submission to the AES selection process. It meets all the
required NIST criteria—128-bit block; 128-, 192-, and 256-bit key; efficient on
various platforms; etc.—and some strenuous design requirements, performance
as well as cryptographic, of our own.

Twofish was designed to meet NIST's design criteria for AES [NIST97b].
Specifically, they are:

- A 128-bit symmetric block cipher.
- Key lengths of 128 bits, 192 bits, and 256 bits.
- No weak keys.
- Efficiency, both on the Intel Pentium Pro and other software and hardware
  platforms.
- Flexible design: e.g., accept additional key lengths; be implementable on
  a wide variety of platforms and applications; and be suitable for a stream
  cipher, hash function, and MAC.
- Simple design, both to facilitate ease of analysis and ease of implementation.

S. Tavares and H. Meijer (Eds.): SAC'98, LNCS 1556, pp. 27–42, 1999.

A central feature of Twofish's security and flexibility is its key schedule. In this paper, we will briefly review the design of Twofish, and then discuss the security features of the key schedule. The remainder of the paper is as follows: First, we discuss the specific design of Twofish. Next, we analyze the Twofish key schedule in some detail. Finally, we point out some open questions with respect to the key schedule. Note that for space reasons, this paper does not include a complete discussion of the Twofish design. Instead, we refer the reader to http://www.counterpane.com.

2   Twofish

Twofish uses a 16-round Feistel-like structure with additional whitening of the input and output. The only non-Feistel elements are the 1-bit rotates. The rotations can be moved into the F function to create a pure Feistel structure, but this requires an additional rotation of the words just before the output whitening step.

The plaintext is split into four 32-bit words. In the input whitening step, these are XORed with four key words. This is followed by sixteen rounds. In each round, the two words on the left are used as input to the g functions. (One of them is rotated by 8 bits first.) The g function consists of four byte-wide key-dependent S-boxes, followed by a linear mixing step based on an MDS matrix. The results of the two g functions are combined using a Pseudo-Hadamard Transform (PHT), and two keywords are added. These two results are then XORed into the words on the right (one of which is rotated left by 1 bit first, the other is rotated right afterwards). The left and right halves are then swapped for the next round. After all the rounds, the swap of the last round is reversed, and the four words are XORed with four more key words to produce the ciphertext.

More formally, the 16 bytes of plaintext $p_0, \ldots, p_{15}$ are first split into 4 words $P_0, \ldots, P_3$ of 32 bits each using the little-endian convention.

$$P_i = \sum_{j=0}^{3} p_{(4i+j)} \cdot 2^{8j} \qquad i = 0, \ldots, 3$$

In the input whitening step, these words are XORed with 4 words of the expanded key.

$$R_{0,i} = P_i \oplus K_i \qquad i = 0, \ldots, 3$$

In each of the 16 rounds, the first two words are used as input to the function F, which also takes the round number as input. The third word is XORed with the first output of F and then rotated right by one bit. The fourth word is rotated left by one bit and then XORed with the second output word of F. Finally, the two halves are exchanged. Thus,

$$(F_{r,0}, F_{r,1}) = F(R_{r,0}, R_{r,1}, r)$$
$$R_{r+1,0} = \mathrm{ROR}(R_{r,2} \oplus F_{r,0}, 1)$$

$$R_{r+1,1} = \mathrm{ROL}(R_{r,3}, 1) \oplus F_{r,1}$$
$$R_{r+1,2} = R_{r,0}$$
$$R_{r+1,3} = R_{r,1}$$

for $r = 0, \ldots, 15$ and where ROR and ROL are functions that rotate their first argument (a 32-bit word) left or right by the number of bits indicated by their second argument.

The output whitening step undoes the 'swap' of the last round, and XORs the data words with 4 words of the expanded key.

$$C_i = R_{16,(i+2) \bmod 4} \oplus K_{i+4} \qquad i = 0, \ldots, 3$$

The four words of ciphertext are then written as 16 bytes $c_0, \ldots, c_{15}$ using the same little-endian conversion used for the plaintext.

$$c_i = \left\lfloor \frac{C_{\lfloor i/4 \rfloor}}{2^{8(i \bmod 4)}} \right\rfloor \bmod 2^8 \qquad i = 0, \ldots, 15$$

2.1 The Function F

The function F is a key-dependent permutation on 64-bit values. It takes three arguments, two input words R_0 and R_1, and the round number r used to select the appropriate subkeys. R_0 is passed through the g function, which yields T_0. R_1 is rotated left by 8 bits and then passed through the g function to yield T_1. The results T_0 and T_1 are then combined in a PHT and two words of the expanded key are added.

$$T_0 = g(R_0)$$
$$T_1 = g(\mathrm{ROL}(R_1, 8))$$
$$F_0 = (T_0 + T_1 + K_{2r+8}) \bmod 2^{32}$$
$$F_1 = (T_0 + 2T_1 + K_{2r+9}) \bmod 2^{32}$$

where (F_0, F_1) is the result of F. We also define the function F' for use in our analysis. F' is identical to the F function, except that it does not add any key blocks to the output. (The PHT is still performed.)

2.2 The Function g

The function g forms the heart of Twofish. The input word X is split into four bytes. Each byte is run through its own key-dependent S-box. Each S-box is bijective, takes 8 bits of input, and produces 8 bits of output. The four results are interpreted as a vector of length 4 over $GF(2^8)$, and multiplied by the 4×4 MDS matrix (using the field $GF(2^8)$ for the computations). The resulting vector is interpreted as a 32-bit word which is the result of g.

$$x_i = \lfloor X/2^{8i} \rfloor \bmod 2^8 \qquad i = 0, \ldots, 3$$

$$y_i = s_i[x_i] \qquad i = 0, \ldots, 3$$

$$\begin{pmatrix} z_0 \\ z_1 \\ z_2 \\ z_3 \end{pmatrix} = \begin{pmatrix} \cdot & \cdots & \cdot \\ & \text{MDS} & \\ \cdot & \cdots & \cdot \end{pmatrix} \cdot \begin{pmatrix} y_0 \\ y_1 \\ y_2 \\ y_3 \end{pmatrix}$$

$$Z = \sum_{i=0}^{3} z_i \cdot 2^{8i}$$

where s_i are the key-dependent S-boxes and Z is the result of g. For this to be well-defined, we need to specify the correspondence between byte values and the field elements of $GF(2^8)$. We represent $GF(2^8)$ as $GF(2)[x]/v(x)$ where $v(x) = x^8 + x^6 + x^5 + x^3 + 1$ is a primitive polynomial of degree 8 over $GF(2)$. The field element $a = \sum_{i=0}^{7} a_i x^i$ with $a_i \in GF(2)$ is identified with the byte value $\sum_{i=0}^{7} a_i 2^i$. This is in some sense the "natural" mapping; addition in $GF(2^8)$ corresponds to a XOR of the bytes.

2.3   The Key Schedule

The key schedule has to provide 40 words of expanded key $K_0, \ldots, K_{39}$, and the 4 key-dependent S-boxes used in the g function. Twofish is defined for keys of length $N = 128$, $N = 192$, and $N = 256$. Keys of any length shorter than 256 bits can be used by padding them with zeroes until the next larger defined key length.

We define $k = N/64$. The key M consists of $8k$ bytes $m_0, \ldots, m_{8k-1}$. The bytes are first converted into $2k$ words of 32 bits each

$$M_i = \sum_{j=0}^{3} m_{(4i+j)} \cdot 2^{8j} \qquad i = 0, \ldots, 2k - 1$$

and then into two word vectors of length k.

$$M_e = (M_0, M_2, \ldots, M_{2k-2})$$
$$M_o = (M_1, M_3, \ldots, M_{2k-1})$$

A third word vector of length k is also derived from the key. This is done by taking the key bytes in groups of 8, interpreting them as a vector over $GF(2^8)$, and multiplying them by a 4×8 matrix derived from a Reed-Solomon code. Each result of 4 bytes is then interpreted as a 32-bit word. These words make up the third vector.

$$\begin{pmatrix} s_{i,0} \\ s_{i,1} \\ s_{i,2} \\ s_{i,3} \end{pmatrix} = \begin{pmatrix} \cdot & \cdots & \cdot \\ & \text{RS} & \\ \cdot & \cdots & \cdot \end{pmatrix} \cdot \begin{pmatrix} m_{8i} \\ m_{8i+1} \\ m_{8i+2} \\ m_{8i+3} \\ m_{8i+4} \\ m_{8i+5} \\ m_{8i+6} \\ m_{8i+7} \end{pmatrix}$$

$$S_i = \sum_{j=0}^{3} s_{i,j} \cdot 2^{8j}$$

for $i = 0, \ldots, k-1$, and

$$S = (S_{k-1}, S_{k-2}, \ldots, S_0)$$

Note that S lists the words in "reverse" order. For the RS matrix multiply, $GF(2^8)$ is represented by $GF(2)[x]/w(x)$, where $w(x) = x^8 + x^6 + x^3 + x^2 + 1$ is another primitive polynomial of degree 8 over $GF(2)$. The mapping between byte values and elements of $GF(2^8)$ uses the same definition as used for the MDS matrix multiply.

Additional Key Lengths Twofish can accept keys of any byte length up to 256 bits. For key sizes that are not defined above, the key is padded at the end with zero bytes to the next larger length that is defined. For example, an 80-bit key $m_0, \ldots, m_9$ would be extended by setting $m_i = 0$ for $i = 10, \ldots, 15$ and treating it as a 128-bit key.

The Function h The function h takes two inputs—a 32-bit word X and a list $L = (L_0, \ldots, L_{k-1})$ of 32-bit words of length k—and produces one word of output. This function works in k stages. In each stage, the four bytes are each passed through a fixed S-box, and XORed with a byte derived from the list. Finally, the bytes are once again passed through a fixed S-box, and the four bytes are multiplied by the MDS matrix just as in g. More formally: we split the words into bytes.

$$l_{i,j} = \lfloor L_i/2^{8j} \rfloor \bmod 2^8$$
$$x_j = \lfloor X/2^{8j} \rfloor \bmod 2^8$$

for $i = 0, \ldots, k-1$ and $j = 0, \ldots, 3$. Then the sequence of substitutions and XORs is applied.

$$y_{k,j} = x_j \qquad j = 0, \ldots, 3$$

If $k = 4$ we have

$$y_{3,0} = q_1[y_{4,0}] \oplus l_{3,0}$$
$$y_{3,1} = q_0[y_{4,1}] \oplus l_{3,1}$$
$$y_{3,2} = q_0[y_{4,2}] \oplus l_{3,2}$$
$$y_{3,3} = q_1[y_{4,3}] \oplus l_{3,3}$$

If $k \geq 3$ we have

$$y_{2,0} = q_1[y_{3,0}] \oplus l_{2,0}$$
$$y_{2,1} = q_1[y_{3,1}] \oplus l_{2,1}$$
$$y_{2,2} = q_0[y_{3,2}] \oplus l_{2,2}$$
$$y_{2,3} = q_0[y_{3,3}] \oplus l_{2,3}$$

In all cases we have

$$y_0 = q_1[q_0[q_0[y_{2,0}] \oplus l_{1,0}] \oplus l_{0,0}]$$
$$y_1 = q_0[q_0[q_1[y_{2,1}] \oplus l_{1,1}] \oplus l_{0,1}]$$
$$y_2 = q_1[q_1[q_0[y_{2,2}] \oplus l_{1,2}] \oplus l_{0,2}]$$
$$y_3 = q_0[q_1[q_1[y_{2,3}] \oplus l_{1,3}] \oplus l_{0,3}]$$

Here, q_0 and q_1 are fixed permutations on 8-bit values that we will discuss shortly. The resulting vector of y_i's is multiplied by the MDS matrix, just as in the g function.

$$\begin{pmatrix} z_0 \\ z_1 \\ z_2 \\ z_3 \end{pmatrix} = \begin{pmatrix} \cdot & \cdots & \cdot \\ \vdots & \text{MDS} & \vdots \\ \cdot & \cdots & \cdot \end{pmatrix} \cdot \begin{pmatrix} y_0 \\ y_1 \\ y_2 \\ y_3 \end{pmatrix}$$

$$Z = \sum_{i=0}^{3} z_i \cdot 2^{8i}$$

where Z is the result of h.

The Key-dependent S-boxes We can now define the S-boxes in the function g by

$$g(X) = h(X, S)$$

That is, for $i = 0, \ldots, 3$, the key-dependent S-box s_i is formed by the mapping from x_i to y_i in the h function, where the list L is equal to the vector S derived from the key.

The Expanded Key Words K_j The words of the expanded key are defined using the h function.

$$\rho = 2^{24} + 2^{16} + 2^8 + 2^0$$
$$A_i = h(2i\rho, M_e)$$
$$B_i = \text{ROL}(h((2i+1)\rho, M_o), 8)$$
$$K_{2i} = (A_i + B_i) \bmod 2^{32}$$
$$K_{2i+1} = \text{ROL}((A_i + 2B_i) \bmod 2^{32}, 9)$$

The constant ρ is used here to duplicate bytes; it has the property that for $i = 0, \ldots, 255$, the word $i\rho$ consists of four equal bytes, each with the value i. The function h is applied to words of this type. For A_i the byte values are $2i$, and the second argument of h is M_e. B_i is computed similarly using $2i+1$ as the byte value and M_o as the second argument, with an extra rotate over 8 bits. The values A_i and B_i are combined in a PHT. One of the results is further rotated by 9 bits. The two results form two words of the expanded key.

The Permutations q_0 and q_1 The permutations q_0 and q_1 are fixed permutations on 8-bit values. They are constructed from four different 4-bit permutations each. We have investigated the resulting 8-bit permutations, q_0 and q_1, extensively, and believe them to be at least no weaker than randomly selected 8-bit permutations.

3 Analysis of The Key Schedule

The key schedule has been designed to provide resistance to attack, while also providing a great deal of flexibility in implementation. For example, after S has been computed from M_e, M_o, all remaining key scheduling can be done "on the fly" during encryption. This allows for very low-memory implementations, and for implementations with excellent key agility. In implementations with more memory, all the subkeys can be precomputed for improved performance. In implementations with still more memory, such as on modern high-end processors with a reasonable RAM cache size, the effects of the key-dependent S-boxes and the MDS matrix multiply can be precomputed, reducing the work per g computation to four table lookups and three XORs.

Note that S is only half the size of the key. This was done so that precomputation of the S-boxes and MDS matrix multiply would be sufficiently fast, and so that low-memory implementations would not have to take too large a performance hit. This means that the g function is slightly different for longer keys than for shorter keys.

3.1 Byte Sequences

The subkeys in Twofish are generated by using the h function, which can be seen as four key-dependent S-boxes followed by an MDS matrix. The input to the S-boxes is basically a counter. In this section we analyze the sequences of outputs that this construction can generate.

All key material is used to define key-dependent S-boxes in h, which are then used to derive subkeys. Each S-box gets a sequence of inputs, $(0, 2, 4, \ldots, 38)$ or $(1, 3, 5, \ldots, 39)$. The S-box generates a corresponding sequence of outputs. The corresponding outputs from the four S-boxes are combined using the MDS matrix multiply to produce the sequence of A_i and B_i words, and those words are processed with the PHT (with a couple of rotations thrown in) to produce a pair of subkey words. Analyzing these byte sequences thus gives us important insights about the whole key schedule.

We can model each byte sequence generated by a key-dependent S-box as a randomly selected non-repeating byte sequence of length 20. This allows us to make many useful predictions about the likelihood of finding keys or pairs of keys with various interesting properties. Because we will be analyzing the key schedule using this assumption in the remainder of this section, we should discuss how reasonable it is to treat this byte sequence as randomly generated.

We have not found any statistical deviations between our key-dependent S-boxes and the random model in any of our extensive statistical tests.

We are looking at sequences of 20 bytes that are all distinct. There are $256!/236!$ of those sequences, which is close to 2^{159}.

3.2   Equivalent S-box Keys

We have verified that there are no equivalent S-box keys that generate the same sequence of 20 bytes. In the random model, the chance of this happening for the $N = 256$ case is about $2^{63} \cdot 2^{-159} = 2^{-96}$. This is the chance of such equivalent S-boxes existing at all. In fact, we recently completed an exhaustive search demonstrating that no pair of key inputs to an S-box produces identical S-box entries.

3.3   Byte Difference Sequences

Let us consider the more general problem of how to get a given 20-byte difference sequence between a pair of S-boxes. Suppose we have two S-boxes, each defined using 32 bits of key material, which are not equal, but which must be chosen to give us a given difference sequence in the XOR of their byte sequences. We can estimate the probability of a pair of 4-byte inputs existing with the desired XOR difference sequence as $2^{63} \cdot 2^{-159} = 2^{-96}$. Note that this is the probability that such a pair of inputs exists, not the probability that a random pair of keys will have this property.

3.4   The A and B Sequences

From the properties of the byte sequences, we can discuss the properties of the A and B sequences generated by each key M.

$$A_i = \mathrm{MDS}(s_0(i, M), s_1(i, M), s_2(i, M), s_3(i, M))$$

Since the MDS matrix multiply is invertible, and since i is different for each round's subkey words generated, we can see that no A or B value can repeat itself.

Similarly, we can see from the construction of h that each key byte affects exactly one S-box used to generate A or B. Changing a single key byte always alters every one of the 20 bytes of output from that S-box, and so always alters every word in the 20-word A or B sequence to which it contributes.

Consider a single byte of output from one of the S-boxes. If we cycle any one of the key bytes that contributes to that S-box through all 256 possible values, the output of the S-box will also cycle through all 256 possible values. If we take four key bytes that contribute to four different S-boxes, and we cycle those four bytes through all possible values, then the result of h will also cycle through all possible values. This proves that A and B are uniformly distributed for all key lengths, assuming the key M is uniformly distributed.

3.5   Difference Sequences in A and B

Let us also consider difference sequences. If we have a specific difference sequence we want to see in A, we are faced with an interesting problem: since the MDS matrix multiply is XOR-linear, each desired output XOR from the matrix multiply allows only one possible input XOR. This means that:

1. A zero output XOR difference in A can occur *only* with a zero output XOR difference in all four of the byte sequences used to build A.
2. Only 1020 possible output differences (out of the 2^{32}) in A_i can occur with a single "active" (altered) S-box. Most differences require all four S-boxes used to form A_i to be active.
3. Each desired output XOR in A requires a specific output XOR in each of the four byte sequences used to form A. This means that getting any desired difference sequence into all 20 A_i values requires getting a desired XOR sequence into all four 20-byte sequences. (Note that if the desired output XOR in A_i is an appropriate value, up to three of the four byte sequences can be identical without much trouble, simply by leaving their key material unchanged.) As mentioned above, this is very unlikely to be possible for a randomly chosen difference pattern in the A sequence. (There are of course difference sequences of A_i's that can occur.)

The above analysis is of course also valid for the B sequence.

3.6   The Sequence (K_{2i}, K_{2i+1})

As A_i and B_i are uniformly distributed (over all keys), so are all the K_i. As all pairs (A_i, B_i) are distinct, all the pairs (K_{2i}, K_{2i+1}) are distinct, although it might happen that $K_i = K_j$ for any pair of i and j.

3.7   Difference Sequences in the Subkeys

Each difference sequence in A and B translate into a difference sequences in (K_{2i}, K_{2i+1}). However, while it is natural to consider A and B difference sequences in terms of XOR differences, subkeys can reasonably be considered either as XOR differences or as differences modulo 2^{32}. Thus, we may discuss difference sequences:

$$D[i, M, M^*] = K_{i,M} - K_{i,M^*}$$
$$X[i, M, M^*] = K_{i,M} \oplus K_{i,M^*}$$

where the difference is computed between the key value M and M^*.

3.8   XOR Differences in the Subkeys

Each round, the subkeys are added to the results of the PHT of two g functions, and the results of those additions are XORed into half of the cipher block. An

XOR difference in the subkeys has a fairly high probability of passing through the addition operation and ending up in the cipher block. (The probability of this is determined by the Hamming weight of the XOR difference, not counting the highest-order bit.) However, to get into the subkeys, a XOR difference must first pass through the first addition.

Consider

$$x + y = z$$
$$(x \oplus \delta_0) + y = z \oplus \delta_1$$

Let k be the number of bits set in δ_0, not counting the highest-order bit. Then, the highest probability value for δ_1 is δ_0, and the probability that this will hold is 2^{-k}. This is true because addition and XOR are very closely related operations. The only difference between the two is the carry between bit positions. If flipping a given bit changes the carry into the next bit position, this alters the output XOR difference. This happens with probability $1/2$ per bit. The situation is more complex for multiple adjacent bits, but the general rule still holds: for every bit in the XOR difference not in the high-order bit position, the probability that the difference will pass through correctly is cut in half.

For the subkey generation, consider an XOR difference, δ_0, in A. This affects two subkey words:

$$K_{2i} = A_i + B_i$$
$$K_{2i+1} = \text{ROL}(A_i + 2B_i, 9)$$

where the additions are modulo 2^{32}. If we assume these XOR differences propagate independently in the two subkeys (which appears to be the case), we see that this leads to an XOR difference of δ_0 in the even subkey word with probability 2^{-k}, and the XOR difference $ROL(\delta_0, 9)$ in the odd subkey with the same probability. The most probable XOR difference in the round's subkey block thus occurs with probabiity 2^{-2k}. A desired XOR difference sequence for all 20 pairs of subkey words is thus quite difficult to get to work when $k \geq 3$, assuming the desired XOR difference sequence can be created in the A sequence at all.

When the XOR difference is in B, the result is slightly more complicated; the most probable XOR difference in a round's pair of subkey words may be either $2^{-(2k-1)}$ or 2^{-2k}, depending on whether or not the XOR difference in B covers the next-to-highest-order bit.

An XOR difference in A or B is easy to analyze in terms of additive differences modulo 2^{32}: an XOR difference with k active bits has 2^k equally likely additive differences. Note that if we have a additive difference in A, we get it in both subkey words, just rotated left nine bits in the odd subkey word. Thus, k-bit XOR differences lead to a given additive difference in a pair of subkey words with probability 2^{-k}. (The rotation does not really complicate things much for the attacker, who knows where the changed bits are.)

Note that when additive subkey differences modulo 2^{32} are used in an attack, they survive badly through the XOR with the plaintext block. We estimate that XOR differences are much more likely to be directly useful in mounting an attack.

3.9   Key-dependent Characteristics and Weak Keys

The concept of a key-dependent characteristic seems to have been introduced in [BB93] in their cryptanalysis of Lucifer, and also appears in [DGV94a] in an analysis of IDEA.[1] The idea is that certain iterative properties of the block cipher useful to an attacker become more effective against the cipher for a specific subset of keys.

A differential attack on Twofish may consider XOR-based differences, additive differences, or both. If an attacker sends XOR differences through the PHT and subkey addition steps, his differential characteristic probabilities will be dependent on the subkey values involved. In general, low-weight subkeys will give an attacker some advantage, but this advantage is relatively small. (Zero bits in the subkeys improve the probabilities of cleanly getting XOR-based differential characteristics through the subkey addition.) Since there appears to be no special way to choose the key to make the subkey sequence especially low weight, we do not believe this kind of key-dependent differential characteristic will have any relevance in attacking Twofish.

A much more interesting issue in terms of key-dependent characteristics is whether the key-dependent S-boxes are ever generated with especially high probability differential or high bias linear characteristics. The statistical analysis presented earlier shows that the best linear and differential characteristics over all possible keys are still quite unlikely.

Note that the structure of both differential and linear attacks in Twofish is such that such attacks appear to generally require good characteristics through at least three of the four key-dependent S-boxes (if not all four), so a single high-probability differential or linear characteristic for one S-box will not create a weakness in the cipher as a whole. Our statistical testing has allowed us to estimate that few or no keys result in a single S-box with a differential characteristic of probability higher than $24/256$ for any length key, and with a linear characteristic with bias higher than $108/256$. These probabilities do not allow for practical differential or linear attacks. Further, for an attacker to mount a differential or linear attack, it appears to be necessary to get very high-probability differential or linear characteristics in all four S-boxes at once.

3.10   Related-key Cryptanalysis

Related-key cryptanalysis [Bih94,KSW96,KSW97] uses a cipher's key schedule to break plaintexts encrypted with related keys. In its most advanced form, differential related-key cryptanalysis, both plaintexts and keys with chosen differentials are used to recover the keys. This type of analysis has had considerable success against ciphers with simplistic key schedules—e.g., GOST and 3-Way [DGV94b]—and is a realistic attack in some circumstances. A conventional attack is usually judged in terms of the number of plaintexts or ciphertexts needed for the attack, and the level of access to the cipher needed to get those texts (e.g.,

---

[1] See [Haw98] for further cryptanalysis of IDEA weak keys.

known plaintext, chosen plaintext, adaptive chosen plaintext); in a related-key attack, we must add the requirement for encryptions to occur under two different, but related, keys.

3.11   Resistance to Related-key Slide Attacks

A "slide" attack occurs in an iterated cipher when the encryption of one block for rounds 1 through n is the same as the encryption of another block for rounds $s+1$ to $s+n$. An attacker can look at two encryptions, and can slide the rounds forward in one of them relative to another. S-1 [Anon95] can be broken with a slide attack [Wag95a]. Travois [Yuv97] has identical round functions, and can also be broken with a slide attack. Conventional slide attacks allow one to break the cipher with only known- or chosen-plaintext queries; however, as we shall see next, there is a generalization to related-key attacks as well.

Related-key slide attacks were first discovered by Biham in his attack on a DES variant [Bih94]. To mount a related-key slide attack on Twofish, an attacker must find a pair of keys M, M^* such that the key-dependent S-boxes in g are unchanged, but the subkey sequences slide down one round. This amounts to finding, for each of the eight byte-permutations used for subkey generation, a change in the keys such that:

$$s_i(j, M) = s_i(j + 2s, M^*)$$

for n values of j. In total, this requires $8n$ of these relations to hold.

Let us look in more detail for a fixed key M. Let $m \in \{5, \ldots, 8\}$ be the number of S-boxes used to compute the round keys that are affected by the difference between M and M^*. Observe that $m \geq 5$ due to the restriction that S cannot change and the properties of the RS matrix that at least 5 inputs must change to keep the output constant. There are at most $\binom{8}{m} 2^{32m-128}$ possible choices of M^*. We have a total of nm 8-bit relations that need to be satisfied. The expected number of M^* that satisfy these relations is thus $\binom{8}{m} \cdot 2^{-8nm+32m-128}$. For $n \geq 4$ this is dominated by the case $m = 5$; we will ignore the other cases for now. So for each M we can expect about 2^{38-40n} keys M^* that support a slide attack for $n \geq 4$. This means that any specific key is unlikely to support a slide attack with $n \geq 4$. Over all possible key pairs, we expect $2^{293-40n}$ pairs M, M^* for which a slide of $n \geq 4$ occurs. Thus, it is unlikely that a pair exists at all with $n \geq 8$.

Resistance to Related-key Differential Attacks A related-key differential attack seeks to mount a differential attack on a block cipher through the key, as well as or instead of through the plaintext/ciphertext port. Against Twofish, such an attack must control the subkey difference sequence for at least the rounds in the middle. For the sake of simplifying discussions of the attack, let us consider an attacker who wants to put a chosen subkey difference into the middle twelve rounds' subkeys. That is, he wants to change M to M^*, and control $D[i, M, M*]$ for $i = 12..35$. At the same time, he needs to keep the g function, and thus the key S, from changing. All else being equal, the longer the key, the more freedom an attacker has to mount a related-key differential attack. We thus will assume the use of 256 bit keys for the remainder of this section. Note that a successful related key attack on 128 or 192 bit keys that gets only zero subkey differences in the rounds whose subkey differences it must control translates directly to an equivalent related key attack on 256 bit keys.

Consider the position of the attacker if he attempts a related-key differential attack with different S keys. This must result in different g outputs for all inputs, since we know that there are no pairs of S values that lead to identical S-boxes. Assuming the pair of S values does not lead to linearly-related S-boxes, it will not be possible to compensate for this change in S with changes in the subkeys in single rounds. The added difficulty is approximately that of adding 24 active S-boxes to the existing related-key attack. For this reason, we believe that any useful related-key attack will require a pair of keys that keeps S unchanged.

The Zero Difference Case The simplest related-key attack to analyze is the one that keeps both S and also the middle twelve rounds' subkeys unchanged. It thus seeks to generate identical A and B sequences for twelve rounds, and thus to keep the individual byte sequences used to derive A and B identical.

The RS code used to derive S from M strictly limits the ways an attacker can change M without altering S. The attacker must try to keep the number of active subkey generating S-boxes as low as possible, since each active S-box is another constraint on his attack. The attacker can keep the number of active S-boxes down to five without altering S, and so this is what he should do. With only the key bytes affecting these five subkey generation S-boxes active, he can alter between one and four bytes in all five S-boxes; the nature of the RS matrix is that if he needs to alter four bytes in any one of these S-boxes, he must alter bytes in all five. In practice, in order to maximize his control over the byte sequences generated by these S-boxes, he must alter four bytes in all five active S-boxes.

To get zero subkey differences, the attacker must get zero differences in the byte sequences generated by all five active S-boxes. Consider a single such byte sequence: The attacker tries to find a pair of four-byte key inputs such that they lead to identical byte sequences in the middle twelve rounds, which means the middle twelve bytes. There are 2^{63} pairs of key inputs from which to choose, and about 2^{95} possible byte sequences available. If the byte sequences behave more-or-less like random functions of the key inputs, this implies that it is extremely unlikely that an attacker can find a pair of key inputs that will get identical byte sequences in these middle twelve rounds. We discuss this kind of analysis of byte sequences in section 3.1. From this analysis, we would not expect to see a pair of keys for even one S-box with more than eight successive bytes unchanged, and we would expect even eight successive bytes of unchanged byte sequence to require control of all four key bytes into the S-box. We would expect a specific pair of key bytes to be required to generate these similar byte sequences.

To extend this to five active S-boxes, we expect there to be, at best, a single pair of values for the twenty active key bytes that leave the middle eight subkeys unchanged.

Other Difference Sequences An attacker who has control of the XOR difference sequences in A_i, B_i does not necessarily have great control over the XOR or modulo 2^{32} difference sequence that appears in the subkeys.

First, we must consider the context of a related-key differential attack. The attacker does not generally know all of the key bytes generating either A_i or B_i. Instead, he knows the XOR difference sequence in A_i and B_i.

Consider an A_i value with an XOR difference of δ. If the Hamming weight of δ is k, not including the high-order bit, then the best estimate for the XOR difference that ends up in the two subkey words for a given round generally has probability about 2^{-2k}. (Control of the A_i, B_i XOR difference sequence does not make controlling the subkey XOR differences substantially easier.)

Consider an A_i value with an XOR difference of δ. If the Hamming weight of δ is k, then the best estimate for the modulo 2^{32} difference of the two subkey words for a given round has probability about 2^{-k}.

This points out one of the difficulties in mounting any kind of successful related-key attack with nonzero A_i, B_i difference sequences. If an attacker can find a difference sequence for A_i, B_i that keeps $k = 3$, and needs to control the subkey differences for twelve rounds, he has a probability of about 2^{-72} of getting the most likely XOR subkey difference sequence, and about 2^{-36} of getting the most likely modulo 2^{32} difference sequence.

Probability of a Successful Attack With One Related-Key Query We consider the use of the RS matrix in deriving S from M to be a powerful defense against related-key differential attacks, because it forces an attacker to keep at least five key generation S-boxes active. Our analysis suggests that any useful control of the subkey difference sequence requires that each active S-box in the attack have all four key bytes changed.

Further, our analysis suggests that, for nearly any useful difference sequence, each active S-box in the attack has a specific pair of defining key bytes it needs to work. At attacker specifying his key relation in terms of bytewise XOR has five pairs of sequences of four key bytes each, which he wants to get. This leaves him with a probability of a pair of keys with his desired relation actually leading to the desired attack of about 2^{-115}, which moves the attack totally outside the realm of practical attacks.

So long as an attacker is unable to improve this, either by finding a way to get useful difference sequences into the subkeys without having so many active key bytes, or by finding a way to mount related-key attacks with different S values for the different keys, we do not believe that any kind of related key differential attack is feasible.

Note the implication of this: Clever ways to control a couple extra rounds' subkey differences are not going to make the attacks feasible, unless they also allow the attacker to use far fewer active key bytes. For reference, note that with one altered key byte per active subkey generation S-box, the attacker ends up with a 2^{-39} probability that a pair of related keys will yield an attack; with two key bytes per active S-box, this increases to 2^{-78}; with three key bytes per active S-box, it increases to 2^{-117}. In practice, this means that any key relation requiring more than one byte of key changed per active S-box appears to be impractical.

3.12   Conclusions

Our analysis suggests that related-key attacks against the full Twofish are not workable. Note, however, that we have spent less time working on resistance to chosen key attacks, such as will be available to an attacker if Twofish is used in the straightforward way to define a hash function. For this reason, we recommend that more analysis be done before Twofish is used in the straightforward way as a hash function, and we note that it appears to be much more secure to use Twofish in this way with 128-bit keys than with 256-bit keys, despite the fact that this also slows the speed of a hash function down by a factor of two.

4   Open Questions

Several questions remain open regarding the strength of the Twofish key schedule. These include for following:

1. We have discussed differential related key attacks within a certain set of assumptions, including the assumption that the subkey generation mechanism has certain more-or-less random properties. We do not have a stronger argument than our intuition and statistical tests that this is the case. A proof or stronger argument in either direction would be of great interest.
2. We have done some analysis (not reflected here for space reasons) on partial chosen key attacks on Twofish. Still remaining are issues raised by the desire to use Twofish in some Davies-Meyer hashing mode. What kind of collision resistance might we expect in this case.
3. We have assumed that the derivation of q_0 and q_1 introduces no weaknesses. Further analysis of this construction, as well as our larger S-box construction methods, would be of interest.
4. We have discussed related-key slide attacks. There are many other ways to reorder the round subkeys. Do any of these ways lead to attacks on the cipher?

References

[Anon95] Anonymous, "this looked like it might be interesting," sci.crypt Usenet posting, 9 Aug 1995.

[BB93]   I. Ben-Aroya and E. Biham, "Differential Cryptanalysis of Lucifer," *Advances in Cryptology — CRYPTO '93 Proceedings*, Springer-Verlag, 1994, pp. 187–199.

[Bih94]   E. Biham, "New Types of Cryptanalytic Attacks Using Related Keys," *Journal of Cryptology*, v. 7, n. 4, 1994, pp. 229–246.

[Bih95]   E. Biham, "On Matsui's Linear Cryptanalysis," *Advances in Cryptology — EUROCRYPT '94 Proceedings*, Springer-Verlag, 1995, pp. 398–412.

[DGV94a] J. Daemen, R. Govaerts, and J. Vandewalle, "Weak Keys for IDEA," *Advances in Cryptology — EUROCRYPT '93 Proceedings*, Springer-Verlag, 1994, pp. 159–167.

[DGV94b] J. Daemen, R. Govaerts, and J. Vandewalle, "A New Approach to Block Cipher Design," *Fast Software Encryption, Cambridge Security Workshop Proceedings*, Springer-Verlag, 1994, pp. 18–32.

[Haw98]   P. Hawkes, "Differential-Linear Weak Key Classes of IDEA," *Advances in Cryptology — EUROCRYPT '98 Proceedings*, Springer-Verlag, 1998, pp. 112–126.

[KSW96]   J. Kelsey, B. Schneier, and D. Wagner, "Key-Schedule Cryptanalysis of IDEA, G-DES, GOST, SAFER, and Triple-DES," *Advances in Cryptology — CRYPTO '96 Proceedings*, Springer-Verlag, 1996, pp. 237–251.

[KSW97]   J. Kelsey, B. Schneier, and D. Wagner, "Related-Key Cryptanalysis of 3-WAY, Biham-DES, CAST, DES-X, NewDES, RC2, and TEA," *Information and Communications Security, First International Conference Proceedings*, Springer-Verlag, 1997, pp. 203–207.

[NIST97a] National Institute of Standards and Technology, "Announcing Development of a Federal Information Standard for Advanced Encryption Standard," *Federal Register*, v. 62, n. 1, 2 Jan 1997, pp. 93–94.

[NIST97b] National Institute of Standards and Technology, "Announcing Request for Candidate Algorithm Nominations for the Advanced Encryption Standard (AES)," *Federal Register*, v. 62, n. 117, 12 Sep 1997, pp. 48051–48058.

[Wag95a] D. Wagner, "Cryptanalysis of S-1," sci.crypt Usenet posting, 27 Aug 1995.

[Yuv97]   G. Yuval, "Reinventing the Travois: Encrytion/MAC in 30 ROM Bytes," *Fast Software Encryption, 4th International Workshop Proceedings*, Springer-Verlag, 1997, pp. 205–209.

Towards Provable Security of Substitution-Permutation Encryption Networks

Zhi-Guo Chen and Stafford E. Tavares

Department of Electrical and Computer Engineering
Queen's University at Kingston, Ontario, Canada, K7L 3N6
{chenz,tavares}@ee.queensu.ca
http://adonis.ee.queensu.ca:8000

Abstract. This paper investigates some security properties of basic substitution-permutation encryption networks (SPNs) by studying the nonlinearity distribution and the XOR table distribution. Based on the idea that mixing small weak transformations results in a large strong cipher, we provide some evidence which shows that a basic SPN converges to a randomly generated s-box with the same dimensions as the SPN after enough number of rounds. We also present a new differential-like attack which is easy to implement and outperforms the classical differential cryptanalysis on the basic SPN structure. In particular, it is shown that 64-bit SPNs with 8 × 8 s-boxes are resistant to our attack after 12 rounds. All of above effort may be regarded as the first step towards provable security for SPN cryptosystems.

Keywords: block cipher, nonlinearity, XOR table, differential attack, provable security.

1 Introduction

Substitution-permutation encryption networks (SPNs) were first suggested by Feistel [2] as a simple and effective implementation of private-key block ciphers (symmetric ciphers) based on the concept of "confusion" and "diffusion" introduced by Shannon [14]. An SPN is constructed by a number of rounds of nonlinear substitutions (s-boxes) followed by bit permutations. Keying the network can be accomplished by XORing the key bits with the data bits before each round of substitutions and after the last round. The key bits associated with each round are derived from the master key according to the key scheduling algorithm. An example of a small SPN with N=16, n=4 and R=3 is illustrated in Fig. 1 where N represents the block size of the SPN consisting of R rounds of $n \times n$ s-boxes.

There are two powerful classes of cryptanalytic attacks that can be mounted against block ciphers such as SPNs. Differential cryptanalysis [1] exploits a highly probable differential characteristic derived from the XOR table of the s-boxes [11]. Linear cryptanalysis [8] depends on the best linear approximation which is directly related to nonlinearity, an important cryptographic property. More details of these attacks on SPNs can be found in [6]. On the other hand, it has been proved that completeness or nondegeneracy [7] can be achieved in the design of SPN cryptosystems. Avalanche characteristics are well studied in [5]

S. Tavares and H. Meijer (Eds.): SAC'98, LNCS 1556, pp. 43–56, 1999.

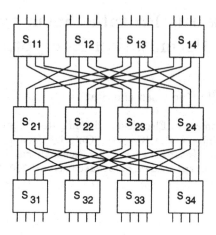

Fig. 1. SPN with N=16, n=4 and R=3

and [18]. All of the above suggests that the basic SPN has many desirable and predictable cryptographic properties useful for the design of cryptosystems.

In this paper, an estimate and upper bound on the nonlinearity distribution of bijective (invertible) s-boxes is presented. Based on the experimental results on nonlinearity and XOR table distribution, we show that the basic SPN resembles random bijective s-boxes of the same size with an increasing number of rounds, i.e., it converges to the ideal cipher. In addition, we present a practical differential-like attack on basic SPNs which exploits the Markov chain model based on the number of active s-boxes [18]. Our attack is effective regardless of the key-scheduling algorithm and more efficient than classical differential cryptanalysis. From the attack, we are able to find some hints on proving the security of SPN cryptosystems.

2   Background

2.1   Nonlinearity

A Boolean function $f(X)$ is an affine function if $f(X) = A \cdot X \oplus b$ where $X, A \in \{0,1\}^n$, $b \in \{0,1\}$; " $\cdot$ " is the dot product and "$\oplus$" is the XOR operation. Affine functions with $b = 0$ are called linear functions. The set of all n-bit affine functions is denoted by $\mathcal{A}_n$. The set of all n-bit Boolean functions is denoted by $\mathcal{F}_n$.

The Hamming weight of a function $f \in \mathcal{F}_n$ is the number of ones in its truth table, denoted by $wt(f)$. The Hamming distance between two functions $f, g \in \mathcal{F}_n$ is defined as $wt(f \oplus g)$. A function $f \in \mathcal{F}_n$ with $wt(f) = 2^{n-1}$ is said to be balanced.

We define the nonlinearity of a Boolean function $f \in \mathcal{F}_n$ as the minimum Hamming distance to all affine functions, denoted by

$$NL(f) = \min_{g \in \mathcal{A}_n} wt(f \oplus g) . \tag{1}$$

Let $\mathcal{O}_n$ denote the set of all non-zero linear combinations of output functions of an $n \times n$ s-box, i.e., $\mathcal{O}_n = \{f \mid f = a_1 \cdot f_1 \oplus \cdots \oplus a_n \cdot f_n\}$ where $a_i \in \{0, 1\}$, all $a_i \neq 0$ and f_i is the i-th output function of the s-box. Then the nonlinearity of the s-box is defined as the minimum nonlinearity of all functions in the set $\mathcal{O}_n$:

$$NL(S) = \min_{f \in \mathcal{O}_n} NL(f) \ . \tag{2}$$

2.2 XOR Table

A dynamic property of an s-box is the XOR difference table. For a given input difference, it provides possible input values to the s-box which generate the corresponding output difference. We define the entry of the XOR difference table for an $n \times n$ s-box with the input difference and the output difference $\Delta X, \Delta Y \in \{0, 1\}^n, \Delta X, \Delta Y \neq 0$ as follows:

$$XOR(\Delta X, \Delta Y) = \{X \mid X \in \{0, 1\}^n, S(X) \oplus S(X \oplus \Delta X) = \Delta Y\} \ . \tag{3}$$

If each entry of the XOR distribution table is replaced by the number of elements in that entry, we consider the number as the XOR value and the new table as the XOR table. The largest value in the XOR table is called the XOR value of the s-box, denoted by XOR^*.

3 Nonlinearity Distribution

An output ciphertext bit of an SPN can be described as a nonlinear function of the input plaintext and the key bits. The nonlinearity of this function depends on the nonlinearity of the s-boxes, the only nonlinear components in the SPN cipher. If the nonlinearities of the s-boxes are very small, the cipher would be subject to a linear attack which makes use of a linear approximation to compute the key bits. Even if the nonlinearities of the s-boxes are made very high, it is still unknown whether the final SPN is highly nonlinear. However, Ritter [13] uses experiments to show that the mixing constructions (permutations in the basic SPN structure) produce nonlinearity levels and distributions similar to those of an ideal cipher, i.e., a keyed look-up table of sufficient size. So what is the nonlinearity distribution? In this section we investigate the nonlinearity distribution of balanced Boolean functions and bijective s-boxes. (We also obtained similar results for random Boolean functions and s-boxes. We leave them out since they are not related to this work.)

Lemma 1. *For a bijective s-box, all non-zero linear combinations of the output functions are balanced.*

Lemma 2. *The probability that the nonlinearity of a randomly selected n-bit balanced Boolean function f is equal to $2i$ is upper bounded by*

$$Pr(NL(f) = 2i) \leq \frac{(2^{n+1} - 2)}{\binom{2^n}{2^{n-1}}} \binom{2^{n-1}}{i}^2 \tag{4}$$

where "=" holds if $i < 2^{n-3}$.

Proof. The total number of balanced affine functions is $2^{n+1} - 2$. For each of them, there are $\binom{2^{n-1}}{i}^2$ balanced functions at Hamming distance $2i$. So the total number of balanced functions with nonlinearity $2i$ is upper bounded by their product. For $i < 2^{n-3}$, the bound is tight because these balanced functions are distinct. The result follows by noting that the total number of balanced functions is $\binom{2^n}{2^{n-1}}$. □

Theorem 3. *For an $n \times n$ bijective s-box, the probability that its nonlinearity is less than or equal to $2i$ is upper bounded by*

$$Pr(NL(S) \leq 2i) \leq \frac{2 (2^n - 1)^2}{\binom{2^n}{2^{n-1}}} \sum_{j=0}^{i} \binom{2^{n-1}}{j}^2 . \tag{5}$$

Proof. Note that

$$Pr(NL(S) \leq 2i) \leq (2^n - 1) Pr(NL(f) \leq 2i) . \tag{6}$$

□

The upper bound in Lemma 2 can be used as an approximation to the non-linearity distribution of a balanced function provided the nonlinearity is not very high. So if we assume that all non-zero linear combinations of all s-box output functions are independent in terms of nonlinearity, then we can get the approximation since

$$Pr(NL(S) \leq 2i) \approx 1 - (1 - Pr(NL(f) \leq 2i))^{2^n - 1} . \tag{7}$$

It can be seen that the approximation and upper bound are close in Table 1. We tested 10^7 random bijective s-boxes of size 8×8. Only probabilities for nonlinearity from 80 to 98 are shown.

Table 1. Experimental result versus theoretical approximation in (7) and upper bound in (5) for the nonlinearity of random bijective 8×8 s-boxes

NL$\leq$	Experiment	Theory	Bound
80	1.83×10^{-4}	1.83×10^{-4}	1.83×10^{-4}
82	8.70×10^{-4}	8.57×10^{-4}	8.57×10^{-4}
84	3.76×10^{-3}	3.74×10^{-3}	3.75×10^{-3}
86	1.52×10^{-2}	1.52×10^{-2}	1.53×10^{-2}
88	5.68×10^{-2}	5.68×10^{-2}	5.85×10^{-2}
90	1.88×10^{-1}	1.89×10^{-1}	2.09×10^{-1}
92	5.03×10^{-1}	5.03×10^{-1}	6.99×10^{-1}
94	8.89×10^{-1}	8.89×10^{-1}	-
96	9.98×10^{-1}	9.99×10^{-1}	-
98	1.00×10^{-0}	1.00×10^{-0}	-

A more complete nonlinearity distribution is plotted in Fig. 2 according to the approximation expression (7). It indicates that s-boxes with nonlinearity greater than 98 are extremely rare. At the low end, the probability also decreases dramatically. So most s-boxes have nonlinearities between 80 and 98. For example, $Pr(NL(S) < 80) = 3.63 \times 10^{-5}$ and $Pr(NL(S) > 98) = 1.01 \times 10^{-8}$. Linear s-boxes (whose nonlinearities are 0) are very unlikely to occur since the probability is about 2.26×10^{-71} which agrees with the previous result in [3]. (After we allow for the fact that [3] uses an old definition of nonlinearity.)

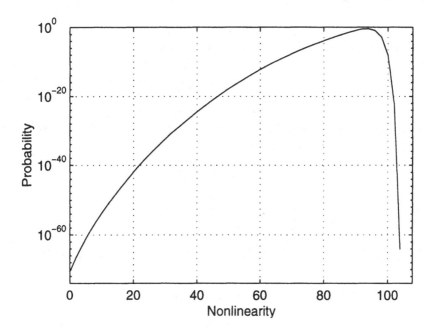

Fig. 2. Nonlinearity distribution for 8×8 s-boxes based on the approximation expression in (7)

4 XOR Distribution Table

The success of differential cryptanalysis relies on the existence of a highly probable differential which is equivalent to the existence of a large value in the XOR table of the SPN cipher, where we view the entire cipher as a big s-box. Unfortunately, it is impractical to examine the properties of the XOR tables of large SPNs (e.g., 64-bit SPNs). One advantage of the basic SPN is its simple "scalable" structure which make it possible to study a smaller version like the 16-bit SPN. Then we may intuitively extrapolate the results to 64-bit SPNs since they are constructed in a similar way. Thus, in this section we consider the 16-bit

SPNs (see Fig. 1) and the corresponding 16×16 bijective s-boxes. We use the chi-square test to show that the XOR table of the basic SPN resembles a large random XOR table as the number of rounds increases. This statistical method is also used as a cryptanalytic attack in [4] and [16].

4.1   Chi-square Test

The first observation is that there are many large entries in the first row of the XOR table of an SPN with a small number of rounds. Based on one experiment, we plot the frequencies of the entries in the first row for SPNs with 4, 5, 6, and 7 rounds, respectively, in Fig. 3.

Although the plot is only one sample and cannot stand for the general case, the distinct differences of the entry distribution show that it is sensitive to the number of rounds. Furthermore, the entry distribution tends to stabilize with increasing number of rounds. It is natural to compare the XOR distribution table of an SPN to that of a random s-box (ideal cipher). We then employ the *chi-square test* to provide a quantitative measure of the difference.

Chi-square test is a standard test for the comparison of two distributions for binned data [12]. The chi-square statistic is defined by

$$\chi^2 = \sum_i \frac{(R_i - S_i)^2}{R_i + S_i} \; . \tag{8}$$

where both R_i and S_i are experimental data. Any term with $R_i = S_i = 0$ is omitted from the sum. In general, a large value of χ^2 indicates a large difference between the two distributions. The above method is described in [12].

4.2   Comparison between SPNs and Random S-boxes

Now we use the chi-square test (8) to examine the entry distribution in the XOR table of SPNs and random s-boxes. As an example, we compare 5-round and 8-round 16-bit SPNs to a random 16×16 s-box with respect to the first row of the XOR tables. We use the random s-box as the reference. The result is shown in Table 2, where the frequencies of the corresponding XOR values and the chi-square values are shown. The threshold for 6 degrees of freedom and 1% significance level is 16.81. Therefore the 5-round SPN is rejected and the 8-round SPN is accepted, which means that an 8-round SPN behaves like a random cipher while a 5-round SPN does not in this particular test.

Although small chi-square values do not always mean that the distribution is ideal (i.e., looks random), consistent large chi-square values indicate a significant difference. Thus for a complete comparison, it is straightforward to compare every row and consider the average of chi-square values. However, our experimental results show that the chi-square values are sensitive to the row index, i.e., the input difference ΔX. It is found that for those input differences which influence a large number of s-boxes in the first round of the SPN, the corresponding row of the XOR table is "closer" to that of the random cipher than those input

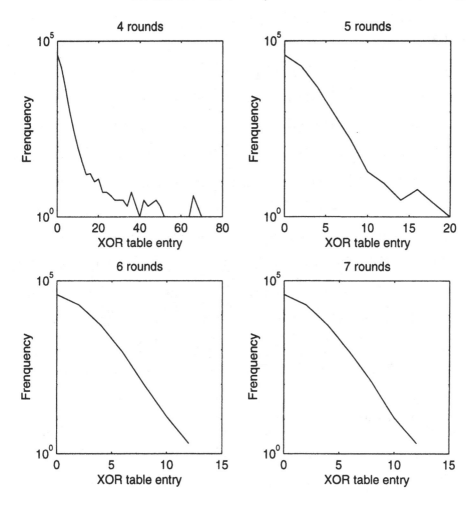

Fig. 3. The entry distribution in the first row of the XOR table for a "sample" 16×16 SPN

differences which influence a small number of s-boxes in the first round of the SPN. For example, Table 3 shows the chi-square values for the input difference of 0001, 0011, 0111, and 1111 (in hexadecimal format) respectively.

Therefore, we should concentrate on the average over those rows where the input difference influences only one s-box in the first round. There are in total $\binom{4}{1}15 = 60$ such rows. Figure 4 illustrates the chi-square test results based on one experiment. The maximum and minimum chi-square values are also presented. The average chi-square value can be regarded as a measure of the distance to the ideal cipher. It can be seen that on average the distribution of the XOR table resembles that of the ideal cipher with increasing number of rounds. And for the 16-bit SPN, we need at least 7 rounds to make the distribution "random". In

Table 2. Chi-square test for SPNs and random s-boxes of size 16 × 16 (the threshold is 16.81)

XOR=	random s-box	5-round SPN	8-round SPN
0	39721	40053	39779
2	19917	19521	19819
4	4972	4908	5002
6	809	861	814
8	107	155	108
10	8	19	13
$\geq$12	2	19	1
χ^2	-	34.43	1.92

Table 3. Chi-square test for different rows of XOR table

Rounds	$\Delta X = 0001$	$\Delta X = 0011$	$\Delta X = 0111$	$\Delta X = 1111$
3	5.53×10^3	1.71×10^3	3.40×10^2	3.33×10^1
4	9.32×10^2	1.31×10^1	7.39×10^0	7.23×10^0
5	4.93×10^1	8.81×10^0	4.06×10^0	4.43×10^0
6	6.46×10^0	4.20×10^0	6.25×10^0	4.03×10^0
7	1.55×10^0	6.07×10^0	2.81×10^0	1.27×10^0
8	4.90×10^0	2.16×10^0	2.14×10^0	7.79×10^0

fact, we usually need more rounds due to the effect of fluctuations. For 7 or more rounds, the average chi-square value is very close to the result of comparing two random s-boxes.

5   Differential-like Attack

Differential cryptanalysis of SPNs is based on the best characteristic instead of the best differential [10]. Heys and Tavares [6] derived upper bounds on the most likely differential characteristic as a function of the maximum XOR value and the number of active s-boxes (i.e., the s-boxes whose inputs are changed in the process of encrypting two plaintexts).

In this section, we present a new differential-like attack on basic SPNs. By modeling the number of active s-boxes in the network using Markov chains [18], we may predict the number of active s-boxes in the second round provided that we make one s-box in the first round (the target s-box) active and know the number of active s-boxes in the last round. This enables us to determine the subkeys of the first round and the subsequent rounds can be attacked similarly.

5.1   Principle of the Attack

Consider an r-round SPN with n_i representing the number of active s-boxes in round i ($1 \leq i \leq r$), the probability of k active s-boxes in round r given one

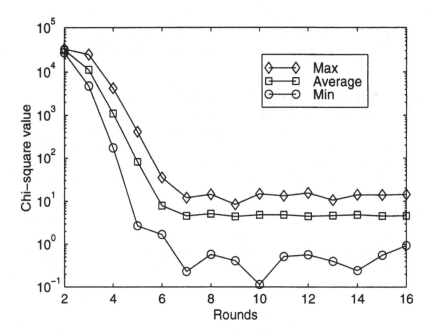

Fig. 4. Chi-square test for SPNs and random s-boxes with respect to the 60 rows of XOR table (with one active s-box in the first round)

active s-box in the first round is denoted by $Pr(n_r = k|n_1 = 1)$. Actually, it is a transition probability of $r - 1$ rounds. Now the selection matrix is defined by $\mathbf{S} = \left[s_{jk}^{(r)} \right]$ where

$$s_{jk}^{(r)} = \frac{Pr(n_r = k; n_2 = j|n_1 = 1)}{Pr(n_r = k|n_1 = 1)} \quad . \tag{9}$$

In other words, $s_{jk}^{(r)}$ is the probability of having j active s-boxes in the second round given that there is one active s-box in the first round and k active s-boxes in the last round. All of the above probabilities can be calculated from the transition matrix of Markov chains [18].

Now from the matrix $\mathbf{S}$ we may predict the number of active s-boxes in the second round by selecting those $s_{jk}^{(r)}$ greater than 50% . If we know how many s-boxes are active in the second round, then we know the output changes of the target s-box. Since the exact inputs to the target s-box are known, we can increment the counters of possible subkeys according to the XOR table of the target s-box. After we examine a number of chosen plaintext pairs, the correct subkey will be counted more often than all the others. The same method is used to derive all subkeys in the first round. If the first round is broken, then we can break the subsequent rounds in the same manner.

We find that it is highly probable that only one s-box in the second round is active if a small number of s-boxes are active in the last round. This also conforms with our intuition. So only the first row of the matrix $\mathbf{S}$ is important. If we define the selection set of the r-round SPN as $\mathcal{T}_r = \{k|s_{1k}^{(r)} > 0.5\}$, then the algorithm for attacking the target s-box in the first round of the r-round SPN is:

1. Encrypt a pair of random plaintexts such that only the target s-box is active. If the number of active s-boxes in the last round is not in the set $\mathcal{T}_r$, then go to 1.
2. Increment the counters of possible subkeys according to those XOR table entries which make only one s-box in the second round active. If there is no such subkey with counter greater than all the others by a threshold value (e.g., 2), then go to 1.
3. Stop. The subkey of the target s-box is found.

The number of chosen plaintext pairs required to determine the subkeys in the first round may be approximated by $N_P = c/P_d$, where c is a constant which may be approximated by $6m$, and m is the number of s-boxes in one round (similar to the results in [1]), and

$$P_d = \sum_{i \in \mathcal{T}_r} Pr\left(n_r = i | n_1 = 1\right) . \tag{10}$$

The threshold value corresponds to the confidence level of success. The higher the value, the more confidence we have that the subkey is correct and the more chosen plaintext pairs we need. The selection set may become empty for SPNs with increasing number of rounds, which suggests that they are immune to our attack.

5.2   Experimental Results

Our analytical results for the 64-bit SPN with randomly selected 8×8 s-boxes are shown in Table 4. Here we define the complexity as the number of chosen plaintext pairs required to determine the first round subkeys according to the calculation of N_P and a choice of $c = 50$. Our attack is effective for up to 11-round SPNs. Moreover, if we guess the 8-bit subkey associated with the target s-box in the first round, we may attack one more round with complexity increased 2^8 times. But this is not practical in that the required plaintext pairs for 12 rounds is approximately equal to the total number of plaintext pairs available for a 64-bit SPN.

A simulation program was run to attack a 64-bit SPN composed of 8×8 s-boxes with maximum XOR table entry $XOR^* = 4$. The experimental results for up to 8 rounds[1] are plotted in Fig. 5 which also shows the theoretical complexity of our new attack and classical differential cryptanalysis. Note that the example seems unfavorable to differential cryptanalysis since we use a set of s-boxes with small XOR^* which are generated from [9] and [17]. The expected XOR^* value

---

[1] The experiment for 8 rounds takes about 20 days on a Sun-ULTRA 1 machine.

Table 4. Differential-like cryptanalysis of a 64-bit SPN with 8×8 s-boxes

Rounds	Complexity (N_P)
2	2^{11}
3	2^{12}
4	2^{14}
5	2^{19}
6	2^{25}
7	2^{30}
8	2^{36}
9	2^{43}
10	2^{47}
11	2^{55}
12	2^{63}

of a randomly selected 8×8 bijective s-box is upper bounded by 16 [11]. But in practice, the value is about 12 and the highly probable characteristic can not always make use of all of these large values. So our attack outperforms the classical differential attack in a practical sense.

5.3 Comments on the Attack

There is an improvement in the implementation of our attack. By carefully choosing such plaintext pairs that make one s-box in the second round active more likely (this can be achieved by inspecting the XOR table of the target s-box), we may enhance the attack by a factor of two. However, the gain is not significant for attacking a large number of rounds.

In fact, our attack exploits the slow avalanche effect of basic SPNs. So the use of s-boxes with a high diffusion order [6] could minimize the impact of the attack. In addition, by replacing the permutation with the linear transformation [6] or multipermutations [15], we could also thwart the attack effectively. However, this introduces a delay which is significant for software implementation.

Both our attack and the classical differential attack are chosen plaintext attacks. The fundamental difference between our attack and the classical attack is that there is a filtering process in our attack.

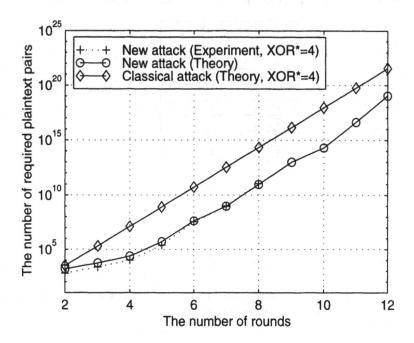

Fig. 5. Comparison of classical differential cryptanalysis and our new attack on a 64-bit SPN with 8×8 s-boxes

Our attack first selects the ciphertexts, then checks those selected, while the classical attack checks every ciphertext trying to derive the key. Hence, our attack can easily take advantage of distributed computing (e.g., over the Internet). Furthermore, it is easy to implement our attack since only minimal preliminary analysis is needed.

In order to get secure SPNs, we need to make the SPNs behave like random big s-boxes. It is necessary to make the one-bit propagation probability less than the random probability so that the SPNs is not distinguishable from the random s-boxes. This results in an estimate of the minimum number of rounds required, denoted by r, as follows:

$$\left(\frac{n}{2^n - 1}\right)^r \leq \frac{n^2}{2^{n^2} - 1} \ . \tag{11}$$

where n is the size of s-boxes. After simplifying the above expression, we can get

$$r > n + \log_2(n) \ . \tag{12}$$

Then it can be seen that when $n = 4$ (i.e., for 16-bit SPNs), we need at least 7 rounds, and when $n = 8$ (i.e., for 64-bit SPNs), we need at least 12 rounds. These agree with our previous results.

6 Conclusion

We have presented an upper bound on the nonlinearity distribution of randomly selected bijective s-boxes which shows that low nonlinearities are very unlikely for large s-boxes. Note that we have only considered bijective s-boxes throughout this work. An SPN may be regarded as a set of large s-boxes indexed by the keys.

Based on the experimental results on XOR table distributions and supported by the results on nonlinearity [13], we have shown that the basic SPN converges to the ideal cipher with an increasing number of rounds. In addition, we have presented a practical differential-like attack on basic SPNs. From the attack, it can be seen that the number of active s-boxes is very important. For a secure SPN, it is necessary to make the number of active s-boxes in the last round independent of the number of active s-boxes in previous rounds. This may be equivalent to $Pr(\Delta Y | \Delta X) = Pr(\Delta Y)$ (where ΔX and ΔY are plaintext and ciphertext differences respectively) which implies that the ciphertext is a random permutation of plaintext. Based on Markov chains [18], it is found that the number of active s-boxes in the last rounds tends to be independent for basic SPNs with an increasing number of rounds. These experiments and analytical estimates may be regarded as some evidence towards provable security for SPN cryptosystems.

References

1. E. Biham and A. Shamir. Differential cryptanalysis of DES-like cryptosystems. *Journal of Cryptology*, vol. 4, no. 1, pp. 3–72, 1991
2. H. Feistel. Cryptography and computer privacy. *Scientific American*, vol. 228, pp. 15–23, 1973
3. J. Gordon and H. Retkin. Are big S-Boxes Best? *Proc. of the Workshop on Cryptography*, LNCS 149, Berlin, pp. 257–262, 1982
4. H. Handschuh and H. Gilbert. χ^2 Cryptanalysis of the SEAL Encryption Algorithm. *Fast Software Encryption*, LNCS 1267, Springer-Verlag, pp. 1–12, 1997
5. H.M. Heys and S.E. Tavares. Avalanche characteristics of substitution-permutation networks. *IEEE Trans. Comp.*, vol. 44, pp. 1131–1139, Sept. 1995
6. H.M. Heys and S.E. Tavares. Substitution Permutation Networks Resistant to Differential and Linear Cryptanalysis. *Journal of Cryptology*, vol. 9, no. 1, pp. 1–19, 1996
7. J.B. Kam and G.I. Davida. Structured design of substitution-permutation encryption networks. *IEEE Trans. Comp.*, C-28, pp. 747–753, 1979
8. M. Matsui. Linear cryptanalysis method for DES cipher. *Proc. of Eurocrypt '93*, Springer-Verlag, Berlin, pp. 386–397, 1994
9. K. Nyberg. Differentially uniform mappings for cryptography. *Advances in Cryptology: Proc. of EUROCRYPT '93*, Springer-Verlag, Berling, pp. 55–64, 1994
10. K. Nyberg and L.R. Knudsen. Provable security against differential cryptanalysis. *Advances in Cryptology: Proc. of CRYPTO '92*, Springer-Verlag, pp. 566–574, 1993
11. L.J. O'Connor. On the distribution of characteristics in bijective mappings. *Proc. of Eurocrypt '93*, Springer-Verlag, Berlin, pp. 360–370, 1994
12. W.H. Press, B.P. Flannery, S.A. Teukplsky and W.T. Vetterling. *Numerical Recipes in C — The Art of Scientific Computing*, Cambridge University Press, pp. 487–494, 1988

13. T. Ritter. Measured nonlinearity in mixing constructions. http://www.io.com/~ritter/
14. C.E. Shannon. Communication theory of secrecy systems. *Bell System Technical Journal*, Vol. 28, pp. 656–715, 1949
15. S. Vaudenay. On the need for multipermutations: cryptanalysis of MD4 and SAFER. *Fast Software Encryption*, LNCS 1008, Springer-Verlag, pp. 286–297, 1995
16. S. Vaudenay. An Experiment on DES - Statistical Cryptanalysis. *Proc. of the 3rd ACM Conference on Computer Security*, ACM Press, pp. 139–147, 1996
17. A.M. Youssef, Z.G. Chen, and S.E. Tavares. Construction of highly nonlinear injective s-boxes with application to CAST-like encryption algorithms. *Proc. of the Canadian Conference on Electrical and Computer Engineering (CCECE '97)*, pp. 330–333, 1997
18. A.M. Youssef. *Analysis and Design of Block Ciphers*. PhD thesis, Queen's University, Kingston, Canada, 1997

An Accurate Evaluation
of Maurer's Universal Test

Jean-Sébastien Coron and David Naccache

[1] Ecole Normale Supérieure, 45 rue d'Ulm
Paris, F-75230, France
coron@clipper.ens.fr
[2] Gemplus Card International,34 rue Guynemer
Issy-les-Moulineaux, F-92447, France
naccache@compuserve.com

Abstract. Maurer's universal test is a very common randomness test, capable of detecting a wide gamut of statistical defects. The algorithm is simple (a few Java code lines), flexible (a variety of parameter combinations can be chosen by the tester) and fast. Although the test is based on sound probabilistic grounds, one of its crucial parts uses the heuristic approximation:

$$c(L, K) \cong 0.7 - \frac{0.8}{L} + \left(1.6 + \frac{12.8}{L}\right) K^{-4/L}$$

In this work we compute the precise value of $c(L, K)$ and show that the inaccuracy due to the heuristic estimate can make the test 2.67 times more permissive than what is theoretically admitted. Moreover, we establish a new asymptotic relation between the test parameter and the source's entropy.

1 Introduction

In statistics, *randomness* refers to these situations where care is taken to see that *each individual has the same chance of being included in the sample group*. In practice, random sampling is not easy : being after a random sample of people, it's not good enough to stand on a street corner and select every fifth person who passes as this would exclude habitual motorists from the sample; call on 50 homes in different areas, and you may end up with only housewives' opinions, their husbands being at work; pin a set of names from a telephone directory, and you exclude *in limine* those who do not have a telephone.

Whilst the use of random samples proves helpful in literally thousands of fields, non-random sampling is fatally disastrous in cryptography. Assessing the randomness of noisy sources is therefore crucial and a variety of tests for doing so exists. Interestingly, most if not all such tests are designed around a common skeleton, called *the monkey paradigm*. Informally, the idea consists in measuring the expectation at which a monkey playing with a typewriter would create a meaningful text. Although one can easily conclude that a complex text (*e.g.* the

S. Tavares and H. Meijer (Eds.): SAC'98, LNCS 1556, pp. 57–71, 1999.
© Springer-Verlag Berlin Heidelberg 1999

IACR's bylaws) has a negligible monkey probability, a simple word such as **cat** is expected to appear more frequently (each $\cong 17,576$ keystrokes) and could be used as a basic (yet very insufficient) randomness test.

However, analyzing *textual features* is much more efficient than pattern-scanning where inter-pattern information is wasted without being re-cycled for deriving additional monkeyness evidence.

Usually, parameters such as the average inter-symbol distance or the length of sequences containing the complete alphabet are measured in a sample and a parameter is calculated from the difference between the measure and its corresponding expectation when a monkey, theorized as a binary symmetric source (BSS), is given control over the keyboard. A BSS is a random source which outputs statistically independent and symmetrically distributed binary random variables. Based on the expected distribution of the BSS' parameter, the test succeeds or fails.

We refer the reader to [2,4] for a systematic treatment of randomness tests and focus the following sections on a particular test, suggested by Maurer in [5].

2   Maurer's universal test

Maurer's universal test [5] takes as input three integers $\{L, Q, K\}$ and a $(Q + K) \times L = N$-bit sample $s^N = [s_1, \ldots, s_N]$ generated by the tested source.

Let B denote the set $\{0,1\}$. Denoting by $b_n(s^N) = [s_{L(n-1)+1}, \ldots, s_{Ln}]$ the n-th L-bit block of s^N, the test function $f_{T_U} : B^N \to \mathbb{R}$ is defined by :

$$f_{T_U}(s^N) = \frac{1}{K} \sum_{n=Q+1}^{Q+K} \log_2 A_n(s^N) \tag{1}$$

where,

$$A_n(s^N) = \begin{cases} n & \text{if } \forall i < n, b_{n-i}(s^N) \neq b_n(s^N) \\ \min\{i : i \geq 1, b_n(s^N) = b_{n-i}(s^N)\} & \text{otherwise.} \end{cases}$$

To tune the test's rejection rate, one must first know the distribution of $f_{T_U}(R^N)$, where R^N denotes a sequence of N bits emitted by a BSS. A sample would then be rejected if the number of standard deviations separating its f_{T_U} from $E[f_{T_U}(R^N)]$ exceeds a reasonable constant[1].

For statistically independent random variables the variance of a sum is the sum of variances but the A_n-terms in (1) are heavily inter-dependent; consequently, [5] introduces a corrective factor $c(L, K)$ by which the standard deviation of f_{T_U} is reduced compared to what it would have been if the A_n-terms were independent :

$$\text{Var}[f_{T_U}(R^N)] = \sigma^2 = c(L, K)^2 \times \frac{\text{Var}[\log_2 A_n(R^N)]}{K} \tag{2}$$

---

[1] the precise value of $E[f_{T_U}(R^N)]$ is computed in [5] and recalled in section 3.3.

A heuristic estimate of $c(L, K)$ is given for practical purposes in [5] :

$$c(L, K) \cong c'(L, K) = 0.7 - \frac{0.8}{L} + \left(1.6 + \frac{12.8}{L}\right) K^{-4/L}$$

In the next section we compute the precise value of $c(L, K)$, under the admissible assumption that $Q \to \infty$ (in practice, Q should be larger than 10×2^L); this enables a much better tuning of the test's rejection rate (according to [5] the precise computation of $c(L, K)$ should have required a considerable if not prohibitive computing effort).

3   An accurate expression of $c(L, K)$

3.1   Preliminary computations

For any set of random variables, we have :

$$\text{Var}[\sum_{i=1}^{n} X_i] = \sum_{i=1}^{n} \text{Var}[X_i] + 2 \sum_{1 \le i < j \le n} \text{Cov}[X_i, X_j] \qquad (3)$$

where $\text{Cov}[X_i, X_j]$ is the covariance of X_i and X_j :

$$\text{Cov}[X_1, X_2] = E[X_1 X_2] - E[X_1] \times E[X_2] \qquad (4)$$

Throughout this paper the notation $a_i = \log_2 A_i$ will be extensively used and, unless specified otherwise, A_i will stand for $A_i(R^N)$.

Formulae (1), (2) and (3) yield :

$$c(L, K)^2 = 1 + \frac{2}{K \times \text{Var}[a_n]} \sum_{1 \le i < j \le K} \text{Cov}[a_{Q+i}, a_{Q+j}]$$

Assuming that $Q \to \infty$ (in practice, $Q > 10 \times 2^L$), the covariance of a_i and a_j is only a function of $k = j - i$ and by the change of variables $k = j - i$ we get :

$$c(L, K)^2 = 1 + \frac{2}{\text{Var}[a_n]} \times \sum_{k=1}^{K-1} (1 - \frac{k}{K}) \times \text{Cov}[a_n, a_{n+k}] \qquad (5)$$

whereas (4) yields :

$$\text{Cov}[a_n, a_{n+k}] = \sum_{i,j \ge 1} \log_2 i \log_2 j \Pr[A_{n+k} = j, A_n = i] - E[a_n]^2 \qquad (6)$$

Considering a source emitting the random variables $U^N = U_1, U_2, \ldots, U_N$, and letting $b_n = b_n(U^N)$, we get :

$$\Pr[A_n(U^N) = i] = \sum_{b \in B^L} \Pr[b_n = b, b_{n-1} \ne b, \ldots, b_{n-i+1} \ne b, b_{n-i} = b]$$

and, when the $b_n(U^N)$-blocks are statistically independent and uniformly distributed,

$$\Pr[A_n(U^N) = i] = \sum_{b \in B^L} \Pr[b_n = b]^2 \times (1 - \Pr[b_n = b])^{i-1}$$

For a BSS we thus have :

$$\Pr[A_n = i] = 2^{-L}(1 - 2^{-L})^{i-1} \quad \text{for } i \geq 1$$

3.2   Expression of $\Pr[A_{n+k} = j, A_n = i]$

Deriving the BSS' $\Pr[A_{n+k} = j, A_n = i]$ for a fixed $i \geq 1$ and variable $j \geq 1$ is somewhat more technical and requires the separate analysis of five distinct cases :

- **Disjoint blocks $1 \leq j \leq k - 1$**

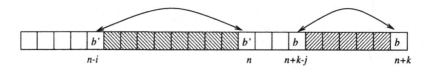

Fig. 1. DISJOINT SEQUENCES.

When $1 \leq j \leq k - 1$, the events $\langle A_{n+k} = j \rangle$ and $\langle A_n = i \rangle$ are independent, as there is no overlap between $[b_{n+k-j} \ldots b_{n+k}]$ and $[b_{n-i} \ldots b_n]$ (figure 1); consequently,

$$\Pr[A_{n+k} = j, A_n = i] = \Pr[A_{n+k} = j] \times \Pr[A_n = i]$$

$$\Pr[A_{n+k} = j, A_n = i] = 2^{-2L}(1 - 2^{-L})^{i+j-2}$$

- **Adjacent blocks $j = k$**

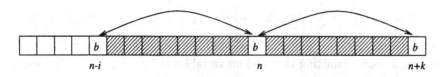

Fig. 2. ADJACENT SEQUENCES.

Letting $b = b_{n+k} = b_n = b_{n-i}$ and letting $\mathcal{E}_{j=k}[b]$ be the event (figure 2) :

$$
\begin{array}{lll}
\mathcal{E}_{j=k}[b] = & & \Pr[\mathcal{E}_{j=k}[b]] = \\
\{b_{n+k} = b, & & \Pr[b_{n+k} = b]\times \\
b_{n+k-1} \neq b, \ldots, b_{n+1} \neq b, & & \Pr[b_{n+k-1} \neq b, \ldots, b_{n+1} \neq b]\times \\
b_n = b, & \Rightarrow & \Pr[b_n = b]\times \\
b_{n-1} \neq b, \ldots, b_{n-i+1} \neq b, & & \Pr[b_{n-1} \neq b, \ldots, b_{n-i+1} \neq b]\times \\
b_{n-i} = b\} & & \Pr[b_{n-i} = b]
\end{array}
$$

we get,

$$
\Pr[\mathcal{E}_{j=k}[b]] = \Pr[b_n = b]^3 \times \Pr[b_n \neq b]^{k+i-2} = 2^{-3L}(1 - 2^{-L})^{k+i-2}
$$

$$
\Pr[A_{n+k} = k, A_n = i] = \sum_{b \in B^L} \Pr[\mathcal{E}_{j=k}[b]]
$$

$$
\Pr[A_{n+k} = k, A_n = i] = 2^{-2L}(1 - 2^{-L})^{i+k-2}
$$

- **Intersecting blocks $k + 1 \leq j \leq k + i - 1$**

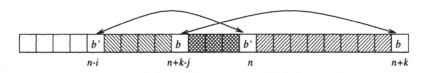

Fig. 3. INTERSECTING SEQUENCES.

For $k+1 \leq j \leq k+i-1$, the sequence $[b_{n+k-j} \ldots b_{n+k}]$ intersects $[b_{n-i} \ldots b_n]$ as illustrated in figure 3. Letting $b = b_{n+k} = b_{n+k-j}$ and $b' = b_n = b_{n-i}$, we get the following configuration, denoted $\mathcal{E}_{k+1 \leq j \leq k+i-1}[b, b']$:

$$
\begin{array}{l}
\mathcal{E}_{k+1 \leq j \leq k+i-1}[b, b'] = \{b_{n+k} = b, \\
\qquad b_{n+k-1} \neq b, \ldots, b_{n+1} \neq b, \\
\qquad b_n = b', \\
\qquad b_{n-1} \notin \{b, b'\}, \ldots, b_{n+k-j+1} \notin \{b, b'\}, \\
\qquad b_{n+k-j} = b, \\
\qquad b_{n+k-j-1} \neq b', \ldots, b_{n-i+1} \neq b', \\
\qquad b_{n-i} = b'\}
\end{array}
$$

whereby :

$$
\Pr[A_{n+k} = j, A_n = i] = \sum_{\substack{b, b' \in B^L \\ b \neq b'}} \Pr[\mathcal{E}_{k+1 \leq j \leq k+i-1}[b, b']]
$$

$$\text{for} \qquad \Pr[b_n = b] = \Pr[b_n = b'] = 2^{-L}$$
$$\Pr[b_n \neq b] = 1 - 2^{-L}$$
$$\Pr[b_n \notin \{b, b'\}] = 1 - 2 \times 2^{-L}$$

and finally :

$$\Pr[A_{n+k} = j, A_n = i] = 2^{-2L}(1 - 2^{-L})^{i+k-2}\left(1 - \frac{1}{2^L - 1}\right)^{j-k-1}$$

- **The forbidden case** $j = k + i$

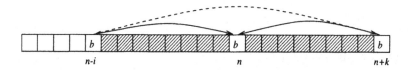

Fig. 4. THE FORBIDDEN CASE.

If $A_n = i$, A_{n+k} can not be equal to $k + i$, as shown in figure 4.

$$\Pr[A_{n+k} = k + i, A_n = i] = 0$$

- **Inclusive blocks** $j \geq k + i + 1$

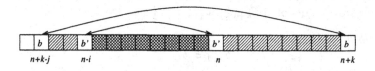

Fig. 5. INCLUSIVE SEQUENCES.

For $j \geq k + i + 1$, the sequence $[b_{n-i} \ldots b_n]$ is included in $[b_{n+k-j} \ldots b_{n+k}]$. As depicted in figure 5, the blocks of $[b_{n+1} \ldots b_{n+k-1}]$ differ from b, those of $[b_{n-i+1} \ldots b_{n-1}]$ differ from both b and b' and those of $[b_{n+k-j+1} \ldots b_{n-i-1}]$ differ from b. Letting $\mathcal{E}_{j \geq k+i+1}[b, b']$ be the event :

$$\mathcal{E}_{j\geq k+i+1}[b, b'] = \{b_{n+k} = b,$$
$$b_{n+k-1} \neq b, \ldots, b_{n+1} \neq b,$$
$$b_n = b',$$
$$b_{n-1} \notin \{b, b'\}, \ldots, b_{n-i+1} \notin \{b, b'\},$$
$$b_{n-i} = b',$$
$$b_{n-i-1} \neq b, \ldots, b_{n+k-j+1} \neq b,$$
$$b_{n+k-j} = b\}$$

$$\Pr[A_{n+k} = j, A_n = i] = \sum_{\substack{b,b' \in B^L \\ b \neq b'}} \Pr[\mathcal{E}_{j\geq k+i+1}[b, b']]$$

we obtain :

$$\Pr[A_{n+k} = j, A_n = i] = 2^{-2L}(1 - 2^{-L})^{j-2} \left(1 - \frac{1}{2^L - 1}\right)^{i-1}$$

3.3   Expression of $c(L, K)$

Let us now define the function :

$$h(z, k) = (1 - z) \sum_{i=1}^{\infty} \log_2(i + k) z^{i-1}$$

For a fixed z, the sequence $\left\{h(z, k)\right\}_{k \in \mathbb{N}}$ has the inductive property :

$$h(z, k) = (1 - z) \log_2(k + 1) + z \times h(z, k + 1) \tag{7}$$

Let

$$u = 1 - 2^{-L} \qquad \text{and} \qquad v = 1 - \frac{1}{2^L - 1}$$

The expected value $E[f_{T_U}(R^N)]$ of the test parameter $f_{T_U}(R^N)$ for a BSS is given by :

$$E[f_{T_U}(R^N)] = E[a_n] = \sum_{i=1}^{\infty} \log_2 i \times \Pr[A_n = i] = h(u, 0)$$

and the variance of a_n is :

$$\text{Var}[a_n] = E[(a_n)^2] - (E[a_n])^2$$
$$= 2^{-L} \sum_{i=1}^{\infty} (\log_2 i)^2 (1 - 2^{-L})^{i-1} - h(u, 0)^2$$

From equation (6) and the expressions of $\Pr[A_{n+k} = j, A_n = i]$, one can derive the following expression :

$$\text{Cov}[a_n, a_{n+k}] = u^k \left(h(u, 0)(h(v, k) - h(u, k)) \right.$$

$$\left. + 2^{-L} \sum_{i=1}^{\infty} \log_2 i \, u^{i-1} v^{i-1} \left(h(u, k + i) - h(v, k + i - 1) \right) \right)$$

and, using equation (5), finally obtain :

$$c(L,K)^2 = 1 - \frac{2}{\text{Var}[a_n]}\left(p(L,1) - p(L,K) - \frac{q(L,1) - q(L,K)}{K}\right)$$

where :

$$p(L,K) = u^{K-1}\sum_{l=1}^{\infty}F(l,L,K)u^{l-1} \quad , \quad q(L,K) = u^{K-1}\sum_{l=1}^{\infty}G(l,L,K)u^{l-1} \quad ,$$

$$F(l,L,K) = u^2\left(h(v,l+K-1) - h(u,l+K)\right)\left(h(v,0) - v^l h(v,l)\right)$$
$$+ u \times h(u,0)\left(h(u,l+K-1) - h(v,l+K-1)\right)$$

and

$$G(l,L,K) = u\left(h(v,l+K-1) - h(u,l+K)\right)$$
$$\left(u\,(l+K)\,(h(v,0) - v^l h(v,l)) - 2^{-L}\sum_{i=1}^{l}i\log_2 i\,v^{i-1}\right)$$
$$+ u\left(l+K-1\right)h(u,0)\left(h(u,l+K-1) - h(v,l+K-1)\right)$$

3.4 Computing $c(L,K)$ in practice

The functions $h(u,k)$, $h(v,k)$, $p(L,K)$ and $q(L,K)$ are all power series in u or v and converge rapidly ($t = 33 \times 2^L$ terms are experimentally sufficient).

To speed things further,

$$\left\{h(u,k)\right\}_{1 \leq k \leq 2t} \quad \text{and} \quad \left\{h(v,k)\right\}_{1 \leq k \leq 2t}$$

could be tabulated to compute $c(L,K)$ in $\mathcal{O}(2^L)$.

For $K \geq t$, we get with an excellent approximation :

$$c(L,K)^2 \cong d(L) + \frac{e(L) \times 2^L}{K} \qquad (8)$$

where $\quad d(L) = 1 - 2\dfrac{p(L,1)}{\text{Var}[a_n]} \quad$ and $\quad e(L) = \dfrac{q(L,1)}{\text{Var}[a_n]} \times 2^{-L+1}$

In most cases approximation (8) is sufficient, as [5] recommends to choose $K \geq 1000 \times 2^L > 33 \times 2^L$.

Although rather complicated to prove (ten pages omitted for lack of space), it is interesting to note that asymptotically :

$$\lim_{L \to \infty}\left(E[f_{T_U}(R^N)] - L\right) = C \stackrel{\Delta}{=} \int_0^{\infty} e^{-\xi}\log_2 \xi \; d\xi \cong -0.8327462$$

$$\lim_{L \to \infty} \text{Var}[a_n] = \frac{\pi^2}{6 \ln^2 2} \cong 3.4237147$$

$$\lim_{L \to \infty} d(L) = 1 - \frac{6}{\pi^2} \cong 0.3920729$$

$$\lim_{L \to \infty} e(L) = \frac{2}{\pi^2}(4 \ln 2 - 1) \cong 0.3592016$$

The distribution of $f_{T_U}(R^N)$ can be approximated by the normal distribution of mean $E[f_{T_U}(R^N)]$ and standard deviation :

$$\sigma = c(L, K)\sqrt{\text{Var}[a_n]/K} \tag{9}$$

$E[f_{T_U}(R^N)]$, $\text{Var}[a_n]$, $d(L)$ and $e(L)$ are listed in table 1 for $3 \leq L \leq 16$ and $L \to \infty$.

L	$E[f_{T_U}(R^N)]$	$\text{Var}[a_n]$	$d(L)$	$e(L)$
3	2.4016068	1.9013347	0.2732725	0.4890883
4	3.3112247	2.3577369	0.3045101	0.4435381
5	4.2534266	2.7045528	0.3296587	0.4137196
6	5.2177052	2.9540324	0.3489769	0.3941338
7	6.1962507	3.1253919	0.3631815	0.3813210
8	7.1836656	3.2386622	0.3732189	0.3730195
9	8.1764248	3.3112009	0.3800637	0.3677118
10	9.1723243	3.3564569	0.3845867	0.3643695
11	10.1700323	3.3840870	0.3874942	0.3622979
12	11.1687649	3.4006541	0.3893189	0.3610336
13	12.1680703	3.4104380	0.3904405	0.3602731
14	13.1676926	3.4161418	0.3911178	0.3598216
15	14.1674884	3.4194304	0.3915202	0.3595571
16	15.1673788	3.4213083	0.3917561	0.3594040
∞	$L - 0.8327462$	3.4237147	0.3920729	0.3592016

Table 1. $E[f_{T_U}(R^N)]$, $\text{Var}[a_n]$, $d(L)$ and $e(L)$ for $3 \leq L \leq 16$ and $L \to \infty$

4   How accurate is Maurer's test ?

Let $c'(L, K)$ be Maurer's approximation for $c(L, K)$, and let σ' be the standard deviation calculated under this approximation.

$$c'(L, K) = 0.7 - \frac{0.8}{L} + \left(1.6 + \frac{12.8}{L}\right) K^{-\frac{4}{L}} \tag{10}$$

$$\sigma' = c'(L, K)\sqrt{\text{Var}[a_n]/K}$$

Letting y' be the approximated number of standard deviations away from the mean allowed for $f_{T_U}(s^N)$, a device is rejected if and only if $f_{T_U}(s^N) < t_1$ or $f_{T_U}(s^N) > t_2$, where t_1 and t_2 are defined by :

$$t_1 = E[f_{T_U}(R^N)] - y'\sigma' \quad \text{and} \quad t_2 = E[f_{T_U}(R^N)] + y'\sigma'$$

y' is chosen such that $\mathcal{N}(-y') = \rho'/2$, where ρ' is the approximated rejection rate. $\mathcal{N}(x)$ is the integral of the normal density function [3] defined as :

$$\mathcal{N}(x) = \frac{1}{\sqrt{2\pi}} \int_{-\infty}^{x} e^{-\xi^2/2} d\xi$$

The actual number of allowed standard deviations is consequently given by $y = y'\,\sigma'/\sigma$, yielding a rejection rate of $\rho = 2\mathcal{N}(-y) = 2\mathcal{N}(-y'\,\sigma'/\sigma)$.

The worst and average *rationes* ρ'/ρ are listed in table 2 for $3 \le L \le 16$ and $1000 \times 2^L \le K \le 4000 \times 2^L$ and $\rho' = 0.001$ (*i.e.* $y' = 3.30$), as suggested in [5]. Figures show that the inaccuracy due to (10) can make the test 2.67 times more permissive than what is theoretically admitted.

The correct thresholds t_1 and t_2 can now be precisely computed using formulae (8), (9) and :

$$t_1 = E[f_{T_U}(R^N)] - y\sigma \quad \text{and} \quad t_2 = E[f_{T_U}(R^N)] + y\sigma$$

where y is chosen such that $\mathcal{N}(-y) = \rho/2$ and ρ is the rejection rate.

L	$\lim\limits_{K \to \infty} c'(L,K)$	$\lim\limits_{K \to \infty} c(L,K)$	worst ρ'/ρ	average ρ'/ρ
3	0.4333333	0.5227547	0.1541921	0.1547350
4	0.5000000	0.5518244	0.3462276	0.3464583
5	0.5400000	0.5741591	0.5058411	0.5097624
6	0.5666667	0.5907426	0.6245271	0.6394724
7	0.5857143	0.6026454	0.7215661	0.7565605
8	0.6000000	0.6109165	0.8118111	0.8775954
9	0.6111111	0.6164930	1.0607613	1.0117992
10	0.6200000	0.6201505	1.2317137	1.1634270
11	0.6272727	0.6224903	1.4245388	1.3337681
12	0.6333333	0.6239543	1.6386583	1.5223726
13	0.6384615	0.6248524	1.8723810	1.7278139
14	0.6428571	0.6253941	2.1234364	1.9481901
15	0.6466667	0.6257157	2.3893840	2.1814850
16	0.6500000	0.6259042	2.6678142	2.4257316

Table 2. A comparison of Maurer's $\{c', \rho'\}$ and the actual $\{c, \rho\}$ values.

5   The entropy conjecture

Maurer's test parameter is closely related to the source's per-bit entropy, which measures the effective key-size of a cryptosystem keyed by the source's output. [5] gives the following result, which applies to every binary ergodic stationary source S with finite memory :

$$\lim_{L\to\infty} \frac{E[f_{T_U}(U_S^N)]}{L} = H_S \tag{11}$$

where H_S is the source's per-bit entropy. Moreover, [5] conjectures that (11) can be further refined as :

$$\lim_{L\to\infty} \left[E[f_{T_U}(U_S^N)] - LH_S \right] \stackrel{c}{=} C \stackrel{\triangle}{=} \int_0^\infty e^{-\xi} \log_2 \xi \, d\xi \cong -0.8327462$$

In this section we show that the conjecture is false and that the correct asymptotic relation between $E[f_{T_U}(U_S^N)]$ and the source's entropy is :

$$\lim_{L\to\infty} \left[E[f_{T_U}(U_S^N)] - \sum_{i=1}^{L} F_i \right] = C$$

where F_i is the entropy of the i-th order approximation of the source, and :

$$\lim_{L\to\infty} F_L = H_S$$

5.1   Statistical model for a random source

Consider a source S emitting a sequence $U_1, U_2, U_3, \ldots$ of binary random variables. S is a *finite memory source* if there exists a positive integer M such that the conditional probability distribution of U_n, given $U_1, \ldots, U_{n-1}$, only depends on the last M emitted bits :

$$P_{U_n|U_1\ldots U_{n-1}}(u_n|u_1 \ldots u_{n-1}) = P_{U_n|U_{n-M}\ldots U_{n-1}}(u_n|u_{n-M} \ldots u_{n-1})$$

for $n > M$ and for every binary sequence $[u_1, \ldots, u_n] \in \{0,1\}^n$. The smallest M is called the *memory* of the source. The probability distribution of U_n is thus determined by the source's *state* $\Sigma_n = [U_{n-M}, \ldots, U_{n-1}]$ at step n. The source is *stationary* if it satisfies :

$$P_{U_n|\Sigma_n}(u|\sigma) = P_{U_1|\Sigma_1}(u|\sigma)$$

for all $n > M$, for $u \in \{0,1\}$ and $\sigma \in \{0,1\}^M$. The state-sequence of a stationary source with memory M forms a finite Markov chain : the source can be in a finite number (actually 2^M) of states σ_i, $0 \le i \le 2^M - 1$, and there is a set of transition probabilities $\Pr[\sigma_j|\sigma_i]$, expressing the odds that if the system is in state σ_i it will next go to state σ_j. For a general treatment of Markov chains, the reader is referred to [1]. In the case of a source with memory M, each of the 2^M states has at most two successor states with non-zero probability, depending on whether a zero or a one is emitted. The transition probabilities are thus determined by the set of conditional probabilities $p_i = \Pr[1|\sigma_i]$, $0 \le i \le 2^M - 1$ of emitting a one from each state σ_i. The entropy of state σ_i is then $H_i = H(p_i)$, where H is the binary entropy function :

$$H(x) = -x \log_2 x - (1 - x) \log_2(1 - x)$$

For the class of *ergodic* Markov processes the probabilities $P_j(N)$ of being in state σ_j after N emitted bits, approach (as $N \to \infty$) an equilibrium P_j which must satisfy the system of 2^M linear equations :

$$\begin{cases} \displaystyle\sum_{j=0}^{2^M-1} P_j = 1 \\[2em] \displaystyle P_j = \sum_{i=0}^{2^M-1} P_i \; \Pr[\sigma_j|\sigma_i] \qquad \text{for } 0 \le j \le 2^M - 2 \end{cases}$$

The source's entropy is then the average of the entropies H_i (of states σ_i) weighted by the state-probabilities P_i :

$$H_S = \sum_i P_i H_i \tag{12}$$

5.2   Asymptotic relation between $E[f_{T_U}(U_S^N)]$ and H_S

The mean of $f_{T_U}(U_S^N)$ for S is given by :

$$E[f_{T_U}(U_S^N)] = \sum_{i \ge 1} \Pr[A_n(U_S^N) = i] \log_2 i \tag{13}$$

with

$$\Pr[A_n(U_S^N) = i] = \sum_{b \in B^L} \Pr[b_n = b, b_{n-1} \ne b, \ldots, b_{n-i+1} \ne b, b_{n-i} = b] \tag{14}$$

Following [6] (theorem 3), the sequences of length L can be looked upon as independent for a sufficiently large L :

$$\Pr[A_n(U_S^N) = i] = \sum_{b \in B^L} \Pr[b]^2 (1 - \Pr[b])^{i-1}$$

and

$$E[f_{T_U}(U_S^N)] = \sum_{b \in B^L} \Pr[b]^2 \sum_{i \ge 1} \log_2 i \; (1 - \Pr[b])^{i-1}$$

Re-using the function $v(r)$ defined in [5],

$$v(r) = r \sum_{i=1}^{\infty} (1 - r)^{i-1} \log_2 i \tag{15}$$

we have

$$E[f_{T_U}(U_S^N)] = \sum_{b \in B^L} \Pr[b] v(\Pr[b])$$

wherefrom one can show that,

$$\lim_{r \to 0}[v(r) + \log_2 r] = \int_0^\infty e^{-\xi} \log_2 \xi \, d\xi \stackrel{\triangle}{=} C \cong -0.8327462 \qquad (16)$$

which yields :

$$\lim_{L \to \infty} \left[E[f_{T_U}(U_S^N)] + \sum_{b \in B^L} \Pr[b] \log_2 \Pr[b] \right] = C \qquad (17)$$

Let G_L be the per-bit entropy of L-bit blocks :

$$G_L = -\frac{1}{L} \sum_{b \in B^L} \Pr[b] \log_2 \Pr[b]$$

then,

$$\lim_{L \to \infty} \left[E[f_{T_U}(U_S^N)] - L \times G_L \right] = C$$

Shannon proved ([6], theorem 5) that

$$\lim_{L \to \infty} G_L = H_S$$

which implies that :

$$\lim_{L \to \infty} \frac{E[f_{T_U}(U_S^N)]}{L} = H_S$$

Let $\Pr[b, j]$ be the probability of a binary sequence b followed by the bit $j \in \{0, 1\}$ and $\Pr[j|b] = \Pr[b, j]/\Pr[b]$ be the conditional probability of bit j after b. Let,

$$F_L = -\sum_{b,j} \Pr[b, j] \log_2 \Pr[j|b] \qquad (18)$$

where the sum is taken over all sequences b of length $L - 1$ and $j \in \{0, 1\}$. We have :

$$F_L = \sum_{b \in B^{L-1}} \Pr[b] H(\Pr[1|b])$$

and, by virtue of Shannon's sixth theorem (op. cit.) :

$$F_L = L \times G_L - (L - 1)G_{L-1} , \qquad G_L = \frac{1}{L} \sum_{i=1}^{L} F_i$$

and

$$\lim_{L \to \infty} F_L = H_S$$

wherefrom

$$\lim_{L \to \infty} \left[E[f_{T_U}(U_S^N)] - \sum_{i=1}^{L} F_i \right] = C$$

5.3   Refuting the entropy conjecture

F_L is in fact the entropy of the L-th order approximation of S [1,6]. Under such an approximation, only the statistics of binary sequences of length L are considered. After a sequence b of length $L-1$ has been emitted, the probability of emitting the bit $j \in \{0,1\}$ is $\Pr[j|b]$. The L-th order approximation of a source is thus a binary stationary source with less than $L-1$ bits of memory, as defined in section 5.1. A source with M bits of memory is then equivalent to its L-th order approximation for $L > M$, and thus $\forall i > M, F_i = H_S$, and :

$$\lim_{L \to \infty} \left[E[f_{T_U}(U_S^N)] - \sum_{i=1}^{M} F_i - (L - M)H_S \right] = C$$

For example, considering a BMS_p (random binary source which emits ones with probability p and zeroes with probability $1 - p$ and for which $M = 0$ and $H_S = H(p)$), we get the following result given in [5] :

$$\lim_{L \to \infty} \left[E[f_{T_U}(U_S^N)] - LH(p) \right] = C$$

The conjecture is nevertheless refuted by considering an STP_p which is a random binary source where a bit is followed by its complement with probability p. An STP_p is thus a source with one bit of memory and two equally-probable states 0 and 1. It follows (12 and 18) that $F_1 = H(1/2) = 1$, $H_S = H(p)$, and :

$$\lim_{L \to \infty} \left[E[f_{T_U}(U_S^N)] - (L - 1)H_S - 1 \right] = C$$

which contradicts Maurer's (7-years old) entropy conjecture :

$$\lim_{L \to \infty} \left[E[f_{T_U}(U_S^N)] - LH_S \right] \overset{C}{=} C$$

6   Further research

Although the universal test is now precisely tuned, a deeper exploration of Maurer's paradigm still seems in order : for instance, it is possible to design a $c(L, K)$-less test by using a newly-sampled random sequence for each $A_n(s^N)$ (since in this setting the $A_n(s^N)$ are truly independent, $c(L, K)$ could be replaced by one). Note however that this approach increases considerably the total length of the random sequence; other theoretically interesting generalizations consist in extending the test to non-binary sources or designing tests for comparing generators to biased references (non-BSS ones).

References

1. R. Ash, *Information theory*, Dover publications, New-York, 1965.
2. D. Knuth, *The art of computer programming, Seminumerical algorithms*, vol. 2, Addison-Wesley publishing company, Reading, pp. 2–160, 1969.
3. R. Langley, *Practical statistics*, Dover publications, New-York, 1968.
4. G. Marsaglia, *Monkey tests for random number generators*, Computers & mathematics with applications, vol. 9, pp. 1–10, 1993.
5. U. Maurer, *A universal statistical test for random bit generators*, Journal of cryptology, vol. 5, no. 2, pp. 89–105, 1992.
6. C. Shannon, *A mathematical theory of communication*, The Bell system technical journal, vol. 27, pp. 379–423, 623–656, July-October, 1948.

Computational Alternatives
to Random Number Generators

David M'Raïhi[1], David Naccache[2], David Pointcheval[3], and Serge Vaudenay[3]

[1] Gemplus Corporation, 3 Lagoon Drive, Suite 300, Redwood City, CA 94065, USA
[2] Gemplus Card International, 34 rue Guynemer, 92447 Issy-les-Moulineaux, France
[3] LIENS – CNRS, École Normale Supérieure, 45 rue d'Ulm, 75230 Paris, France.
{david.pointcheval,serge.vaudenay}@ens.fr
URL: http://www.dmi.ens.fr/~{pointche, vaudenay}

Abstract. In this paper, we present a simple method for generating random-based signatures when random number generators are either unavailable or of suspected quality (malicious or accidental).

By opposition to all past state-machine models, we assume that the signer is a memoryless automaton that starts from some internal state, receives a message, outputs its signature and returns *precisely* to the same initial state; therefore, the new technique *formally* converts randomized signatures into deterministic ones.

Finally, we show how to translate the random oracle concept required in security proofs into a realistic set of tamper-resistance assumptions.

1   Introduction

Most digital signature algorithms rely on random sources which stability and quality crucially influence security: a typical example is El-Gamal's scheme [9] where the secret key is protected by the collision-freedom of the source.

Although biasing tamper-resistant generators is difficult[1], discrete components can be easily short-circuited or replaced by fraudulent emulators.

Unfortunately, for pure technological reasons, combining a micro-controller and a noise generator on the same die is not a trivial engineering exercise and most of today's smart-cards do not have real random number generators (traditional substitutes to random sources are keyed state-machines that receive a query, output a pseudo-random number, update their internal state and halt until the next query: a typical example is the BBS generator presented in [4]).

In this paper, we present an alternative approach that converts randomized signature schemes into deterministic ones: in our construction, the signer is a memoryless automaton that starts from some internal state, receives a message, outputs its signature and returns *precisely* to the same initial state.

Being very broad, we will illustrate our approach with Schnorr's signature scheme [22] before extending the idea to other randomized cryptosystems.

---

[1] such designs are usually buried in the lowest silicon layers and protected by a continuous scanning for sudden statistical defects, extreme temperatures, unusual voltage levels, clock bursts and physical exposure.

S. Tavares and H. Meijer (Eds.): SAC'98, LNCS 1556, pp. 72–80, 1999.
© Springer-Verlag Berlin Heidelberg 1999

2   Digital signatures

In EUROCRYPT'96, Pointcheval and Stern [20] proved the security of an El-Gamal variant where the hash-function has been replaced by a random oracle. However, since hash functions are fully specified (non-random) objects, the factual significance of this result was somewhat unclear. The following sections will show how to put this concept to work in practice.

In short, we follow Pointcheval and Stern's idea of using random oracles[2] but distinguish two fundamental implementations of such oracles (private and public), depending on their use.

Recall, *pro memoria*, that a digital signature scheme is defined by a distribution **generate** over a key-space, a (possibly probabilistic) signature algorithm **sign** depending on a secret key and a verification algorithm **verify** depending on the public key (see Goldwasser et al. [11]).

We also assume that **sign** has access to a private oracle f (which is a part of its private key) while **verify** has access to the public oracle h that commonly formalizes the hash function transforming the signed message into a digest.

Definition 1. *Let $\Sigma^h = ($ **generate**, **sign**^{h}, **verify**$^h)$ denote a signature scheme depending on a uniformly-distributed random oracle h. Σ is (n, t, ϵ)-secure against existential-forgery adaptive-attacks if no probabilistic Turing machine, allowed to make up to n queries to h and **sign** can forge, with probability greater than ϵ and within t state-transitions (time), a pair $\{m, \sigma\}$, accepted by **verify**.*

More formally, for any (n, t)-limited probabilistic Turing machine $\mathcal{A}$ that outputs valid signatures or fails, we have:

$$\Pr_{\omega, h}\left[\mathcal{A}^{h, \text{sign}}(\omega) \text{ succeeds}\right] \leq \epsilon$$

where ω is the random tape.

Figure 1 presents such a bi-oracle variant of Schnorr's scheme: h is a public (common) oracle while f is a secret oracle (looked upon as a part of the signer's private key); note that this variant's **verify** is strictly identical to Schnorr's original one.

Definition 2. *Let $\mathcal{H} = (h_K)_{K \in \mathcal{K}}$: $A \to B$ be a family of hash-functions, from a finite set A to a finite set B, where the key K follows a distribution $\mathcal{K}$. $\mathcal{H}$ is an (n, ϵ)-pseudo-random hash-family if no probabilistic Turing machine $\mathcal{A}$ can distinguish h_K from a random oracle in less than t state-transitions and n queries, with an advantage greater than ϵ.*

In other words, we require that for all n-limited $\mathcal{A}$:

$$\left| \Pr_{\omega, K}\left[\mathcal{A}^{h_K}(\omega) \text{ accepts}\right] - \Pr_{\omega, h}\left[\mathcal{A}^h(\omega) \text{ accepts}\right] \right| \leq \epsilon$$

---

[2] although, as showed recently, there is no guarantee that a provably secure scheme in the random oracle model will still be secure in reality [5].

System parameters:	k, security parameter
	p and q primes, $q|(p-1)$
	$g \in \mathbb{Z}_p^*$ of order q
	$h : \{0,1\}^* \to \mathbb{Z}_q$
Key generation:	generate(1^k)
	secret: $x \in_R \mathbb{Z}_q$ and $f : \{0,1\}^* \to \mathbb{Z}_q$
	public: $y = g^x \bmod p$
Signature generation:	sign$(m) := \{e, s\}$
	$u = f(m, p, q, g, y)$
	$r = g^u \bmod p$
	$e = h(m, r)$
	$s = u - xe \bmod q$
Signature verification:	verify$(m; e, s)$
	$r = g^s y^e \bmod p$
	check that $e = h(m, r)$

Fig. 1. A deterministic variant of Schnorr's scheme.

where ω is the random tape and h is a random mapping from A to B.

So far, this criterion has been used in block-cipher design but never in conjunction with hash functions. Actually, Luby and Rackoff [16] proved that a truly random 3-round, ℓ-bit message Feistel-cipher is $(n, n^2/2^{\ell/2})$-pseudo-random and safe until $n \cong 2^{\ell/4}$ messages have been encrypted (this argument was brought as an evidence for DES' security).

Note that (n, ϵ)-pseudo-randomness was recently shown to be close to the notion of n-wise decorrelation bias, investigated by Vaudenay in [24].

This construction can be adapted to pseudo-random hash-functions as follows: we first show how to construct a pseudo-random hash-function from a huge random string and then simplify the model by de-randomizing the string and shrinking it to what is strictly necessary for providing provable security. Further reduction will still be possible, at the cost of additional pseudo-randomness assumptions.

Theorem 1. *Let B be the set of ℓ-bit strings and $A = B^2$. Let us define two B-to-B functions, denoted F and G, from an $\ell \times 2^{\ell+1}$-bit key $K = \{F, G\}$. Let $h_K(x, y) = y \oplus G(x \oplus F(y))$. The family $(h_K)_K$ is $(n, n^2/2^{\ell+1})$-pseudo-random.*

Proof. The considered family is nothing but a truncated two-round Feistel construction and the proof is adapted from [16, 19] and [17]. The core of the proof consists in finding a meaningful lower bound for the probability that n different $\{x_i, y_i\}$'s produce n given z_i's. More precisely, the *ratio* between this probability and its value for a truly random function needs to be greater than $1 - \epsilon$. Letting $T = x \oplus F(y)$, we have:

$$\Pr[h_K(x_i y_i) = z_i; i = 1, \ldots, n] \geq \Pr[h_K(x_i y_i) = z_i \text{ and } T_i \text{ pairwise different}]$$

$$\geq \left(\frac{1}{2^\ell}\right)^n \left(1 - \frac{n(n-1)}{2} \min_{i,j} \Pr[T_i = T_j]\right)$$

and for any $i \neq j$ (since $x_i y_i \neq x_j y_j$), we either have $y_i \neq y_j \Rightarrow \Pr[T_i = T_j] = 1/2^\ell$, or $y_i = y_j$ and $x_i \neq x_j$ which implies $\Pr[T_i = T_j] = 1$.

Consequently:

$$\Pr[h_K(x_i y_i) = z_i; i = 1, \ldots, n] \geq \left(\frac{1}{2^\ell}\right)^n \left(1 - \frac{n(n-1)}{2} \frac{1}{2^\ell}\right) \Rightarrow \epsilon = \frac{n^2}{2^{\ell-1}}.$$

Considering a probabilistic distinguisher $\mathcal{A}^O$ using a random tape ω, we get:

$$\Pr_{\omega,K}[\mathcal{A}^{h_K}(\omega) \text{ accepts}] = \sum_{\substack{\text{accepting} \\ x_1 y_1 z_1 \cdots x_n y_n z_n}} \Pr_{\omega,K}[x_1 y_1 z_1 \ldots x_n y_n z_n]$$

$$= \sum_{x_i y_i z_i} \Pr_{\omega}[x_i y_i z_i / x_i y_i \overset{O}{\to} z_i] \Pr_K[h_K(x_i y_i) = z_i]$$

$$\geq (1 - \epsilon) \sum_{x_i y_i z_i} \Pr_{\omega}[x_i y_i z_i / x_i y_i \overset{O}{\to} z_i] \Pr_O[O(x_i y_i) = z_i]$$

$$= (1 - \epsilon) \Pr_{\omega,O}[\mathcal{A}^O(\omega) \text{ accepts}]$$

and

$$\Pr_{\omega,K}[\mathcal{A}^{h_K}(\omega) \text{ accepts}] - \Pr_{\omega,O}[\mathcal{A}^O(\omega) \text{ accepts}] \geq -\epsilon$$

which yields an advantage smaller than ϵ by symmetry (*i.e.* by considering another distinguisher that accepts if and only if $\mathcal{A}$ rejects). $\qquad\square$

Note that this construction can be improved by replacing F by a random linear function: if $K = \{a, G\}$ where a is an ℓ-bit string and G an $n\ell$-bit string defining a random polynomial of degree $n - 1$, we define $h_K(x) = y \oplus G(x \oplus a \times y)$ where $a \times y$ is the product in $GF(2^\ell)$ (this uses Carter-Wegman's xor-universal hash function [6]).

More practically, we can use standard hash-functions such as:

$$h_K(x) = \text{HMAC-SHA}(K, x)$$

at the cost of adding the function's pseudo-randomness hypothesis [2,3] to the (already assumed) hardness of the discrete logarithm problem.

To adapt random oracle-secure signatures to everyday's life, we regard $(h_K)_K$ as a pseudo-random keyed hash-family and require an indistinguishability between elements of this family and random functions. In engineering terms, this *precisely* corresponds to encapsulating the hash function in a tamper-resistant device.

Theorem 2. *Let $\mathcal{H}$ be a (n, ϵ_1)-pseudo-random hash-family. If the signature scheme Σ^h is (n, t, ϵ_2)-secure against adaptive-attacks for existential-forgery, where h is a uniformly-distributed random-oracle, then $\Sigma^{\mathcal{H}}$ is $(n, t, \epsilon_1 + \epsilon_2)$-secure as well.*

Proof. Let $\mathcal{A}^{\mathcal{H},\text{sign}}$ be a Turing machine capable of forging signatures for h_K with a probability greater than $\epsilon_1 + \epsilon_2$. h_K is distinguished from h by applying $\mathcal{A}$ and considering whether it succeeds or fails. Since $\mathcal{A}^{h,\text{sign}}$ can not forge signatures with a probability greater than ϵ_2, the advantage is greater than ϵ_1, which contradicts the hypothesis.   □

3   Implementation

An interesting corollary of theorem 2 is that if n hashings take more than t seconds, then K can be chosen randomly by a trusted authority, with some temporal validity. In this setting, long-term signatures become very similar to time-stamping [13, 1].

Another consequence is that random oracle security-proofs are no longer theoretical arguments with no practical justification as they become, *de facto*, a step towards practical and provably-secure schemes using pseudo-random hash families; however, the key has to remain secret, which forces the implementer to distinguish two types of oracles:

- A public random oracle h, that could be implemented as keyed pseudo-random hash function protected in a all tamper-resistant devices (signers and verifiers).
- A private random oracle f, which in practice could also be any pseudo-random hash-function keyed with a secret (unique to each signature device) generated by **generate**.

An efficient variant of Schnorr's scheme, provably-secure in the standard model under the tamper-resistance assumption, the existence of one-way functions and the DLP's hardness is depicted in figure 2.

The main motivation behind our design is to provide a memoryless pseudo-random generator, making the dynamic information related to the state of the generator avoidable. In essence, the advocated methodology is very cheap in terms of entropy as one can re-use the already existing key-material for generating randomness.

Surprisingly, the security of realistic random-oracle implementations is enhanced by using *intentionally* slow devices:

- use a slow implementation (*e.g.* 0.1 seconds per query) of a $(2^{40}, 1/2000)$-pseudo-random hash-family.
- consider an attacker having access to 1000 such devices during 2 years ($\cong 2^{26}$ seconds).
- consider Schnorr's scheme, which is $(n, t, 2^{20}nt/T_{\text{DL}})$-secure in the random oracle model, where T_{DL} denotes the inherent complexity of the DLP [21].

For example, $\{|p| = 512, |q| = 256\}$-discrete logarithms can not be computed in less than 2^{98} seconds ($\cong$ a 10,000-processor machine performing 1,000 modular multiplications per processor per second, executing Shank's baby-step giant-step algorithm [23]) and theorem 2 guarantees that within two years, no attacker can

System parameters:	k, security parameter
	p and q primes, $q\|(p-1)$
	$g \in \mathbb{Z}_p^*$ of order q
	$(h_v : \{0,1\}^* \to \mathbb{Z}_q)_{v \in \mathcal{K}}$ pseudo-random hash-family
	$v \in_R \mathcal{K}$ secret key
	(same in all tamper-resistant devices)
Key generation:	generate(1^k)
	secret: $x \in_R \mathbb{Z}_q$ and $z \in_R \mathcal{K}$
	public: $y = g^x \bmod p$
Signature generation:	sign(m) := $\{e, s\}$
	$u = h_z(m, p, q, g, y)$
	$r = g^u \bmod p$
	$e = h_v(m, r)$
	$s = u - xe \bmod q$
Signature verification:	verify($m; e, s$)
	$r = g^s y^e \bmod p$
	check that $e = h_v(m, r)$

Fig. 2. A provably-secure deterministic Schnorr variant.

succeed an existential-forgery under an adaptive-attack with probability greater than $1/1000$.

This proves that realistic low-cost implementation and provable security can survive in harmony. Should a card be compromised, the overall system security will simply become equivalent to Schnorr's original scheme.

Finally, we would like to put forward a variant (see figure 3) which is not provably-secure but presents the attractive property of being *fully* deterministic (a given message m, will always yield the same signature):

Lemma 1. *Let* $\{r_1, s_1\}$ *and* $\{r_2, s_2\}$ *be two Schnorr signatures, generated by the same signer using algorithm 2 then* $\{r_1, s_1\} = \{r_2, s_2\} \Leftrightarrow m_1 = m_2$.

Proof. If $m_1 = m_2 = m$ then $r_1 = r_2 = g^{h(x, m, p, q, g, y)} = r \bmod p$, $e_1 = e_2 = h(m, r) = e \bmod q$ and $s_1 = h(x, m, p, q, g, y) - xe \bmod q = s_2 = s$, therefore $\{r_1, s_1\} = \{r_2, s_2\}$.

To prove the converse, observe that if $r_1 = r_2 = r$ then $g^{u_1} = g^{u_2} \bmod p$ meaning that $u_1 = u_2 = u$. Furthermore, $s_1 = u - xe_1 = u - xe_2 = s_2 \bmod q$ implies that $e_1 = h(m_1, r) = h(m_2, r) = e_2 \bmod q$; consequently, unless we found a collision, $m_1 = m_2$. □

Industrial motivation: This feature is a cheap protection against direct physical attacks on the signer's noise-generator (corrupting the source to obtain twice an identical u).

System parameters:	k, security parameter
	p and q prime numbers such that $q\|(p-1)$
	$g \in \mathbb{Z}_p^*$ of order q
	h, hash function
Key generation:	generate(1^k)
	secret: $x \in_R \mathbb{Z}_q$
	public: $y = g^x \bmod p$
Signature generation:	sign(m) := $\{e, s\}$
	$u = h(x, m, p, q, g, y) \bmod q$
	$r = g^u \bmod p$
	$e = h(m, r) \bmod q$
	$s = u - xe \bmod q$
Signature verification:	verify($m; e, s$)
	$r = g^s y^e \bmod p$
	check that $e = h(m, r) \bmod q$

Fig. 3. A practical deterministic Schnorr variant.

4   Deterministic versions of other schemes

The idea described in the previous sections can be trivially applied to other signature schemes such as [10] or [12]. Suffice it to say that one should replace each session's random number by a digest of the keys (secret and public) and the signed message.

Blind signatures [8] (a popular building-block of most e-cash schemes) can be easily transformed as well: in the usual RSA setting the user computes $w = h(k, m, e, n)$ (where k is a short secret-key) and sends $m' = w^e m \bmod n$ to the authority who replies with $s' = w^{ed} m^d \bmod n$ that the user un-blinds by a modular division ($s = s'/w = m^d \bmod n$).

The "blinding" technique can also be used to prevent timing-attacks [15], but it requires again a random blinding factor [14].

More fundamentally, our technique completely *eliminates* a well-known attack on Mc Eleice's cryptosystem [18] where, by asking the sender to re-encrypt logarithmically many messages, one can filter-out the error vectors (e, chosen randomly by the sender at each encryption) through simple majority votes.

We refer the reader to section III.1.4.A.C of [7] for more detailed description of this attack (that disappears by replacing e by a hash-value of m and the receiver's public-keys).

References

1. D. Bayer, S. Haber, and W. S. Stornetta. Improving the Efficiency and Reliability of Digital Time-Stamping. *Sequences II, Methods in Communication, Security and Computer Science*, pages 329–334, 1993.

2. M. Bellare, R. Canetti, and H. Krawczyk. Keying Hash Functions for Message Authentication. In *Crypto '96*, LNCS 1109. Springer-Verlag, 1996.

3. M. Bellare, R. Canetti, and H. Krawczyk. Message Authentication using Hash Functions: The HMAC construction. *RSA Laboratories' Cryptobytes*, 2(1), Spring 1996.

4. L. Blum, M. Blum, and M. Shub. A Simple Unpredictable Random Number Generator. *SIAM Journal on computing*, 15:364–383, 1986.

5. R. Canetti, O. Goldreich, and S. Halevi. The Random Oracles Methodology, Revisited. In *Proc. of the 30th STOC*. ACM Press, 1998.

6. L. Carter and M. Wegman. Universal Hash Functions. *Journal of Computer and System Sciences*, 18:143–154, 1979.

7. F. Chabaud. *Recherche de Performance dans l'Algorithmique des Corps Finis, Applications à la Cryptographie*. PhD thesis, École Polytechnique, 1996.

8. D. Chaum. Blind Signatures for Untraceable Payments. In *Crypto '82*, pages 199–203. Plenum, NY, 1983.

9. T. El Gamal. A Public Key Cryptosystem and a Signature Scheme Based on Discrete Logarithms. In *IEEE Transactions on Information Theory*, volume IT–31, no. 4, pages 469–472, July 1985.

10. A. Fiat and A. Shamir. How to Prove Yourself: practical solutions of identification and signature problems. In *Crypto '86*, LNCS 263, pages 186–194. Springer-Verlag, 1987.

11. S. Goldwasser, S. Micali, and R. Rivest. A Digital Signature Scheme Secure Against Adaptive Chosen-Message Attacks. *SIAM Journal of Computing*, 17(2):281–308, April 1988.

12. L. C. Guillou and J.-J. Quisquater. A Practical Zero-Knowledge Protocol Fitted to Security Microprocessor Minimizing Both Transmission and Memory. In *Eurocrypt '88*, LNCS 330, pages 123–128. Springer-Verlag, 1988.

13. S. Haber and W. S. Stornetta. How to Timestamp a Digital Document. *Journal of Cryptology*, 3:99–111, 1991.

14. B. Kaliski. Timing Attacks on Cryptosystems. *RSA Laboratories' Bulletin*, 2, January 1996.

15. P. C. Kocher. Timing Attacks on Implementations of Diffie-Hellman, RSA, DSS, and Other Systems. In *Crypto '96*, LNCS 1109, pages 104–113. Springer-Verlag, 1996.

16. M. Luby and Ch. Rackoff. How to Construct Pseudorandom Permutations from Pseudorandom Functions. *SIAM Journal of Computing*, 17(2):373–386, 1988.

17. U. M. Maurer. A Simplified and Generalized Treatment of Luby-Rackoff Pseudorandom Permutation Generators. In *Eurocrypt '92*, LNCS 658, pages 239–255. Springer-Verlag, 1993.

18. R. J. McEliece. A Public-Key Cryptosystem Based on Algebraic Coding Theory. *DSN progress report*, 42-44:114–116, 1978. Jet Propulsion Laboratories, CALTECH.

19. J. Patarin. *Étude des Générateurs de Permutations Pseudo-aléatoires Basés sur le Schéma du DES*. PhD thesis, Université de Paris VI, November 1991.

20. D. Pointcheval and J. Stern. Security Proofs for Signature Schemes. In *Eurocrypt '96*, LNCS 1070, pages 387–398. Springer-Verlag, 1996.
21. D. Pointcheval and J. Stern. Security Arguments for Digital Signatures and Blind Signatures. *Journal of Cryptology*, 1998. To appear.
22. C. P. Schnorr. Efficient Identification and Signatures for Smart Cards. In *Crypto '89*, LNCS 435, pages 235–251. Springer-Verlag, 1990.
23. D. Shanks. Class number, a theory of factorization, and genera. In *Proceedings of the symposium on Pure Mathematics*, volume 20, pages 415–440. AMS, 1971.
24. S. Vaudenay. Provable Security for Block Ciphers by Decorrelation. In *STACS '98*, LNCS 1373, pages 249–275. Springer-Verlag, 1998.

Storage-Efficient Finite Field Basis Conversion

Burton S. Kaliski Jr.[1] and Yiqun Lisa Yin[2]

[1] RSA Laboratories, 20 Crosby Drive, Bedford, MA 01730, **burt@rsa.com**
[2] RSA Laboratories, 2955 Campus Drive, San Mateo, CA 94402, **yiqun@rsa.com**

Abstract. The problem of finite field basis conversion is to convert from the representation of a field element in one basis to the representation of the element in another basis. This paper presents new algorithms for the problem that require much less storage than previous solutions. For the finite field $GF(2^m)$, for example, the storage requirement of the new algorithms is only $O(m)$ bits, compared to $O(m^2)$ for previous solutions. With the new algorithms, it is possible to extend an implementation in one basis to support other bases with little additional cost, thereby providing the desired interoperability in many cryptographic applications.

1 Introduction

Finite field arithmetic is becoming increasingly important in today's computer systems, particularly for cryptographic operations. Among the more common finite fields in cryptography are odd-characteristic finite fields of degree 1, conventionally known as $GF(p)$ arithmetic or arithmetic modulo a prime, and even-characteristic finite fields of degree greater than 1, conventionally known as $GF(2^m)$ arithmetic, where m is the degree. Arithmetic in $GF(2^m)$ (or any finite field of degree greater than 1) can be further classified according to the choice of basis for representing elements of the finite field; two common choices are a polynomial basis and a normal basis.

For a variety of reasons, including cost, performance, and compatibility with other applications, implementations of $GF(2^m)$ arithmetic vary in their choice of basis. The variation in choice affects interoperability, since field elements represented in one basis cannot be operated on directly in another basis. The problem of interoperability limits the applicability of implementations to cryptographic communication. As an example, if two parties wish to communicate with cryptographic operations and each implements finite field arithmetic in a different basis, then at least one party must do some conversions, typically before or after communicating a field element or at certain points in the cryptographic operations. Otherwise, the results of the cryptographic operations will be different.

It is well known that it is possible to convert between two choices of basis for a finite field; the general method involves a matrix multiplication. However, the matrix is often too large. For instance, the change-of-basis matrix for $GF(2^m)$ arithmetic will have m^2 entries, requiring several thousand bytes or more of storage in typical applications (e.g., $m \approx 160$). While such a matrix may be

S. Tavares and H. Meijer (Eds.): SAC'98, LNCS 1556, pp. 81–93, 1999.
© Springer-Verlag Berlin Heidelberg 1999

reasonable to store in a software implementation, it is likely to be a significant burden in a low-cost hardware implementation.

We describe in this paper new algorithms for basis conversion for normal and polynomial bases that require much less storage than previous solutions. Our algorithms are also very efficient in that they involve primarily finite-field operations, rather than, for instance, matrix multiplications. This has the advantage of benefiting from the optimizations that are presumably already available for finite-field operations.

With our algorithms, it is possible to extend an implementation in one basis so that it supports other choices of basis, with only a small additional cost in terms of circuitry, program size, or storage, relative to typical cryptographic applications of finite-field arithmetic, such as elliptic curve cryptosystems. Our work applies both to even-characteristic and odd-characteristic finite fields of degree greater than one, though even-characteristic arithmetic is the most likely application of our work, since it is more common in practice. We also suggest how to generalize our algorithms to other choices of basis than polynomial and normal bases.

2   Background

In this section, we introduce some basic notation and definitions. Let $GF(q)$ denote the finite field with q elements where $q = p^r$ for some prime p and integer $r \geq 1$. The *characteristic* of the field is the prime p. For even-characteristic fields, we have $p = 2$. Throughout the paper, we use $GF(q^m)$ to denote the finite field defined over the ground field $GF(q)$; the degree of $GF(q^m)$ over $GF(q)$ is m.

A *basis* for the finite field $GF(q^m)$ is a set of m elements $\omega_0, \ldots, \omega_{m-1} \in GF(q^m)$ such that every element of the finite field can be represented uniquely as a linear combination of basis elements. That is, given an element $\epsilon \in GF(q^m)$, we can write

$$\epsilon = \sum_{i=0}^{m-1} B[i]\omega_i$$

where $B[0], \ldots, B[m-1] \in GF(q)$ are the *coefficients*. The row vector $B = (B[0], \ldots, B[m-1])$ is called the *representation* of the element ϵ in the basis $\omega_0, \ldots, \omega_{m-1}$. Once the basis is chosen, rules for field operations (such as addition, multiplication, inversion) can be derived.

Elements of a finite field can be represented in a variety of ways, depending on the choice of basis for the representation. Two common choices of basis are a polynomial basis and a normal basis. In a polynomial basis, the basis elements are successive powers of an element γ (called the generator), that is, $\omega_i = \gamma^i$. The element γ must satisfy certain properties, namely that the powers $\gamma^0, \ldots, \gamma^{m-1}$ are linearly independent. In a normal basis, the basis elements are successive exponentiations of an element γ (again called the generator), that is, $\omega_i = \gamma^{q^i}$. In this case, the successive exponentiations must be linearly independent. Each basis has its own advantages and disadvantages in terms of implementation, and some discussions can be found in [4].

The *basis conversion* or *change-of-basis* problem is to compute the representation of an element of a finite field in one basis, given its representation in another basis. The general solution to the problem is to apply the change-of-basis matrix relating the two bases. Suppose that we are converting from the representation B of ϵ in the basis $\omega_0, \ldots, \omega_{m-1}$ to another basis. Let W_i be the representation of the element ω_i in the second basis, and let M, the change-of-basis matrix, be the $m \times m$ matrix whose columns are $W_0, \ldots, W_{m-1}$. It follows that the representation A of the element ϵ in the second basis can be computed as the matrix-vector product $A^T = M B^T$ where we view A and B as row vectors of dimension m. A change-of-basis matrix is invertible, and we can convert in the reverse direction by computing $B^T = M^{-1} A^T$.

The change-of-basis-matrix solution is straightforward and effective. But it is limited by two factors. First, the matrix M is potentially quite large, consisting of m^2 coefficients. Moreover, if we wish to convert in both directions, we must store the matrix M^{-1} as well, or else compute it, which could be time-consuming. Second, the operations involved in computing the matrix-vector product, while involving coefficients in the ground field, are not necessarily implementable with operations in either basis. Thus, the conversion process may not be as efficient as we would like.

Another approach to conversion is to multiply by elements of a dual basis (see Page 58 of [6]), but the storage requirement will again be quite large, if the entire dual basis is stored.

Our objective is to overcome the difficulties of the approaches just described. We wish to convert from one basis to another without involving a large amount of storage or requiring a large number of operations. And, we would like to take advantage of the built-in efficiency of finite field operations in one basis, rather than implementing new operations, such as matrix multiplications. We will call the basis in which finite field operations are primarily performed the *internal basis*. The other basis will be called the *external basis*. The conversion operation from the external basis to the internal basis will be called an *import* operation, and the reverse an *export* operation.

The specific problems to be solved are thus as follows.

- *Import problem.* Given an internal basis and an external basis for a finite field $GF(q^m)$ and the representation B of a field element in the external basis (the *external representation*), determine the corresponding representation A of the same field element in the internal basis (the *internal representation*) primarily with internal-basis operations, and with minimal storage.
- *Export problem.* Given an internal basis and an external basis for a finite field $GF(q^m)$ and the internal representation A of a field element, determine the corresponding external representation B of the same field element primarily with internal-basis operations, and with minimal storage.

The more general problem of converting from one basis to another with operations in a third basis is readily solved by importing to and re-exporting from the third basis; thus, our algorithms for converting to and from an internal basis will suffice for the more general problem.

3 Conversion Algorithms

In this section, we present four conversion algorithms for the import and export problem. We will focus our discussion on the case that both bases are defined over the same ground field $GF(q)$, and that the coefficients in the ground field are represented the same way in both bases. Section 4 addresses the case that the bases are defined over different ground fields, or that the coefficients are represented differently.

We require that the external basis is a polynomial basis or a normal basis, so that elements in the external basis have either the form

$$\epsilon = \sum_{i=0}^{m-1} B[i]\gamma^i \tag{1}$$

or the form

$$\epsilon = \sum_{i=0}^{m-1} B[i]\gamma^{q^i} \tag{2}$$

where γ is the generator of the external basis and $B[0], \ldots, B[m-1] \in GF(q)$ are the coefficients of the external representation. However, as discussed in Section 4, similar algorithms may be constructed for other choices of basis. In addition, we assume that the internal representation G of the generator γ is given, which is reasonable in most practical settings.[1]

We make no assumptions on the internal basis, other than that it is defined over the same ground field $GF(q)$ and that the representation of the coefficients in the ground field is the same. Our algorithms involve the same sequence of operations whether the internal basis is a polynomial basis, a normal basis, or some other type of basis. Thus, as examples, our algorithms can convert from a polynomial basis to a normal basis, from a normal basis to a polynomial basis, from a polynomial basis with one generator to a polynomial basis with another generator, or from a normal basis with one generator to a normal basis to another generator, to give a few possibilities.

We assume that addition, subtraction, and multiplication operations are readily available in the internal basis. A special case of multiplication, which can be more efficient, is scalar multiplication, where one of the operands is a coefficient, i.e., an element of the ground field.

In the following, we denote by I the internal representation of the identity element.

---

[1] If not, it can be computed given information about the internal and external bases, such as the minimal polynomial of the generator. There may be several acceptable internal representations of the generator, and hence several equivalent internal representations of a given element. For interoperability we need only that conversion into and out of the internal basis involve the same choice of G.

3.1   Basic Techniques for Conversion

Before presenting our conversion algorithms, we first describe some useful techniques which will serve as the building blocks of the conversion algorithms.

Algorithms for importing from an external basis can be constructed based on a direct computation of Equations (1) or (2). Since the internal representation G of the generator γ is given, we can easily convert each external basis element into its internal representation using only operations in the internal basis. In Sections 3.2 and 3.3, we give alternatives to these algorithms which are amenable to further performance improvements.

Algorithms for exporting to an external basis, however, cannot be constructed in the same direct manner. The major obstacle comes from the following fact: It is not obvious how to convert each internal basis element into its external representation using *only* operations in the internal basis, even given the external representation of the generator. So instead of converting the basis element, we will use some new techniques, and they are described in the following three lemmas.

Lemma 1. *Suppose the external basis is a polynomial basis with a generator γ. Let B be the external representation of an element ϵ, and let B' be the external representation of the element $\epsilon\gamma^{-1}$. Then for all indexes $0 < i < m - 1$,*

$$B'[i] = B[i+1]$$

provided that $B[0] = 0$.

Lemma 3.1 shows that if the external basis is a polynomial basis, then multiplication by the inverse γ^{-1} of the generator γ shifts the coefficients down, provided that the coefficient at index 0 is initially 0. The result leads to an algorithm for exporting to a polynomial basis: compute the coefficient $B[0]$, subtract $B[0]$, multiply by G^{-1}, and repeat, computing successive coefficients of B.

Related to this, multiplication by the generator γ shifts coefficients up, provided that $B[m-1] = 0$. Rotation of the coefficients in either direction is also possible, though we will not need it for our algorithms.

Lemma 2. *Suppose the external basis is a normal basis. Let B be the external representation of an element ϵ, and let B' be the external representation of the element ϵ^q. Then for all indexes $0 \le i \le m - 1$,*

$$B'[i] = B[(i-1) \bmod m].$$

Lemma 3.2 shows that if the external basis is a normal basis, then raising to the power q rotates the coefficients up. The result leads to an algorithm for exporting to a normal basis: compute the coefficient $B[m-1]$, raise to the power q, and repeat.

We still need a technique for obtaining the coefficient $B[0]$ or $B[m-1]$. From the fact that the coefficients of the internal and external representations are related by a change-of-basis matrix M $B^T = M^{-1}A^T$, we know that a coefficient $B[i]$ can be obtained by a linear combination

$$B[i] = \sum_{j=0}^{m-1} M^{-1}[i,j]A[j]$$

where the values $M^{-1}[i,j] \in GF(q)$ are elements of the matrix M^{-1}. We can thus obtain a coefficient $B[i]$ by operations over the ground field $GF(q)$. We may also compute the coefficient $B[i]$ with only internal-basis operations, as will be shown in the next lemma.

In preparation, we will need to consider the multiplication matrices for the internal basis. Let $\omega_0, \ldots, \omega_{m-1}$ be the internal basis. The multiplication matrix for the coefficient at index k, denoted K_k, is the $m \times m$ matrix whose $[i,j]$th element, $0 \le i,j < m$, is the coefficient at index k of the representation in the internal basis of the product $\omega_i\omega_j$. In other words, the matrices are defined so that for all i, j, $0 \le i, j < m$,

$$\omega_i\omega_j = \sum_{k=0}^{m-1} K_k[i,j]\omega_k.$$

The multiplication matrices is invertible. It follows from this definition that the coefficient at index 0 of the product of two internal representations R and S, which we may write as $(R \times S)[0]$, is equal to RK_0S^T.

Lemma 3. *Let $s_0, \ldots, s_{m-1}$ be elements of $GF(q)$, and let K_0 be the multiplication matrix for the coefficient at index 0 in the internal basis. Then for any internal representation A of an element,*

$$\sum_{i=0}^{m-1} s_j A[j] = (A \times V)[0],$$

where V is defined as $V^T = K_0^{-1}[s_0 \cdots s_{m-1}]^T$.

Proof. Since the multiplication matrix K_0 is invertible, the element V exists. By definition of multiplication, we have $(A \times V)[0] = AK_0V^T$. It follows directly that $(A \times V)[0]$ equals the desired linear function.

Lemma 3.3 shows that any linear combination of coefficients of the internal representation of an element may be computed with internal-basis operations. To generalize the result, we denote by V_i the value such that

$$B[i] = (A \times V_i)[0],$$

i.e., the one where the values $s_0, \ldots, s_{m-1}$ are the matrix row $M^{-1}[i,0], \ldots,$ $M^{-1}[i,m-1]$. Like the internal representation G of the generator of the external basis, a value V_i is particular to an external basis; a different set of values V_i would be needed for each external basis with which one might want to convert.

In what follows, we present four conversion algorithms, for importing and exporting with external polynomial and normal bases. For each algorithm, we measure the number of full multiplications and scalar multiplications involved and the amount of storage required for constants and intermediate results. As noted previously, there are direct import algorithms based on direct computation of Equations (1) and (2); we give different versions that allow further performance improvements.

As we will see, each basic conversion algorithm requires the storage of only one or two constants (vectors of length m). So for $GF(q^m)$ the total storage requirement is $O(m \log q)$ bits, compared to $O(m^2 \log q)$ bits for previous solutions.

3.2 Importing from a Polynomial Basis

Algorithm IMPORTPOLY converts from a polynomial-basis representation for $GF(q^m)$ over $GF(q)$ to an internal representation over the same ground field, primarily with internal-basis operations.

Input: $B[0], \ldots, B[m-1]$, the external representation to be converted
Output: A, the corresponding internal representation
Constants: G, the internal representation of the generator of the external basis

```
proc IMPORTPOLY
A ← 0
for i ← m − 1 down to 0 do
    A ← A × G
    A ← A + B[i] × I
endfor
```

The algorithm processes one coefficient per iteration, scanning from highest index to lowest, accumulating powers of G up to G^i for each $B[i]$ term. It involves m full multiplications (and may also involve m scalar multiplications, depending on the form of I), and requires storage for one constant.

We can view the above algorithm as computing the internal representation A according to the following formula.

$$A = B[0] \times I + G \times (B[1] \times I + G \times (B[2] \times I + \cdots + G \times (B[m-1] \times I) \cdots)).$$

So IMPORTPOLY bears some similarity to the method of evaluating a polynomial $f(x)$ at a given value x^* using Horner's rule [2]. More specifically, $f(x^*)$ can be evaluated with $O(m)$ operations by rewriting $f(x)$ as follows.

$$f(x^*) = a_0 + x^*(a_1 + x^*(a_2 + \cdots + x^*(a_{m-1})) \cdots)).$$

There are some distinctions between the two methods. The inputs to basis conversion are the coefficients $B[0], \ldots, B[m-1]$, and the generator G is fixed. In contrast, the input to polynomial evaluation is the value x^*, and the coefficients $a_0, \ldots, a_{m-1}$ are fixed.

It is possible to reduce the number of iterations of the loop and thereby improve performance by processing more than one coefficient per iteration. One of the ways to do this is, in the case that m is even, to change the loop to

> **for** $i \leftarrow m/2 - 1$ **down to** 0 **do**
> $\quad A \leftarrow A \times G$
> $\quad A \leftarrow A + B[i + m/2] \times G^{m/2} + B[i] \times I$
> **endfor**

where $G^{m/2}$ is an additional constant. Although the number of scalar multiplications remains the same, the number of full multiplications can be reduced by up to a factor of k, if k coefficients are processed per iteration. In addition, all k coefficients can be processed in parallel.

Another improvement is to unroll the first iteration and start with $A \leftarrow B[m-1] \times I$.

3.3   Importing from a Normal Basis

Algorithm IMPORTNORMAL converts from a normal-basis representation for $GF(q^m)$ over $GF(q)$ to an internal representation over the same ground field, primarily with internal-basis operations.

Input: $B[0], \ldots, B[m-1]$, the external representation to be converted
Output: A, the corresponding internal representation
Constants: G, the internal representation of the generator of the external basis

> **proc** IMPORTNORMAL
> $A \leftarrow 0$
> **for** $i \leftarrow m - 1$ **down to** 0 **do**
> $\quad A \leftarrow A^q$
> $\quad A \leftarrow A + B[i] \times G$
> **endfor**

The algorithm processes one coefficient per iteration, scanning from highest index to lowest, accumulating powers of G up to G^{q^i} for each $B[i]$ term. (We make use of the fact that $(A + B[i] \times G)^q = A^q + B[i] \times G^q$.) It involves m exponentiations to the power q and m scalar multiplications, and requires storage for one constant, in addition to the intermediate results for exponentiation. (The exponentiation will typically involve about $1.5 \log_2 q$ multiplications and require storage for one intermediate result, though better performance is possible if the internal basis is a normal basis. For $q = 2$, the exponentiation will be just a squaring.)

As with IMPORTPOLY, it is possible to improve performance by unrolling the first iteration or by processing more than one coefficient per iteration.

3.4 Exporting to a Polynomial Basis

Algorithm EXPORTPOLY converts from an internal representation for $GF(q^m)$ over $GF(q)$ to an external polynomial-basis representation over the same ground field, primarily with internal-basis operations.

Input: A, the internal representation to be converted
Output: $B[0], \ldots, B[m-1]$, the corresponding external representation
Constants:
    G^{-1}, the internal representation of the inverse of the generator of the external basis
    V_0, the value such that $(A \times V_0)[0] = B[0]$ (see Lemma 3.3)

> **proc** EXPORTPOLY
> $A \leftarrow A \times V_0$
> **for** $i \leftarrow 0$ **to** $m-1$ **do**
> $B[i] \leftarrow A[0]$
> $A \leftarrow A - B[i] \times V_0$
> $A \leftarrow A \times G^{-1}$
> **endfor**

The algorithm computes one coefficient per iteration, applying the observations previously given, with the additional enhancement of premultiplying by the value V_0.[2] The algorithm involves $m + 1$ full multiplications and m scalar multiplications, and requires storage for two constants. The input A is modified by the algorithm.

As with the previous algorithms, more than one coefficient can be processed per iteration. An additional constant such as $V_{m/2}/V_0$ is required, where $V_{m/2}$ is the value such that $(A \times V_{m/2})[0] = B[m/2]$; the coefficient $B[i + m/2]$ would be computed as

$$T \leftarrow A \times V_{m/2}/V_0$$
$$B[i + m/2] \leftarrow T[0]$$

The number of multiplications is not reduced in this case, due to the method for computing the coefficient $B[i+m/2]$. The improvement would be more significant if the coefficients were computed as a direct linear function of A, provided that such computation were efficient.

Another improvement is to unroll the last iteration and end with $B[m-1] = A[0]$.

---

[2] This is the reason that the correction step involves subtracting the value $B[i] \times V_0$ rather than $B[i] \times I$. The alternative to premultiplying A by V_0 is to multiply it by V_0 during each iteration before computing the coefficient $B[i]$; but this involves an additional multiplication per iteration.

3.5   Exporting to a Normal Basis

Algorithm EXPORTNORMAL converts from an internal representation for $GF(q^m)$ over $GF(q)$ to a normal-basis representation over the same ground field, primarily with internal-basis operations.

Input: A, the internal representation to be converted
Output: $B[0], \ldots, B[m-1]$, the corresponding external representation
Constants: V_{m-1}, the value such that if $(A \times V_{m-1})[0] = B[m-1]$ (see Lemma 3.3)

> proc EXPORTNORMAL
> for $i \leftarrow m-1$ down to 0 do
> $T \leftarrow A \times V_{m-1}$
> $B[i] \leftarrow T[0]$
> $A \leftarrow A^q$
> endfor

The algorithm computes one coefficient per iteration, applying the observations previously given. The algorithm involves m exponentiations to the power q and m full multiplications, and requires storage for one constant and one intermediate result, T, in addition to the intermediate results for exponentiation. The input A, though modified by the algorithm, returns to its initial value.

As with EXPORTPOLY, it is possible to improve performance by unrolling the last iteration or by processing more than one coefficient per iteration.

4   Extensions

The algorithms presented so far all assume that the ground field is the same for the internal and the external basis and have the same representation.

If the ground fields are the same but have different representations, the individual coefficients can be converted through techniques similar to those for the entire representation.

If the ground fields have different representations, however, we can convert individual "subcoefficients" of each coefficient, where the subcoefficients are elements of $GF(p)$. In the import algorithms, we would add terms of the form $B[i][j] \times H_j$ (or $B[i][j] \times GH_j$) to A, where $B[i][j]$ is a subcoefficient and H_j is the internal representation of an element of the ground-field basis. In the export algorithms, we would multiply by values like $V_{0,j}$ where $(A \times V_{0,j}) = B[0][j]$. The storage requirements for these methods would depend on the degree of the ground field over $GF(p)$, and would be modest in many cases. More storage-efficient algorithms are possible, however, involving techniques different than those described here.

The algorithms presented here are examples of a more general class of conversion algorithms, where successive coefficients are processed as an internal representation is "shifted" or "rotated" in terms of its corresponding external-basis representation. For the algorithms here, such "external" shifting or rotation is accomplished by such operations as exponentiation, multiplication by G, or multiplication by G^{-1} (combined with subtraction of the $B[0]$ coefficient), depending on the algorithm. The import algorithms interleave shifting with insertion of coefficients into the internal representation, by addition of a the term $B[i] \times I$ or $B[i] \times G$; the export algorithms interleave shifting with extraction of the coefficients, by a computation of the form $(A \times V_0)[0]$.

Algorithms of the same class may be constructed for any choices of basis for which an efficient external shifting or rotation operation can be defined.

5   Applications

Many public-key cryptosystems are based on operations in large finite mathematical groups, and the security of these cryptosystems relies on the computational intractability of computing discrete logarithms in the underlying groups. The group operations usually consist of arithmetic in finite fields, in particular $GF(p)$ and $GF(2^m)$. In this section, we focus on the application of our conversion algorithms to elliptic curve cryptosystems over $GF(2^m)$ [5]. The general principles extend to other applications as well.

At a high level, an elliptic curve over $GF(2^m)$ is a set of points which form a group with respect to certain rules for adding two points. Such an addition consists of a series of field operations in $GF(2^m)$. In a generic implementation using projective coordinates, adding two distinct points needs 10 multiplications and 3 squarings, and doubling a point needs 5 multiplications and 5 squarings. [3]

Elliptic curve cryptosystems over $GF(2^m)$ that are of particular interest today are analogs of conventional discrete logarithm cryptosystems. Such elliptic curve schemes (e.g., EC Diffie-Hellman and EC DSA [1][3]) usually involve one or two *elliptic curve scalar multiplications* of the form $Q = kP$ where P and Q are points and k is an integer of length about m. Again in a generic implementation, such an operation can be done with m doublings of a point and $m/2$ additions of two distinct points, giving a total of $(3 \times m/2 + 5 \times m) = 6.5m$ field squarings and $(10 \times m/2 + 5 \times m) = 10m$ field multiplications.

To illustrate the cost of conversion, we consider a general situation in which two parties implement some elliptic curve scheme over $GF(2^m)$ with different choices of basis. In such a situation, each elliptic curve scalar multiplication in the scheme would require at most two conversions by one of the parties, one before and one after the operation. [4] The following table compares the cost of two conversions (back and forth) with the cost of one elliptic curve scalar multiplication.

---

[3] In general, the number of operations depends on the particular formulas and constraints on the parameters of the curve. The number given here is based on Annex A "Number-theoretic algorithms" in [3].

[4] For example, suppose Alice has a polynomial basis and Bob a normal basis, and they want to agree on a shared key using EC Diffie-Hellman. After obtaining Bob's public key P, Alice would import P from the normal basis to the polynomial basis, compute $Q = kP$ in the polynomial basis, and export Q from the polynomial basis back to the normal basis. Alternatively, Bob could perform the conversions.

operation	multiplications	squarings
IMPORTPOLY + EXPORTPOLY	$2m + 1$	0
IMPORTNORMAL + EXPORTNORMAL	m	$2m$
EC scalar multiplication	$10m$	$6.5m$

Based on the above table, we can estimate the extra cost of conversion compared with one elliptic curve scalar multiplication. When the external basis is a polynomial basis, the extra cost (IMPORTPOLY + EXPORTPOLY) is around $2/(10 + 6.5 \times 0.5) = 15\%$ for an internal polynomial basis (assuming squaring is twice as fast as multiplication for an internal polynomial basis) and about $2/10 = 20\%$ for an internal normal basis (since squarings are essentially free in an internal normal basis). Similarly, when the external basis is a normal basis, the extra cost (IMPORTNORMAL + EXPORTNORMAL)is about $(1 + 2 \times 0.5)/(10 + 6.5 \times 0.5) = 15\%$ for an internal polynomial basis and about $1/10 = 10\%$ for an internal normal basis. To summarize, the extra cost of conversion is between 10% and 20%, depending on the implementation.

Overall, we conclude that our conversion algorithms incur only a small extra cost in an elliptic curve cryptosystem, and that the memory requirement is quite reasonable: only a few additional constants or intermediate values need to be stored. The cost can be reduced still further by additional optimizations such as processing more than one coefficient at a time, with the only additional requirement being the storage of a small number of additional elements.

6   Conclusions

We have described in this paper several new algorithms for basis conversion that require much less storage than previous solutions. Our algorithms are also very efficient in that they involve primarily finite-field operations. This has the advantage of benefiting from the optimizations that are presumably already available for finite-field operations.

As examples, our algorithms can convert from a polynomial basis to a normal basis and from a normal basis to a polynomial basis. Related algorithms can convert to and from other choices of basis for which an efficient "external shifting" operation can be defined. Our algorithms are particularly applicable to even-characteristic finite fields, which are typical in cryptography.

The variation in choice of basis for representing finite fields has affected interoperability, especially of cryptosystems. With our algorithms, it is possible to extend an implementation in one basis so that it supports other choices of basis at only a small additional cost, thereby providing the desired interoperability and extending the set of parties that can communicate with cryptographic operations.

Acknowledgment

We would like to thank the IEEE P1363 working group for motivating this problem, Leonid Reyzin for helpful discussions, and the anonymous referees for their comments. We would also like to thank Rob Lambert for bringing Horner's Rule to our attention.

References

1. ANSI X9.62: *The Elliptic Curve Digital Signature Algorithm (ECDSA)*, draft, November 1997.
2. T.H. Cormen, C.E. Leiserson, and R.L. Rivest. *Introduction to Algorithms.* The MIT Press, 1990.
3. IEEE P1363: *Standard Specifications for Public-Key Cryptography*, Draft version 3, May 1998. **http://stdsbbs.ieee.org/groups/1363/draft.html**.
4. A. Menezes, I. Blake, X. Gao, R. Mullin, S. Vanstone, and T. Yaghoobian. *Applications of Finite Fields.* Kluwer Academic Publishers, 1993.
5. A. Menezes. *Elliptic Curve Public Key Cryptosystems.* Kluwer Academic Publishers, 1993.
6. R. Lidl and H. Niederreiter. *Finite Fields,* volume 20 of *Encyclopedia of Mathematics and Its Applications.* Addison-Wesley, 1983.

Verifiable Partial Sharing of Integer Factors

Wenbo Mao

Hewlett-Packard Laboratories
Filton Road, Stoke Gifford, Bristol BS34 8QZ, United Kingdom
wm@hplb.hpl.hp.com

Abstract. It is not known to date how to partially share the factors of an integer (e.g., an RSA modulus) with verifiability. We construct such a scheme on exploitation of a significantly lowered complexity for factoring $n = pq$ using a non-trivial factor of $\phi(n)$.

1 Introduction

Partial key escrow purports to add a great deal of difficulty to mass privacy intrusion which is possible in ordinary key escrow with abusive authorities while preserving the property of an ordinary escrowed cryptosystem in targeting small number of criminals. In partial key escrow, a portion of an individual's private key with an agreed and proved size will not be in escrow. Key recovery requires a non-trivial effort to determine the missing part. A partial key escrow scheme must render that the determination of the missing key part will only be possible *after* recovery of the key part which is in escrow (usually with a set of distributed agents who are collectively trusted to share the escrowed key part). If the missing part can be determined before, or without taking, a prescribed key recovery procedure, then off-line pre-computations can be employed for finding the missing part and this can be done in a massive scale with many or all users targeted. This constitutes a so-called prematured key recovery attack: the missing key part is not really missing and the whole private key of each user can be made available right after recovery of the escrowed key part. The effect of partial key escrow is thereby nullified and the scenario of mass privacy intrusion can still be assumed just as the case of an ordinary key escrow scheme. In their recent work "Verifiable partial key escrow", Bellare and Goldwasser [1] discussed scenarios of prematured key recovery attacks.

Thus, a necessary step in verifiable partial key escrow is for a key owner to prove that a private key contains a hidden number which will not be in escrow and has an agreed size. To discover this number requires first to recover the escrowed part of the private key, and only after that recovery can an exhaustive search procedure be lunched to determine the missing number. The cost of the search will be a well-understood problem given the proved size of the missing number.

The previous verifiable partial key escrow scheme of Bellare and Goldwasser [1] was proposed for discrete logarithm based cryptosystems. In that realization,

S. Tavares and H. Meijer (Eds.): SAC'98, LNCS 1556, pp. 94–105, 1999.

a key $y = g^x$ is cut into two parts $y_1 y_2 = g^{x_1} g^{x_2} = g^{x_1 + x_2}$ where x_1 is a partial private key and is proved in escrow, x_2 is the remaining private component which will not be in escrow but has an agreed bit size to be proved by the key owner during the key escrow time. Using a bit commitment scheme these proofs can be done without revealing the partial private and public keys. Key recovery will have to be via recovering x_1, y_1 then $y_2 = y/y_1$ followed by searching x_2 from y_2. Note that neither of the partial public keys should be revealed before recovering x_1, or else searching x_2 can take place before x_1 is recovered.

It is not known to date how to partially escrow integer factors (e.g., prime factors of an RSA modulus) with verifiability. Full key escrow for integer factoring based cryptosystems can take the approach of escrowing a prime factor of an integer [12] (the scheme in [12] achieves public verifiability). It is however not straightforward to perceive a partial integer-factor sharing scheme along that approach. A major problem is to establish a precise cost for key recovery (e.g., a 2^{40}-level time cost, which is non-trivial but expensively manageable by a well resourced agent, and has been chosen as a suitable workload for the partial key escrow scheme for discrete-log based cryptosystems [1].) Chopping down an 80-bit block[1] from a prime factor and throwing it away will unlikely be a correct rendering because, in integer factoring based cryptosystems, a prime factor should have a size significantly larger than 80 bits, and so the remaining escrowed part of the factor will be sufficiently large to allow exploitation of polynomial-time factoring methods (e.g., Coppersmith [6]) to factor the integer in question. On the other hand, throwing away a much larger number block may render a key unrecoverable.

We will construct a scheme for verifiable partial sharing of the factors of an integer. The central idea in our realization is an observation (one of the cryptanalysis results in [13]) on a significantly lowered and precisely measurable time complexity for factoring $n = pq$ if a factor of $\phi(n)$ (the Euler phi function) is known. This factor will be proved to have an agreed size and will be in escrow in verifiable secret sharing. We will reason that without recovery of this factor, integer factoring will be as the same difficult as the original factoring problem.

In the remainder of the paper, Section 2 describes a lowered complexity for factoring n with a known factor of $\phi(n)$, Section 3 constructs the proposed scheme; Section 4 analyzes the security and performance of the scheme; finally, Section 5 concludes the work.

2 Integer Factoring with an Additional Knowledge

Let $n = pq$ for p and q being distinct primes. Then

$$n + 1 = (p - 1)(q - 1) + (p + q). \tag{1}$$

---

[1] Searching a number of this size requires 2^{40} operations using known best algorithms based on a square-root reduction; we will discuss this reduction in Section 2.

Let r be a factor of $(p-1)(q-1)$, but not a factor of $p-1$ and $q-1$ (otherwise r^2 will be a factor of $(p-1)(q-1)$). Then

$$p + q \equiv n + 1 \, (\text{mod} \, r). \tag{2}$$

When r is smaller than $p + q$, congruence (2) means the equation

$$p + q = (n + 1 \bmod r) + kr, \tag{3}$$

for an unknown k. If $p + q$ becomes known, factoring n into p and q follows a simple calculation. Equation (3) shows that if r is known, finding the quantity $p + q$ is equivalent to finding the unknown k, and hence the difficulty to factor n is equivalent to that to find k.

Let $|a|$ denote the bit size of the integer a in the binary representation. For $|p + q| > |r|$, noting $(n + 1 \bmod r) < r$, we have

$$|k| + |r| - 1 \le |p + q| \le |k| + |r| + 1,$$

or

$$|k| \approx |p + q| - |r|.$$

So when r is known, the time needed to determine $p + q$, or to factor n, is bounded above by performing

$$2^{|p+q|-|r|} \tag{4}$$

computations. The algorithm is to search k in equation (3).

The bound in (4) can further be lowered significantly. Note that most elements in the multiplicative group of integers modulo n Z_n^* have orders larger than $p + q$ unless factoring n is easy. This means that arbitrarily picking $u \in Z_n^*$, and letting v be the least number satisfying

$$u^v \bmod n = 1,$$

then in an overwhelming probability,

$$v > p + q.$$

Combining (1) and (3), we will have

$$n + 1 = (p - 1)(q - 1) + (n + 1 \bmod r) + kr. \tag{5}$$

Raising u to the both sides of (5), writing $w = u^r \bmod n$, and noting $u^{(p-1)(q-1)} \equiv 1 \, (\text{mod} \, n)$, we have

$$u^{n+1-(n+1 \bmod r)} \equiv w^k \, (\text{mod} \, n). \tag{6}$$

We point out that the quantity k in (6) is exactly that in (3) because $kr < p + q$ which should be smaller than the order of u (namely, transforming (3) to (6), the number k will not be reduced in modulo the order of u). With r (hence

w) known, extracting k from the equation (6) using Shank's baby-step giant-step algorithm (see e.g., [5], Section 5.4.1) will only need about

$$2^{\frac{|k|}{2}}$$

operations (multiplication in Z_n^*) and the same order in the space complexity. This is a much lowered bound from (4) as it is the positive square root of (4). To this end we have proven the following statement.

Lemma 1. *Let $n = pq$ for p, q being two distinct primes, and r be a known factor of $\phi(n) = (p-1)(q-1)$ with $|r| < |p+q|$. Then time and space for factoring n can use $2^{\frac{|p+q|-|r|}{2}}$ as an upper bound.* □

A number of previous integer factoring based cryptosystems make use of a disclosed sizable factor of $\phi(n)$ where n is the product two secret primes [8, 9, 11]. Each of these systems includes moduli settings that allow feasible factorization using our method. For instance, a modulus setting in [8] satisfies $\frac{|p+q|-|r|}{2} = 35$.

Remark   Lemma 1 states an upper bound, which means to factor $n = pq$ with a known $r|(p-1)(q-1)$ needs *at most* $2^{\frac{|p+q|-|r|}{2}}$ operations. However, this complexity measurement reaches the lowest known to date on exploitation of computing small discrete logarithms. Any algorithm using fewer operations will provide a breakthrough improvement on solving the (small) discrete logarithm problem. The same argument applies to the verifiable partial key escrow scheme of Bellare and Goldwasser.

If r is unknown, then equation (3), hence equation (6), are not usable. For n being the product of large primes p and q, factorization is so far known to be an infeasible problem. The verifiable partial integer-factor sharing scheme to be specified in the next section will make use of this big time-complexity difference between factoring n with a sizeable factor of $(p-1)(q-1)$, and doing it without.

3   The Proposed Scheme

A user (Alice) shall construct her public key $n = pq$ satisfying that p and q are primes of the same size, $(p-1)(q-1)$ has a factor r with $|p+q| - |r| = 80$ and r is not a factor of both $p-1$ and $q-1$.

Alice shall then prove in zero-knowledge the following things: (i) n is the product of two hidden prime factors p and q of the same size; (ii) a hidden number r is a factor of $(p-1)(q-1)$ but not that of both $p-1$ and $q-1$, satisfying $\frac{|n|}{2} - |r| = 80$; (iii) verifiable secret sharing of r with a set of agents (called shareholders). After all of these have been done, we know from the previous section that if r is recovered by co-operative shareholders, n can be factored by finding k in (6) with 2^{40} operations using the known fastest algorithms in computing small discrete logarithms. On the other hand, if r is not available, n

will be in the same position as an ordinary RSA modulus, and given the size of n, no computationally feasible algorithm is known to factor it.

To this end we have fully described the working principle of a verifiable partial integer-factor sharing scheme. In the remaining of this section we shall construct computational zero-knowledge protocols for proof of the required structure of n, and for verifiable secret sharing of r. These constructions will use various previous results as building blocks, which can be found, respectively, in the work of Chaum and Pedersen [4] (for proof of discrete logarithm equality), Damgård [7] (for proof of integer size), Pedersen [14] (for verifiable threshold secret sharing), and Mao [12] (for proof of correct integer arithmetic).

These protocols will make use of a public cyclic group with the following construction. Let P be a large prime such that $Q = (P - 1)/2$ be also prime. Let $f \in Z_P^*$ be a fixed element of order Q. The public group is the multiplicative group generated by f. We assume that it is computationally infeasible to compute discrete logarithms to the base f. Once setup, the numbers f and P (hence $Q = (P - 1)/2$) will be announced for use by the system wide entities.

3.1   Proof of Correct Integer Arithmetic

We shall apply $y = f^x \bmod P$ as the one-way function needed to prove the correct integer arithmetic. (In the sequel we will omit the presentation of modulo operations whenever the omission does not cause confusion.)

For integers a and b, Alice can use the above one-way function to prove $c = ab$ without revealing a, b and c. She shall commit to the values a, b and c by sending to a verifier (Bob) their one-way images $(A, B, C) = (f^a, f^b, f^{ab})$ and prove to him that the pre-image of C is the product of those of A and B.

However note that for the one-way function f^x used, the proved multiplication relationship depicted above is in terms of modulo $ord(f)$ (here $ord(f) = Q$ is the order of the element f), and in general this relationship does not demonstrate that $\log_f(C) = \log_f(A) \log_f(B)$ is also true in the space of integers. Nevertheless, if Alice can show the bit sizes of the respective discrete logarithms (i.e., $|\log_f(A)|$, $|\log_f(B)|$ and $|\log_f(C)|$), then the following lemma guarantees the correct multiplication.

Lemma 2. *Let* $ab = c \,(\bmod Q)$ *and* $|c| + 2 < |Q|$. *If* $|a| + |b| \le |c| + 1$ *then* $ab = c$.

Proof. Suppose $ab \ne c$. Then $ab = c + \ell Q$ for some integer $\ell \ne 0$. Noticing $0 < c < Q$, so

$$|a| + |b| \ge |ab| = |c + \ell Q| \ge |Q| - 1 > |c| + 1,$$

contradicting the condition $|a| + |b| \le |c| + 1$.   □

Thus $y = f^x$ forms a suitable one-way function to be used for proof of the correct product of integers in its pre-image space, provided the bit sizes of the pre-images are shown. Given f^x, there exists efficient protocols to show $|x|$ without revealing x ([7]). In the next subsection we will specify a simplified variation of the proof which protects x up to computational zero-knowledge. Applying the bit-size proof protocol, we can specify a protocol (predicate) $Product(A, B, C)$ below. The predicate will return 1 (Bob accepts) if $\log_f(C) = \log_f(A) \log_f(B)$, or return 0 (Bob rejects) if otherwise.

Protocol $Product(f^a, f^b, f^{ab})$
(a run will be abandoned with 0 returned if any party finds any error in any checking step, otherwise it returns 1 upon termination.)
Alice sends to Bob: $|a|, |b|, |ab|$, and demonstrates the following evidence:

i) $\log_f(f^a) = \log_{f^b}(f^{ab})$ and $\log_f(f^b) = \log_{f^a}(f^{ab})$;
ii) $|a| + |b| \leq |ab| < |Q| - 2$.

In $Product$, showing (i) can use the protocol of Chaum and Pedersen [4], and showing (ii) can use a protocol $Bit\text{-}Size$ to be specified in the next subsection. Note that only the bit sizes regarding the first two input values need to be shown because, having shown the equations in step (i), the size regarding the third value is always the summation of those regarding the former two.

We will analyze the security and performance of this protocol in Section 4.

3.2   Proof of Integer Size

The basic technique is due to Damgård [7]. We specify a simplified version based on the discrete logarithm problem. In our variation, the number in question is protected in computational (statistical) zero-knowledge.

Let I be the interval $[a, b] (= \{x | a \leq x \leq b\})$, $e = b - a$, and $I \pm e = [a - e, b + e]$. In Protocol $Bit\text{-}Size$ specified below, Alice can convince Bob that the discrete logarithm of the input value to the agreed base is in the interval $I \pm e$.

Protocol $Bit\text{-}Size(f^s)$
Execute the following k times:

1. Alice picks $0 < t_1 < e$ at uniformly random, and sets $t_2 := t_1 - e$; she sends to Bob the unordered pair $C_1 := f^{t_1}$, $C_2 := f^{t_2}$;
2. Bob selects $b = 0$ or $b = 1$ at uniformly random, and sends b to Alice;
3. Alice sets $u_1 := t_1$, $u_2 := t_2$ for $b = 0$, $u_1 := t_1 + s$, $u_2 := t_2 + s$ for $b = 1$, and sends u_1, u_2 to Bob;
4. Bob checks the following (with $i = 1, 2$):
     for $b = 0$, $-e < u_i < e$ and $C_i = f^{u_i}$;
     for $b = 1$, $a - e < u_i < b + e$ and $C_i f^s = f^{u_i}$;

Set, for instance, $I := [2^{\ell-1}, 2^\ell]$. Then e will be $2^{\ell-1}$, and $Bit\text{-}Size$ will prove that the discrete logarithm of the input value does not exceed $\ell + 1$ binary bits.

3.3   Proof of the Prime Factors' Structure

Let $n = pq$ be the user Alice's public key for an integer factoring based cryptosystem. The proposed scheme will require Alice to set the primes p and q with the following structure

$$p = 2p's + 1, \quad q = 2q't + 1, \tag{7}$$

Here p', q', s, t are distinct odd numbers, any two of them are relatively prime, and their sizes satisfy

$$|p'| - |s| = |q'| - |t| = 80. \tag{8}$$

Then set

$$r = 4st. \tag{9}$$

As a procedure for key setup, Alice should first choose the numbers p', q', s, t at random with the required sizes, oddity and relative-prime relationship. She then samples if p and q in (7) are prime. The procedure repeats until p and q are prime. For p', q', s, t being odd, both p and q will be congruent to 3 modulo 4, rendering n to be a Blum integer [2]. We will need this property for proof of two prime structure of n. It is advisable that p' and q' be chosen as primes, resulting in p and q as the so-called strong primes (for n in a secure size, p' and q' will be large). This follows a desirable moduli setting for integer factoring based cryptosystems.

From (7) and (8) we know

$$|p| - 2|s| = |p'| - |s| = 80 = |q'| - |t| = |q| - 2|t|.$$

Noting $|p| = |q|$, so the above implies $|s| = |t|$, and

$$|p + q| - |r| \approx |p| + 1 - |4st| \approx |p| + 1 - (2 + 2|s|) \in \{79, 80, 81\}.$$

Once the key values are fixed, Alice shall publish

$$A = f^p, \quad B = f^q, \quad C = f^r, \quad D = f^{\frac{(p-1)(q-1)}{r}}.$$

Using these published values, Alice can prove to Bob the following two facts:

i) $n = \log_f(A) \log_f(B)$, by running $Product(A, B, f^n)$;

ii) $\log_f(C)$ divides $(\log_f(A) - 1)(\log_f(B) - 1)$, by running $Product(C, D, \frac{f^{n+1}}{AB})$ (note here the dis-log of the third input is $(\log_f(A) - 1)(\log_f(B) - 1)$).

During the proof that runs $Product$, Alice has also demonstrated the bit size values $|\log_f(A)|$, $|\log_f(B)|$ and $|\log_f(C)|$. Bob should check that

$$|\log_f(A)| - |\log_f(C)| \approx \frac{|n|}{2} - |\log_f(C)| \in \{79, 80, 81\}.$$

Finally Alice should prove that n consists of two prime factors only. This can be achieved by applying the protocol of Van de Graaf and Peralta [10] for proof of Blum integers (we have constructed n to be a Blum integer), and the protocol of Boyar et al [3] for proof of square-free integers.

3.4   Verifiable Secret Sharing of r

For self-containment, we include Pedersen's threshold verifiable secret sharing scheme [14] for sharing the secret r among a multi number of shareholders with threshold recoverability.

Let the system use m shareholders, Using Shamir's t ($< m$) out of m threshold secret sharing method ([15]), Alice can interpolate a t-degree polynomial $r(x)$

$$r(x) = \sum_{i=0}^{t-1} c_i x^i \bmod Q,$$

where the coefficients $c_1, c_2, \cdots, c_{t-1}$ are randomly chosen from Z_Q^*, and $c_0 = r$. The polynomial satisfies $r(0) = r$. She shall send the secret shares $r(i)$ ($i = 1, 2, \cdots, m$) to each of the m shareholders, respectively (via secret channels), and publishes

$$f^{r(i)} \bmod P, \text{ for } i = 0, 1, \cdots, m,$$

and

$$f^{c_j} \bmod P \text{ for } j = 0, 1, \cdots, t-1.$$

Each shareholder can verify

$$f^{r(i)} \equiv \prod_{j=0}^{t-1} (f^{c_j})^{i^j} \pmod{P} \text{ for } i = 1, 2 \cdots, m.$$

Assume that at least t of the m shareholders are honest by performing correct checking. Then the polynomial $r(x)$ is now secretly shared among them. When key recovery is needed, they can use their secret shares to interpolate $r(x)$ and recover $r = r(0)$.

We should point out that this sub-protocol is not a necessary component in the proposed scheme. There exists other schemes to prove a correct threshold encryption of a discrete logarithm (e.g., [12] achieves verifiable secret sharing with a public verifiability; that is, secret sharing can be done without the presence of the m shareholders and without assuming t of them to be honest). We have chosen Pedersen's scheme for simplicity in presentation.

4   Analysis

We now provide security and performance analyzes on the proposed verifiable partial integer-factor sharing scheme.

4.1   Security

The security of the proposed scheme consists of the correctness and the privacy. The former concerns whether Alice can successfully cheat Bob to accept her proof using a key in which the private component is either not recoverable, or recoverable at a much higher cost than that of the procedure specified in Section 2. The latter concerns whether Bob can gain any knowledge, as a result of verifying Alice's proof, leading to discovery of Alice's private key without going through the prescribed key recovery procedure.

Correctness

In Section 2 we have established the precise cost for key recovery. The remark following Lemma 1 further emphasizes that should there exist an algorithm that can find the missing key part with fewer than 2^{40} operations, that algorithm will also form a breakthrough improvement from all methods we know to date for computing small discrete logarithms. The remaining correctness issue is on that of the sub-protocols which realize the mathematics established in Section 2.

The two sub-protocols that construct *Product* have well understood correctness properties [4, 7]. The cheating probability for proof of a Diffie-Hellman triple is $1/Q$ where Q is the order of f, and that for *Bit-Size* is $1/2^k$, here k is the number of iterations in the it. Using $k = 40$ will achieve a sufficiently low chance for successful cheating by Alice. So we can use $1/2^{40}$ as the cheating probability for *Product* (since $1/Q \ll 1/2^{40}$ for usual sizes of Q).

Next, the correctness of the protocol for proof of two-prime structure is also well established [10, 3], with error probability $1/2^k$ for k iterations of message verification. Again, k can be set to 40.

We note that in the proofs for the structural validity of n (i.e., $n = pq$, the primality of p, q, the relation $r|(p-1)(q-1)$, and the required sizes of these numbers), the verification job does not involve handling any secret. Therefore it can be carried out by anybody and can be repeated if necessary. So Bob cannot collude with Alice without risking to be caught.

Finally, the correctness of the protocol for verifiable sharing of r [14] has a two-sided error. The error probability for Alice's successful cheating is $1/Q$ which is negligible as Q is sufficiently large. The other side of the error is determined by the number of dishonest shareholders. The number r can be correctly recovered if no fewer than t out of the m shareholders are honest (i.e., they follow the protocol). Usually the threshold value t is set to $\lfloor \frac{m}{2} \rfloor$.

Privacy

Although each of the sub-protocols used in the construction of the main scheme has its own proved privacy, we cannot assume that when they run in combination they will still provide a good privacy for the main scheme. Thus, we shall study the privacy of the main scheme in the following manner: (i) the values that have been made public in the main scheme do not themselves leak useful information for finding the prime factors, and (ii) when viewed together, they will not produce a formula usable for finding the prime factors. We emphasize that the quality of (ii) is essential as we must be sure that the published values will not damage the original privacy of each sub-protocol used in the main scheme.

In addition to the public key n, in the main scheme Alice has published the following values:

$$A = f^p, \quad B = f^q, \quad C = f^r, \quad D = f^{\frac{(p-1)(q-1)}{r}}, \quad |p| = |q| = \frac{|n|}{2}, \quad |r| \approx \frac{|n|}{2} - 80.$$

First of all it is obvious that there is no danger for disclosing the bit size information of the secret numbers. So we need only to consider the first four numbers: A, B, C and D.

We note that (A, B, f^n) and $(C, D, \frac{f^{n+1}}{AB})$ forming two Diffie-Hellman triples. This fact, however, will not help Bob to find the discrete logarithms of A, B, C, and D, following the security basis of the Diffie-Hellman problem. Analogously, any arithmetic of these numbers will not produce anything easier than the original Diffie-Hellman problem.

We should remind that, even for n having a widely-believed least size of 512 bits, the discrete logarithms of the published values above will all be sufficiently large. The least one is r whose size satisfies

$$|r| \geq 256 - 81 = 175,$$

which is still sufficiently large and immune to the square-root attack aimed for extracting it from C.

Next we shall review if any of the published numbers will damage the original privacy of each sub-protocol used. We can safely begin without concerning any impact on the two sub-protocols used in proof of two-prime structure of n because those protocols involves calculations in a different group: Z_n^* rather than Z_p^*.

The two sub-protocols that construct *Product* require Alice to hide the private input values (which are discrete logs of the input to *Product* or to *Bit-Size*) by using a genuine random number in each round of message exchange (these random numbers form the prover's random input; review [4] for proof of Diffie-Hellman triple, and Section 3.2 for proof of integer size). In both sub-protocols, the prover's random input will have similar sizes as those of the private input. In fact, the sub-protocol for proof of Diffie-Hellman triple is perfect (computational) zero-knowledge [4], and that for proof of bit size is statistical zero-knowledge [7]. As a result, *Product* can be simulated using the standard way for demonstrating an honest verifier zero-knowledge protocol.

Finally we point out that Pedersen's protocol for verifiable sharing of r (the discrete log of C) has a well understood perfect privacy following the privacy of Shamir's secret sharing scheme based on polynomial interpolation [14, 15].

4.2 Performance

The main scheme consists of (i) verifiable secret sharing of r, (ii) two instances of running *Product*, and (iii) proof of two-prime structure of n.

Since (i) can be viewed as common in verifiable secret sharing schemes, we shall only analyze the performance of (ii) and (iii) as an additional cost for achieving verifiable partial sharing of integer factors.

The major cost in a run of *Product* is to prove the size of two integers (twice running *Bit-Size*), adding a trivial cost for proof of discrete logarithm equality (a three-move protocol). For the two runs of *Product*, total data to be exchanged in *Product* will be bounded above by $(4k+2)|P|$ (binary bits) where k is the number of iterations in *Bit-Size*. The number of computations needed to be performed by both the prover and the verifier can be expressed by $4k|P|$ (multiplications modulo P).

Next, in the proof of two-prime structure of n, the protocol of Van de Graaf and Peralta [10] involves agreeing k pairs of random numbers in Z_n^* and k bits as random signs. For each pair of the numbers agreed, the two parties will evaluate a Jacobi symbol which is at a level of performing a few multiplications in Z_n^*. So the data to be exchanged here will be bounded above by $(2k+1)|n|$ (binary bits), where k can be as the same as that in *Bit-Size*. A similar estimate apply to the protocol for proof of square-free numbers [3]. The number of computations need to be performed by both the prover and the verifier can be expressed by $2k|n|$ (multiplications modulo n). We can use $|P|$ to up-bound $|n|$.

Combining the results in the two paragraphs above, we conclude in following statement for the performance of the proposed scheme.

Lemma 3. *With a verifiable secret sharing scheme for sharing the discrete logarithm of an integer, partial sharing of the prime factors of an integer of size up to $|P|-3$ (binary bits) can be achieved by further transmitting $(8k+3)|P|$ binary bits, with both the prover and the verifier computing $6k|P|$ multiplications.* □

If $|P|$ is set to 1540, then the maximum size of the integers that can dealt with by the proposed scheme will be 1536 (binary bits) $(= 3 \times 512)$. Considering setting $k = 40$ will allow a sufficiently small probability of $1/2^{40}$ for the prover successfully cheating. Then, the total number of bits to be transmitted will be 497,420 (62 kilobytes). A slightly smaller number of multiplications (modulo P and modulo n) will need to be performed by both the prover and the verifier.

5   Conclusion

We have constructed a verifiable partial integer-factor sharing scheme and shown that the scheme is secure and practically efficient. The working principle of the scheme is an observation on a significantly lowered time complexity for factoring $n = pq$ using a known factor of $\phi(n)$. This is of independent interest in that the lowered complexity bound should be regarded as a piece of must-know knowledge for designing protocols or cryptosystems based on disclosing a factor of $\phi(n)$ of a non-trivial size.

Acknowledgments

I wish to thank the reviewers of the SAC'98 for helpful comments.

References

1. Bellare, M. and S. Goldwasser. Verifiable partial key escrow. Proceedings of 4th ACM Conference on Computer and Communications Security. ACM Press. April 1997. pp. 78–91.

2. Blum, M. Coin flipping by telephone: a protocol for solving impossible problems. Proceedings of 24th IEEE Computer Conference (CompCon), 1982. pp. 133–137.

3. Boyar, J., K. Friedl and C. Lund. Practical zero-knowledge proofs: Giving hints and using deficiencies. Advances in Cryptology: Proceedings of EUROCRYPT 89 (J.-J. Quisquater and J. Vandewalle, eds.), Lecture Notes in Computer Science, Springer-Verlag, 434 (1990) pp 155–172.

4. Chaum, D. and T. P. Pedersen. Wallet databases with observers. Advances in Cryptology: Proceedings of CRYPTO 92 (E.F. Brickell, ed.), Lecture Notes in Computer Science Springer-Verlag, 740 (1993) pp. 89–105.

5. Cohen, H. *A Course in Computational Algebraic Number Theory.* Springer-Verlag Graduate Texts in Mathematics 138 (1993).

6. Coppersmith, D. Finding a small root of a bivariate integer equation; factoring with high bits known. Advances in Cryptology: Proceedings of EUROCRYPT 96 (U. Maurer, ed.), Lecture Notes in Computer Science, Springer-Verlag, 1070 (1996) pp. 178–189.

7. Damgård, I.B. Practical and provably secure release of a secret and exchange of signatures. Advances in Cryptology: Proceedings of EUROCRYPT 93 (T. Helleseth, ed.), Lecture Notes in Computer Science, Springer-Verlag, 765 (1994) pp. 201–217.

8. Girault, M. An identity-based identification scheme based on discrete logarithms modulo a composite number. Advances in Cryptology: Proceedings of EUROCRYPT 90 (I.B. Damgård, ed.), Lecture Notes in Computer Science, Springer-Verlag, 473 (1991) pp. 481–486.

9. Girault, M. and J.C. Paillès. An identity-based scheme providing zero-knowledge authentication and authenticated key-exchange. First European Symposium on Research in Computer Security – ESORICS 90 (1990) pp. 173–184.

10. Van de Graaf, J. and R. Peralta. A simple and secure way to show the validity of your public key. Advances in Cryptology: Proceedings of CRYPTO 87 (E. Pomerence, ed.), Lecture Notes in Computer Science, Springer-Verlag, 293 (1988) pp. 128–134.

11. Kim, S. J., S. J. Park and D. H. Won. Convertible group signatures. Advances in Cryptology: Proceedings of ASIACRYPT 96 (K. Kim, T. Matsumoto, eds.), Lecture Notes in Computer Science, Springer-Verlag, 1163 (1996) pp. 310–321.

12. Mao, W. Necessity and realization of universally verifiable secret sharing. 1998 IEEE Symposium on Security and Privacy, IEEE Computer Society (1998) pp. 208–214.

13. Mao, W. and C.H. Lim. Cryptanalysis of prime order subgroup of Z_n^*. Advances in Cryptology: Proceedings of ASIACRYPT 98, Lecture Notes in Computer Science, Springer-Verlag (to appear: November 1998).

14. Pedersen, T. Non-interactive and information-theoretic secure verifiable secret sharing. Advances in Cryptology: Proceedings of CRYPTO 91 (J. Feigenbaum, ed.), Lecture Notes in Computer Science, Springer-Verlag, 576 (1992) pp. 129–120.

15. Shamir, A. How to share a secret. Communications of the ACM, Vol 22 (1979) pp 612–613.

Higher Order Differential Attack
Using Chosen Higher Order Differences

Shiho Moriai[1], Takeshi Shimoyama[2], and Toshinobu Kaneko[2,3]

[1] NTT Laboratories
1-1 Hikari-no-oka, Yokosuka, 239-0847 Japan
shiho@isl.ntt.co.jp
[2] TAO
1-1-32 Shin-urashima-cho, Kanagawa-ku, Yokohama, 221-0031 Japan
shimo@yokohama.tao.go.jp
[3] Science University of Tokyo
2641 Yamazaki, Noda, Chiba, 278-8510 Japan
kaneko@ee.noda.sut.ac.jp

Abstract. This paper introduces an improved higher order differential attack using chosen higher order differences. We can find a lower order of the higher order differential by choosing higher order differences. It follows that the designers of a block cipher can evaluate the lower bound of the number of chosen plaintexts and the complexity required for the higher order differential attack. We demonstrate an improved higher order differential attack of a CAST cipher with 5 rounds using chosen higher order differences with fewer chosen plaintexts and less complexity. Concretely, we show that a CAST cipher with 5 rounds is breakable with 2^{16} plaintexts and $< 2^{24}$ times the computation of the round function, which half the values reported in Fast Software Encryption Workshop'98. We also show that it is breakable with 2^{13} plaintexts and about 2^{44} times the computation of the round function, which are $\frac{1}{16}$-th of those reported in Fast Software Encryption Workshop'97.

1 Introduction

Higher order differential attack is a powerful algebraic cryptanalysis. It is useful for attacking ciphers which can be represented as Boolean polynomials of low degrees. After Lai mentioned the cryptographic significance of derivatives of Boolean functions in [9], Knudsen used this notion to attack ciphers that were secure against conventional differential attacks [8]. At FSE'97 Jakobsen and Knudsen [7] gave an extension of Knudsen's attacks and broke ciphers using quadratic functions such as the cipher $\mathcal{KN}$ [12]. They were provably secure ciphers against differential and linear cryptanalysis. Furthermore, at Information Security Workshop'97 [13], we reduced the complexity required for the higher order differential attack of the cipher $\mathcal{KN}$ by solving the attack equation algebraically. At Fast Software Encryption Workshop'98 [11], we generalized it and applied it to a CAST cipher.

S. Tavares and H. Meijer (Eds.): SAC'98, LNCS 1556, pp. 106–117, 1999.
© Springer-Verlag Berlin Heidelberg 1999

This paper introduces an improved higher order differential attack using chosen higher order differences. The higher order differential attack exploits the fact that a higher order differential (e.g., the d-th order differential) of an intermediate data is constant, or independent of the key. In this paper we call the order "d" *the order of the higher order differential.*

In the known higher order differential attack [7, Theorem 1], the order of the higher order differential was found from the algebraic degree of Boolean polynomials of the ciphertexts. That is, if a ciphertext bit is represented by a Boolean polynomial of plaintext bits of degree d, then $d + 1$ is the order of the higher order differential, since the $(d + 1)$-th order differential of the ciphertexts becomes 0. Furthermore, in the higher order differential attack described in [11], it was shown that if all subkeys are combined using operation XOR in a Feistel cipher, the order of the higher order differential is equal to the algebraic degree of the Boolean polynomials of ciphertexts. That is, if a ciphertext bit is represented by a Boolean polynomial of plaintexts of degree d, then d is the order of the higher order differential, since the d-th order differential of the ciphertexts becomes 1. It is known that the order of the higher order differential determines the required number of plaintexts and ciphertexts pairs (p/c pairs) and complexity. Therefore, it is important to find the lowest order of the higher order differential to estimate the security of a cipher against the higher order differential attack.

This paper shows that we can find the lower order of the higher order differential by choosing higher order differences. For example, we demonstrate the higher order differential attack of a CAST cipher using chosen higher order differences with fewer chosen p/c pairs and less complexity. Concretely, we show that a CAST cipher with 5 rounds is breakable with 2^{16} plaintexts and $< 2^{24}$ times the computation of the round function, which half the values reported in Fast Software Encryption Workshop'98 [11]. We also show that it is breakable with 2^{13} plaintexts and about 2^{44} times the computation of the round function, which are $\frac{1}{16}$-th of those reported in Fast Software Encryption Workshop'97 [7]. The reason why we apply the improved higher order differential attack to a CAST cipher with 5 rounds is that we wan to to show how much improvement is achieved by choosing higher order differences. This attack is also applicable to other block ciphers. A similar improved higher order differential attack of a 5-round MISTY without FL functions is shown in [14].

2   Higher Order Differential Attack

This section gives an outline of the higher order differential attack. Fuller descriptions of the attack are presented in the references [7, 9, 11].

Let the target be a Feistel cipher with block size 64 bits and r rounds. We assume that the right half 32-bit of the plaintext is fixed at any value. We denote the left half 32-bit of the plaintext by $x = (x_{31}, \ldots, x_0) \in GF(2)^{32}$, the ciphertext by $y = (y_L, y_R)$, $y_L, y_R \in GF(2)^{32}$, and the key by $k = (k_{l-1}, \ldots, k_0)$, where the key length is l bits. Let X and K be sets of variables s.t. $X = \{x_{31}, \ldots, x_0\}$ and

$K = \{k_{l-1}, \ldots, k_0\}$. Let $k^{(i)} = (k_{31}^{(i)}, \ldots, k_0^{(i)})$ be the i-th round key. Throughout this paper, the subscript 0 indicates the least significant bit of the data.

When the key k is fixed, an intermediate bit in the encryption process denoted by $z \in GF(2)$ can be represented as a Boolean polynomial with respect to X whose coefficients are Boolean polynomials with respect to K, i.e., $z = g[k](x)$, where

$$g[k](x) = \sum c_{i_{31}, \ldots, i_0}(k) \cdot x_{31}^{i_{31}} \cdots x_0^{i_0}.$$

Note that $c_{i_{31}, \ldots, i_0}(k)$ is the coefficient of $x_{31}^{i_{31}} \cdots x_0^{i_0}$, and $i_{31}, \ldots, i_0$ is 0 or 1.

Definition 1. *We define the i-th order differential of $g[k](x)$ with respect to X, denoted by $\Delta_{(a_i, \ldots, a_1)}^{(i)} g[k](x)$, as follows;*

$$\Delta_{(a)}^{(1)} g[k](x) = g[k](x) + g[k](x + a),$$

$$\Delta_{(a_i, \ldots, a_1)}^{(i)} g[k](x) = \Delta_{(a_i)}^{(1)} \left(\Delta_{(a_{i-1}, \ldots, a_1)}^{(i-1)} g[k](x) \right),$$

where $a \in GF(2)^{32}$, and $\{a_i, \ldots, a_1\} \subseteq GF(2)^{32}$ are linearly independent. Let "+" denote bitwise XOR. In this paper, since we consider only the higher order differential with respect to X, we omit "with respect to X."

Definition 2. *On $\Delta_{(a_i, \ldots, a_1)}^{(i)} g[k](x)$, which is the i-th order differential of $g[k](x)$, we define i as the order of the higher order differential. Furthermore, we define $\{a_i, a_{i-1}, \ldots, a_1\}$ as the i-th order differences. The i-th order differences consist of i-tupple differences in $GF(2)^{32}$.*

The following theorems are known on the higher order differential of Boolean functions [9, 13].

Theorem 1. [9] *The following equation holds for any $b \in GF(2)^{32}$ and $\{a_i, \ldots, a_1\} \subseteq GF(2)^{32}$.*

$$\Delta_{(a_i, \ldots, a_1)}^{(i)} g[k](b) = \sum_{x \in V^{(i)}[a_i, \ldots, a_1]} g[k](x + b).$$

Note that $V^{(i)}[a_i, \ldots, a_1]$ denotes the i-dimensional subspace spanned by $\{a_i, \ldots, a_1\}$. In other words, it is the set of all 2^i possible linear combinations of $a_i, \ldots, a_1$, where each a_i is in $GF(2)^{32}$ and linearly independent.

Theorem 2. [13] *Let d be a natural number. Let $\{a_{d+1}, \ldots, a_1\} \subseteq GF(2)^{32}$ be linearly independent. If the degree of $g[k](x)$ with respect to X is d, then we have the following equations.*

$$\Delta_{(a_d, \ldots, a_1)}^{(d)} g[k](x) \in R[K], \quad \text{and}$$

$$\Delta_{(a_{d+1}, \ldots, a_1)}^{(d+1)} g[k](x) = 0,$$

where $R[K]$ is the Boolean polynomial ring of K.

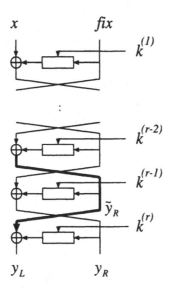

Fig. 1. A higher order differential attack of Feistel ciphers

Attack Procedure. In the improved higher order differential attack described in this paper, the last round key $k^{(r)}$ is recovered as follows (See also Fig.1.).

Step 1. Find the lowest order d s.t. for the d-th order differences $^\exists\{a_d, \ldots, a_1\}$ $\in GF(2)^{32}$ and $^\exists b \in GF(2)^{32}$, the d-th order differential of $\tilde{y}_R(b)$, i.e., $\Delta^{(d)}_{(a_d, \ldots, a_1)} \tilde{y}_R(b)$, is independent of the key.

Step 2. Construct attack equation (1) and solve it with respect to the last round key $k^{(r)}$.

$$\Delta^{(d)}_{(a_d, \ldots, a_1)} F[k^{(r)}](y_R(b)) + \Delta^{(d)}_{(a_d, \ldots, a_1)} y_L(b) = \Delta^{(d)}_{(a_d, \ldots, a_1)} \tilde{y}_R(b)$$

$$\iff \sum_{x \in V^{(d)}[a_d, \ldots, a_1]+b} F[k^{(r)}](y_R(x)) + \sum_{x \in V^{(d)}[a_d, \ldots, a_1]+b} y_L(x) = \sum_{x \in V^{(d)}[a_d, \ldots, a_1]+b} \tilde{y}_R(x) \quad (1)$$

One way to find the last round key $k^{(r)}$ is exhaustive search [7], where the attacker tries all 2^{32} possible candidates of $k^{(r)}$ and finds the correct key. Another way is the algebraic solution [13, 11], where the attacker transforms the algebraic equations into the system of linear equations and solves it, for example.

3 CAST Ciphers

This section describes CAST ciphers, which we use for demonstrating the improved higher order differential attack.

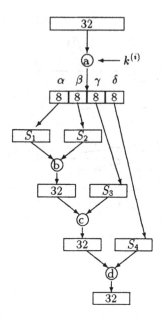

Fig. 2. Round function of CAST ciphers

CAST ciphers are a family of symmetric ciphers constructed using the CAST design procedure proposed by Adams and Tavares [1]. The CAST design procedure describes that they appear to have good resistance to differential cryptanalysis [5], linear cryptanalysis [10], and related-key cryptanalysis [4].

CAST ciphers are based on the framework of the Feistel cipher. The round function is specified as follows (See also Fig.2.). A 32-bit data half is input to the function along with a subkey $k^{(i)}$. These two quantities are combined using operation "a" and the 32-bit result is split into four 8-bit pieces. Each piece is input to a different 8×32 S-box (S_1, S_2, S_3, and S_4). S-boxes S_1 and S_2 are combined using operation "b"; the result is combined with S_3 using operation "c"; this second result is combined with S_4 using operation "d". The final 32-bit result is the output of the round function.

The CAST design procedure allows a wide variety of possible round functions: 4 S-boxes and 4 operations (a,b,c, and d). As for S-boxes, reference [3] suggested constructing the S-boxes from bent functions. Later, on reference [6] CAST ciphers with random S-boxes was proposed. In our attack, we use the S-boxes based on bent functions proposed for CAST-128. CAST-128 is a famous example CAST cipher used in several commercial applications, e.g., PGP5.0. As for operations, a simple way to define the round function is to specify that all operations are XORs, which is addition on $GF(2)$, although other operations may be used instead. Actually, according to reference [1], some care in the choice of operation "a" can conceivably give intrinsic immunity to differential and linear cryptanalysis.

As for the number of rounds, it seems that the CAST design procedure doesn't specify a concrete number. For example, CAST-128 has 12 or 16 rounds [2]. There are also several key schedules for CAST ciphers, but for the purpose of our attack the key schedule makes no difference.

4   Higher Order Differential Attacks of a 5-round CAST Cipher using Chosen Higher Order Differences

This section demonstrates an improved higher order differential attack using chosen higher order differences. The target is the CAST cipher with 5 rounds described in Section 3. The improvement consists in Step 1 in the attack procedure in Section 2.

4.1   How to find the lowest order of the higher order differential

Using degree of Boolean polynomials In the previously known higher order differential attack, the order of the higher order differential was derived using the degree of Boolean polynomials of $\tilde{y}_R(x)$ with respect to X.

The way to find the order of the higher order differential of the CAST cipher with 5 rounds is as follows. We begin by considering the degree of the round function. Let S_1, S_2, S_3, and S_4 be the functions of S-boxes: $GF(2)^8 \rightarrow GF(2)^{32}$. It is shown in [11] that for every S-box all output bits can be represented by Boolean polynomials of input bits of degree 4. Considering the structure of the round function (See Fig.2), all output bits of the round function can be represented by Boolean polynomials of input bits of degree 4, since we assume that operations "a", "b", "c", and "d" are XORs [11].

If we fix the right half of the plaintext at $0 \in GF(2)^{32}$, the right half of the output of the 4-th round $\tilde{y}_R(x) \in GF(2)^{32}$ can be represented as Eq. (2).

$$\tilde{y}_R(x) = f(f(x + f(k^{(1)}) + k^{(2)}) + k^{(3)}) + x + f(k^{(1)}), \qquad (2)$$

where $f : GF(2)^{32} \rightarrow GF(2)^{32}$ is the round function. Since f can be represented by Boolean polynomials of input bits of degree 4, the degree of $f(x+f(k^{(1)})+k^{(2)})$ with respect to $X = \{x_{31}, \ldots, x_0\}$ is 4, and the terms of the 4-th degree have the coefficient in $GF(2)^{32}$, which means that it is independent of the key. Hence, the degree of $\tilde{y}_R(x)$ with respect to X is at most 16, and the terms of the 16-th degree included in Eq. (2) have the coefficient in $GF(2)^{32}$, which means that it is independent of the key. Therefore, the 16-th order differential of $\tilde{y}_R(x)$ is constant for any linearly independent 16-th order difference $\{a_{16}, a_{15}, \ldots, a_1\}$.

$$\Delta^{(16)}_{(a_{16}, \ldots, a_1)} \tilde{y}_R(x) = \sum_{x \in V^{(16)}[a_{16}, \ldots, a_1]} \tilde{y}_R(x) \;\; = c \in GF(2)^{32} \;\; (= \text{const.})$$

Using chosen higher order differences In this section, we show that if we choose higher order differences, some higher order differential attacks of the CAST cipher with 5 rounds are possible where the order of the higher order differential is less than 16. Whether a higher order differential attack is possible when the order of the higher order differential is less than 16 depends on whether a higher order differential of $\tilde{y}_R(x)$ is independent of the key for the order of the higher order differential less than 16.

[WHEN THE ORDER OF THE HIGHER ORDER DIFFERENTIAL IS 15] Let us consider the 15-th order differential of $\tilde{y}_R(x)$. First, we prove the following lemma.

Lemma 1. *If we choose $\{e_{14}, e_{13}, \ldots, e_0\}$ for the 15-th order differences $\{a_{15}, a_{14}, \ldots, a_1\}$, each bit of the 15-th order differential of $\tilde{y}_R(x)$ is constant or linear with respect to a key bit. Note that e_i is defined as:*

$$e_i = (0, \ldots, \overset{i}{1}, \ldots, 0) \in GF(2)^{32}.$$

Proof. The 15-th order differential of $\tilde{y}_R(x)$ for 15-th order differences $\{e_{14}, e_{13}, \ldots, e_0\}$ is the same as the 15-th order differential of $\tilde{y}_R(x)$ with respect to $\{x_{14}, x_{13}, \ldots, x_0\}$. Therefore, the 15-th order differential of $\tilde{y}_R(x)$ doesn't have the terms that don't include all variables of $\{x_{14}, x_{13}, \ldots, x_0\}$. All the terms of $\tilde{y}_R(x)$ that remain in the 15-th order differential are as follows;

$$\text{degree-15: } c_1(k)x_{14}x_{13}\cdots x_0 \quad \text{and} \tag{3}$$
$$\text{degree-16: } c_2(k)x_{15}x_{14}\cdots\cdots x_0. \tag{4}$$

First, we show why the terms such as $c_3(k)x_{16}x_{14}\cdots x_0$ don't remain. Let X_1, X_2, X_3, and X_4 be sets of variables as follows:

$$X_1 = \{x_{31}, x_{30}, \ldots, x_{24}\},$$
$$X_2 = \{x_{23}, x_{22}, \ldots, x_{16}\},$$
$$X_3 = \{x_{15}, x_{14}, \ldots, x_8\},$$
$$X_4 = \{x_7, x_6, \ldots, x_0\}.$$

The terms of degree 16 included in $\tilde{y}_R(x)$ is the product of four terms of degree 4 with respect to X in the output of the 2-nd round function, $f(x + f(k^{(1)}) + k^{(2)})$. The terms in the output of the 2-nd round function consist of terms with respect to only X_1, terms with respect to only X_2, terms with respect to only X_3, terms with respect to only X_4, and constant terms depending on k. Therefore, the terms of degree 16 that contain variables $x_{16} \subset X_2$, $\{x_{15}, \ldots, x_8\} \subset X_3$, and $\{x_7, \ldots, x_0\} \subset X_4$ don't remain in the 15-th order differential of $\tilde{y}_R(x)$ for 15-th order differences $\{e_{14}, e_{13}, \ldots, e_0\}$.

Second, consider the coefficient of the degree-15 term $c_1(k)x_{14}x_{13}\cdots x_0$ (Eq. (3)). We begin by considering the terms included in the output of the 2-nd round function. The output of the 2-nd round function includes the terms in the following, as one example, since the input of the 2-nd round function is

$x + f(k^{(1)}) + k^{(2)}$. Let $f_i : GF(2)^{32} \to GF(2)$ be the function which outputs the i-th bit of the output of f.

degree-4:
$$(x_3 + f_3(k^{(1)}) + k_3^{(2)})(x_2 + f_2(k^{(1)}) + k_2^{(2)})(x_1 + f_1(k^{(1)}) + k_1^{(2)})(x_0 + f_0(k^{(1)}) + k_0^{(2)}) \quad (5)$$

degree-3:
$$(x_2 + f_2(k^{(1)}) + k_2^{(2)})(x_1 + f_1(k^{(1)}) + k_1^{(2)})(x_0 + f_0(k^{(1)}) + k_0^{(2)}) \quad (6)$$

The coefficients of the terms of degree-4 with respect to X included in the output of the 2-nd round function are 1, if they exist, since they come from terms such as Eq. (5). The coefficients of the terms of degree-3 with respect to X are the sum of the coefficients of the terms expanded from terms such as Eq. (5) and terms such as Eq. (6). The coefficient from the former is linear with respect to a key bit, and the coefficient from the latter is 1. Since the terms of degree 15 with respect to X included in $\tilde{y}_R(x)$ are the products of three terms of degree 4 and one term of degree 3 included in the output of the 2-nd round function, from the discussion above, the coefficients of the terms of degree 15 are represented as

$$\alpha_1 \times k_i + \alpha_0,$$

where k_i is a key bit and $\alpha_1, \alpha_0 \in GF(2)$.

From similar considerations, the coefficients of the terms of degree 16 included in $\tilde{y}_R(x)$ are 0 or 1.

In conclusion, considering

- the coefficient of the degree-15 term $c_1(k)$ is $\alpha_1 k_i + \alpha_0$, where $\alpha_1, \alpha_0 \in GF(2)$, and
- the coefficient of the degree-16 term $c_2(k)$ is 0 or 1

which remain in the Boolean polynomial of each bit of $\Delta_{(e_{14}, \ldots, e_0)}^{(15)} \tilde{y}_R(x)$, each bit of $\Delta_{(e_{14}, \ldots, e_0)}^{(15)} \tilde{y}_R(x)$ is one of the following:

$$\{x_{15} + k_i + 1, x_{15} + k_i, x_{15} + 1, x_{15}, k_i + 1, k_i, 1, 0\}.$$

Moreover, when the degree-16 term $x_{15}x_{14} \cdots x_0$ exists, α_1 is always 1, since the input of the 2-nd round function is $x + f(k^{(1)}) + k^{(2)}$. Therefore, each bit of $\Delta_{(e_{14}, \ldots, e_0)}^{(15)} \tilde{y}_R(x)$ is one of the following:

$$\begin{cases} x_{15} + k_i + 1 \\ x_{15} + k_i \\ 1 \\ 0 \end{cases} \quad (7)$$

This proves that if we choose $\{e_{14}, \ldots, e_0\}$ for the 15-th order differences, each bit of the 15-th order differential of $\tilde{y}_R(x)$ is constant or linear with respect to a key bit. □

From Eq. (7) in the proof of Lemma 1, the following corollary is proved.

Table 1. The number of constant bits of the 15-th order differential of $\tilde{y}_R(x)$ for some chosen differences

differences	$\{E_1, E_2\} \setminus e_i$	$\{E_1, E_3\} \setminus e_i$	$\{E_1, E_4\} \setminus e_i$
# of constant bits	15	13	12
differences	$\{E_2, E_3\} \setminus e_i$	$\{E_2, E_4\} \setminus e_i$	$\{E_3, E_4\} \setminus e_i$
# of constant bits	11	14	9

Corollary 1. *if we choose $\{e_{14}, \ldots, e_0\}$ for the 15-th order differences, some bits of the 15-th order differential of $\tilde{y}_R(x)$ are constant, and the XORed values of any two bits of the other bits are also constant.*

A similar corollary is proved for the following 15-th order differences $\{a_{15}, \ldots, a_1\}$.

$$\{a_{15}, \ldots, a_1\} = $$

$$\left\{ \bigcup_{\text{Two sets of } E_1, E_2, E_3, \text{ and } E_4} \{E_1, E_2, E_3, E_4\} \right\} \setminus (\text{one of chosen } e_i\text{'s}) \quad (8)$$

where $E_1 = \{e_{31}, e_{30}, \ldots e_{24}\}$, $E_2 = \{e_{23}, e_{22}, \ldots e_{16}\}$, $E_3 = \{e_{15}, e_{14}, \ldots e_8\}$, $E_4 = \{e_7, e_6, \ldots e_0\}$, and "\" denotes the exclusion from the set.

Experimental verification. We computed the 15-th order differential of $\tilde{y}_R(x)$ for all 15-th order differences represented by Eq. (8) by computer experiments. Table 1 shows the number of constant bits in the 15-th order differential of $\tilde{y}_R(x)$. Note that $\{E_1, E_2\} \setminus e_i$ denotes the set $\{E_1, E_2\}$ excluding an arbitrary difference e_i in $\{E_1, E_2\}$. The number of constant bits and the bit-positions don't depend on the excluded e_i. This is obvious from the fact that the constant doesn't depend on x_i.

[WHEN THE ORDER OF THE HIGHER ORDER DIFFERENTIAL IS 14] For the 14-th order differential of $\tilde{y}_R(x)$, it is shown that if we choose $\{e_{13}, e_{12}, \ldots, e_0\}$ for the 14-th order differences, each bit of the 14-th order differential of $\tilde{y}_R(x)$ is quadratic with respect to key bits. Table 2 shows that the degree with respect to the key of each bit of the 14-th order differential of $\tilde{y}_R(x)$ for some chosen differences. Note that the 14-th order differences $\{a_{14}, a_{13}, \ldots, a_1\}$ are chosen from $\{e_{31}, e_{30}, \ldots, e_{16}\}$. The column "differences" in Table 2 holds the XORed values $a_{14} + a_{13} + \cdots + a_1$. Table 2 shows that some bits of the 14-th order differential of $\tilde{y}_R(x)$ are constant, or independent of the key, a fact that is exploited in the improved higher order differential.

[WHEN THE ORDER OF THE HIGHER ORDER DIFFERENTIAL IS 13] For the 13-th order differential of $\tilde{y}_R(x)$, it is shown that if we choose $\{e_{12}, e_{11}, \ldots, e_0\}$ for the 13-th order differences, each bit of the 13-th order differential of $\tilde{y}_R(x)$ is degree

Table 2. The degree of each bit of the 14-th order differential of $\tilde{y}_R(x)$

differences	bit position of the 14-th order differential of $\tilde{y}_R(x)$ 31 30 29 28 27 26 25 24 23 22 21 20 19 18 17 16 15 14 13 12 11 10 9 8 7 6 5 4 3 2 1 0
111111111111100	1 2 2 2 2 2 2 2 2 0 0 1 2 2 2 1 2 0 1 2 2 1 1220110121
111111111111010	1 2 2 2 2 2 2 2 2 1 0 1 2 2 2 1 2 1 0 2 2 1 1220111122
111111111110110	1 2 2 2 2 2 2 2 2 0 0 1 2 2 2 1 2 1 1 2 2 1 1221011120
111111111101110	1 2 2 2 2 2 2 2 2 0 1 1 2 2 2 1 2 0 0 2 2 1 1220111121

Table 3. The bit-position of constant bits of the 13-th order differential of $\tilde{y}_R(x)$

differences	bit position of the 13-th order differential of $\tilde{y}_R(x)$ 31 30 29 28 27 26 25 24 23 22 21 20 19 18 17 16 15 14 13 12 11 10 9 8 7 6 5 4 3 2 1 0
111111111111000	0
111111111100011	0
111111000111111	0　　　　　　　　　　　　　0

3 or less with respect to key bits. Table 3 shows that the bit positions where the 13-th order differential of $\tilde{y}_R(x)$ is constant for some chosen differences. Note that the 13-th order differences $\{a_{13}, a_{11}, \ldots, a_1\}$ are chosen from 13 differences of $\{e_{31}, e_{30}, \ldots, e_{16}\}$. The column "differences" in Table 3 holds the XORed values $a_{13} + a_{12} + \cdots + a_1$. Similarly, Table 3 shows that some bits of the 13-th order differential of $\tilde{y}_R(x)$ are constant, or independent of the key, a fact that is exploited in the improved higher order differential.

4.2   Construct the attack equation and find the last round key $k^{(5)}$

In this section we construct attack equation (1) using the higher order differences found in Step 1 (Section 4.1), and find the last round key $k^{(5)}$. If we find the last round key $k^{(5)}$ by checking all possible 2^{32} candidates exhaustively as shown in [7], higher order differential attacks are possible where the 13-th order differences given in Section 4.1 are used. The required number of chosen p/c pairs is only 2^{13}, and the required complexity is about $\frac{1}{2} \times 2^{13} \times 2^{32} = 2^{44}$ times the computation of round function on average (see new result (II) in Table 4).

If we find the last round key $k^{(5)}$ by solving attack equation (1) algebraically as shown in [13, 11], the required complexity can be reduced, though the required number of chosen p/c pairs increases slightly. Hereafter, let $k^{(5)} = (k_{31}, k_{30}, \ldots, k_0)$ denote the last round key, and define the set of variables $K^{(5)}$ as $K^{(5)} = \{k_{31}, k_{30}, \ldots, k_0\}$. According to reference [11], the degree of attack equation (1) is 3 with respect to $K^{(5)}$ and we have to solve algebraic equations of degree 3 with 32 unknown variables. If we transform it to a system of linear equations regarding all monomials on $k^{(5)}$ in attack equation (1) as independent unknown variables, all variables in $K^{(5)}$ can be determined uniquely. The number of unknown variables is 368 [11, Section 4.2], and we have to prepare 368 equations.

Table 4. Required # of chosen p/c pairs and complexity for attacking a 5-round CAST

Attacks	# of p/c pairs	complexity
Jakobsen & Knudsen [7]	2^{17}	2^{48}
Moriai, Shimoyama & Kaneko [11]	2^{17}	2^{25}
New result (I)	2^{16}	2^{24}
New result (II)	2^{13}	2^{44}

If we use one of the 15-th order differences given in Section 4.1, we can obtain 32 equations (equations for 32 bits) using 2^{15} chosen p/c pairs. For the remaining $368 - 32 = 336$ equations, we can choose 15 different 2^{14} chosen p/c pairs from the same 2^{15} p/c pairs as above, but it seems difficult to prepare 336 equations according to Table 2. Therefore, we use arbitrary 16-th order differences and obtain 32 equations using 2^{16} chosen p/c pairs, and for the remaining 336 equations, we obtain them using some 15-th order differences given in Section 4.1, which we can choose from the 2^{16} p/c pairs above. In this case, the required number of chosen p/c pairs is only 2^{16}, and the required complexity is less than 2^{24} times the computation of round function (see new result (I) in Table 4). Deriving the required complexity is explained in reference [11]. Table 4 shows new results on the number of p/c pairs and the complexity required for attacking a 5-round CAST cipher and compares them with previous results.

5   Conclusion

This paper introduced an improved higher order differential attack using chosen higher order differences. We demonstrated a higher order differential attack of a CAST cipher with 5 rounds using chosen higher order differences with fewer chosen p/c pairs and less complexity than the previous results. It is open whether the attack can be extended beyond 5 rounds. The target cipher is an example of a family of symmetric ciphers constructed using the CAST design procedure. CAST-128, which is used in several commercial applications, e.g., PGP5.0, has a stronger round function and more rounds, hence the improved higher order differential attack seems difficult to mount against CAST-128.

We're working on how to find the lowest order of the higher order differential, which will lead to provably security against higher order differential attacks.

References

1. C.M.Adams, "Constructing Symmetric Ciphers Using the CAST Design Procedure," Designs, Codes and Cryptography, Volume 12, Number 3, November, pp.283–316, Kluwer Academic Publishers, 1997.
2. C.M.Adams, "The CAST-128 Encryption Algorithm," Request for Comments (RFC) 2144, Network Working Group, Internet Engineering Task Force, May, 1997.

3. C.M.Adams and S.E.Tavares, "Designing S-boxes for ciphers resistant to differential cryptanalysis," In Proceedings of the 3rd symposium on State and Progress of Research in Cryptography, pp.181–190, 1993.

4. E.Biham, "New Types of Cryptanalytic Attacks Using Related Keys," Advances in Cryptology – EUROCRYPT'93, Lecture Notes in Computer Science 765, pp.398–409, Springer-Verlag, 1994.

5. E.Biham and A.Shamir, "Differential Cryptanalysis of DES-like Cryptosystems," Journal of Cryptology, Volume 4, Number 1, pp.3–72, Springer-Verlag, 1991.

6. H.M.Heys and S.E.Tavares, "On the security of the CAST encryption algorithm," Canadian Conference on Electrical and Computer Engineering, pp.332–335, 1994.

7. T.Jakobsen and L.R.Knudsen, "The Interpolation Attack on Block Ciphers," Fast Software Encryption, FSE'97, Lecture Notes in Computer Science 1267, pp.28–40, Springer-Verlag, 1997.

8. L.R.Knudsen, "Truncated and Higher Order Differentials," Fast Software Encryption – Second International Workshop, Lecture Note in Computer Science 1008, pp.196–211, Springer-Verlag, 1995.

9. X.Lai, "Higher Order Derivatives and Differential Cryptanalysis," Communications and Cryptography, pp.227–233, Kluwer Academic Publishers, 1994.

10. M.Matsui, "Linear Cryptanalysis Method for DES Cipher," Advances in Cryptology – EUROCRYPT'93, Lecture Notes in Computer Science 765, pp.386–397, Springer-Verlag, 1994.

11. S.Moriai, T.Shimoyama, and T.Kaneko, "Higher Order Differential Attack of a CAST Cipher," Fast Software Encryption, FSE'98, Lecture Notes in Computer Science 1372, pp.17–31, Springer-Verlag, 1998.

12. K.Nyberg and L.R.Knudsen, "Provable Security Against a Differential Attack," Journal of Cryptology, Vol.8, No.1, pp.27–37, Springer-Verlag, 1995.

13. T.Shimoyama, S.Moriai, and T.Kaneko, "Improving the Higher Order Differential Attack and Cryptanalysis of the $\mathcal{KN}$ Cipher," Information Security, First International Workshop, ISW'97, Lecture Notes in Computer Science 1396, pp.32–42, Springer-Verlag, 1998.

14. H.Tanaka, K. Hisamatsu, and Toshinobu Kaneko, "Higher Order Differential Attack of MISTY without FL functions," Technical Report of IEICE, ISEC98-5, The Institute of Electronics, Information and Communication Engineers, 1998. (in Japanese)

On Maximum Non-averaged
Differential Probability

Kazumaro Aoki

NTT Laboratories
1-1 Hikarinooka, Yokosuka-shi, Kanagawa-ken, 239-0847 Japan
`maro@isl.ntt.co.jp`

Abstract. Maximum average of differential probability is one of the security measures used to evaluate block ciphers such as the MISTY cipher. Here *average* means the average for all keys. Thus, there are keys which yield larger maximum differential probability even if the maximum average of differential probability is sufficiently small.

This paper presents the cases in which the maximum differential probability is larger than the maximum average of differential probability for some keys, and we try to determine the maximum differential probability considering the key effect.

Keywords: Differential cryptanalysis, linear cryptanalysis, differential, maximum average of differential probability, linear hull, maximum average of linear probability, maximum non-averaged differential probability, maximum non-averaged linear probability, DES-like cipher

1 Introduction

The security of symmetric key block ciphers against differential cryptanalysis [2] can be evaluated using several security measures. The maximum average of differential probability [5] is one such measure. We can regard that differential cryptanalysis fails for a block cipher if the maximum average of differential probability for the block cipher is sufficiently small. Thus, designers of a block cipher should guarantee that the maximum average of differential probability is sufficiently small. Some block ciphers were shown to have maximum averages of differential probability that were sufficiently small by Knudsen et al. [8, 7].

It is important to note that *average* of the maximum average of differential probability means the average over all keys. That is, even if the maximum average of differential probability is sufficiently small, the block cipher may be insecure for some keys. Canteaut evaluated the maximum differential probability for all keys not just the *average* of all keys for some types of DES-like ciphers [3]. However, the proof of her main theorem was flawed.

This paper points out a flaw in the proof of [3], and extends the theorems that have correct proof to linear cryptanalysis [6]. Our conclusion is that inequalities similar to [1] hold if the number of rounds is less than 3. Moreover, we report experimental results. The results show that more rigorous inequalities may be proved on non-averaged differential probability for a specific F-function.

S. Tavares and H. Meijer (Eds.): SAC'98, LNCS 1556, pp. 118–130, 1999.
© Springer-Verlag Berlin Heidelberg 1999

2 Preliminaries

We define the following.

1. $\text{Prob}[A|B] = 0$ if B is an empty set.
2. Operations not specified here are as given in $\text{GF}(2)^i$.
3. Suffixes L and R are the left and right half of the variable letter regarded as bit string, respectively.

We define the following notations.

$\text{dom } f$	the domain of function f
$a \bullet b$	Even parity of bitwise logical AND operation of bit strings a and b
$\Delta_f(a,b)$	$\{x \mid f(x) + f(x+a) = b\}$
$\delta_f(a,b)$	$^\#\Delta_f(a,b)$
$\delta_f^{\max}$	$\max\limits_{a \neq 0, b} \delta_f(a,b)$
ΔX	$X + X^*$
$\Lambda_f(a,b)$	$\{x \mid x \bullet a = f(x) \bullet b\}$
$\lambda_f(a,b)$	$2^\# \Lambda_f(a,b) - {}^\# \text{dom } f$
$\lambda_f^{\max}$	$\max\limits_{a, b \neq 0} \lambda_f(a,b)^2$
(a,b)	Concatenation of bit strings a and b
$\exp_b(e)$	b^e
$S(X)$	$\{f : X \to X \mid f\text{: bijective}\}$
$C(f)$	equivalence class of f
$T/\sim$	quotient set of set T with the equivalence relation $\sim$, i.e. $\{C(f) \mid f \in T\}$
$\mathcal{P}(T)$	power set of T

We define the precedence of operations as the following.

$$\Delta \succ \bullet \succ +$$

We define the cipher $E_{k_1, k_2, \dots, k_r}$, the analysis target, as follows.

1. $L(i), R(i)$ (n-bit)
2. $Y(0)$: plaintext ($2n$-bit)
3. $(L(i), R(i)) = Y(i)$
4. $\begin{cases} Z(i) = F_{k_{i+1}}(R(i)) \\ L(i+1) = R(i) \qquad \text{for } 0 \leq i < r \\ R(i+1) = L(i) + Z(i) \end{cases}$

We define F-function as $F_{k_{i+1}}(R(i)) = f(R(i) + k_{i+1})$, and f as bijective.
Moreover, we define

$$\delta_E^{\max} = \max_{k_1, k_2, \dots, k_r} \delta_{E_{k_1, k_2, \dots, k_r}}^{\max}$$

$$\lambda_E^{\max} = \max_{k_1, k_2, \dots, k_r} \lambda_{E_{k_1, k_2, \dots, k_r}}^{\max}$$

for the block cipher E.

The following lemma is useful for evaluating linear probability.

Lemma 1. *Following equation holds for n-input Boolean function* $f : GF(2)^n \to GF(2)$.

$$\frac{1}{2^n} \sum_{x \in GF(2)^n} \exp_{-1}(f(x)) = 2 \operatorname*{Prob}_x[f(x) = 0] - 1$$

3   Previous Results

3.1   Averaged Case

Nyberg and Knudsen showed a bound of the maximum average of differential probability for $r \geq 3$ [8], and Matsui pointed out that a similar inequality holds in linear cryptanalysis using duality [7]; later Aoki and Ohta showed the strict bounds for bijective F-Function [1].[1]

Lemma 2 ([8, 7, 1]).

$$\max_{\alpha \neq 0, \beta} \operatorname*{Average}_{k_1, k_2, \ldots, k_r} \delta_{E_{k_1, k_2, \ldots, k_r}}(\alpha, \beta) \leq (\delta_f^{\max})^2 \quad \textit{if } r \geq 3$$

$$\max_{\alpha, \beta \neq 0} \operatorname*{Average}_{k_1, k_2, \ldots, k_r} \lambda_{E_{k_1, k_2, \ldots, k_r}}(\alpha, \beta) \leq \lambda_f^{\max} \quad \textit{if } r \geq 3$$

3.2   Non-Averaged Case

Canteaut showed some results of differential probability as dependent on keys [3].

Lemma 3 (2-round differential probability).

$$\operatorname*{Prob}_{Y(0), Y^*(0)}[\Delta Y(2) = \beta | \Delta Y(0) = \alpha] = \frac{\delta_f(\alpha_R, \alpha_L + \beta_L)\delta_f(\beta_L, \alpha_R + \beta_R)}{2^{2n}}$$

Lemma 4 (3-round differential probability with trivial 1-st round).

$$\operatorname*{Prob}_{Y(0), Y^*(0)}[\Delta Y(3) = \beta | \Delta Y(0) = (\alpha_L, 0)] = \frac{\delta_f(\alpha_L, \beta_L)\delta_f(\beta_L, \alpha_L + \beta_R)}{2^{2n}}$$

Lemma 5 (3-round differential probability with trivial 2-nd round).

$$\operatorname*{Prob}_{Y(0), Y^*(0)}[\Delta Y(3) = (\alpha_R, \beta_R) | \Delta Y(0) = \alpha] = \frac{\delta_f(\alpha_R, \alpha_L)\delta_f(\alpha_R, \beta_R)}{2^{2n}}$$

---

[1] Kaneko et al. showed similar inequalities for general F-function [4].

4   Extension to Linear Cryptanalysis

We can prove the lemmas in Sect. 3.2 for the case of linear probability using Lemma 1. We show the lemmas here and prove them in Appendices.

Lemma 6 (2-round linear probability).

$$2 \operatorname*{Prob}_{Y(0)}[Y(0) \bullet \alpha = Y(2) \bullet \beta] - 1$$

$$= \exp_{-1}(k_1 \bullet (\alpha_R + \beta_R) + k_2 \bullet (\alpha_L + \beta_L)) \times \frac{\lambda_f(\alpha_R + \beta_R, \alpha_L)\lambda_f(\alpha_L + \beta_L, \beta_R)}{2^{2n}}$$

Lemma 7 (3-round linear probability with trivial 3-rd round).

$$2 \operatorname*{Prob}_{Y(0)}[Y(0) \bullet \alpha = Y(3) \bullet (\beta_L, 0)] - 1$$

$$= \exp_{-1}(k_1 \bullet (\alpha_R + \beta_L) + k_2 \bullet \alpha_L) \times \frac{\lambda_f(\alpha_R + \beta_L, \alpha_L)\lambda_f(\alpha_L, \beta_L)}{2^{2n}}$$

Lemma 8 (3-round linear probability with trivial 2-nd round).

$$2 \operatorname*{Prob}_{Y(0)}[Y(0) \bullet \alpha = Y(3) \bullet (\beta_L, \alpha_L)] - 1$$

$$= \exp_{-1}(k_1 \bullet \alpha_R + k_3 \bullet \beta_L) \times \frac{\lambda_f(\alpha_R, \alpha_L)\lambda_f(\beta_L, \alpha_L)}{2^{2n}}$$

5   Computer Evaluations

5.1   Extension Trial to General Rounds

Canteaut evaluated DES-like ciphers with general rounds similar to Sect. 3.2 [3]. Unfortunately, the results contain errors. First, she described

$$P[\Delta Y(3) = (\beta_L, \beta_R) | \Delta Y(0) = (\alpha_L, \alpha_R), K_1 = k_1, K_2 = k_2, K_3 = k_3]$$
$$= \sum_d P[\Delta Z(2) = \beta_R + d | \Delta R(2) = \beta_L, \Delta Y(0) = (\alpha_L, \alpha_R),$$
$$K_1 = k_1, K_2 = k_2, K_3 = k_3]$$
$$\times \frac{\delta_f(\alpha_R, d + \alpha_L)\delta_f(d, \beta_L + \alpha_R)}{2^{2n}}$$

in [3, Theorem 1]. The conditional part of the probability formula of the right side of this equality misses $\Delta Y(2) = (d, \beta_L)$. So, this equality does not hold.[2] In addition, we believe that the induction part of the proof that she did not describe has the same error.

---

[2] Her probability formulas of [3] does not contain information on random variables, so her proofs are hard to understand. We did not understand the correctness of the proof of Propositions 2 and 3 in [3]. So, the proof of Propositions 2 and 3 may be flawed, however, we confirmed that the statements of these propositions are correct.

5.2   Preliminaries

We tried to prove [3, Theorem 1], but computer based approaches seemed feasible.

We used computers to evaluate the differential probabilities for all f in the case that n is small and for randomly generated f in the case that n is not small.

The evaluations consider all round keys and all bijective functions for f-function. Thus the evaluations are enormously complex.

Following lemmas are effective for decreasing the complexity of computer evaluations. Proofs are in Appendices.

Lemma 9. *We can obtain the same value which is an element of a set T of a measure $h : \mathcal{P}(\mathrm{GF}(2)^{2n} \times \mathrm{GF}(2)^{2n}) \to T$ by adjusting other keys for r-round cipher $Y(r) = E(Y(0))$ even if we change one of any even round key and one of any odd round key in an arbitrary manner, where the measure $h(\{(Y(0), Y(r))|Y(r) = E(Y(0))\})$ satisfies the following equation.*

$$\forall \mu, \nu \; [h(\{(Y(0) + \mu, Y(r) + \nu)|Y(r) = E(Y(0))\})$$
$$= h(\{(Y(0), Y(r))|Y(r) = E(Y(0))\})] \tag{1}$$

Corollary 1. $\delta_E^{\max}$ *is independent of* (k_1, k_2).

Corollary 2. $\lambda_E^{\max}$ *is independent of* (k_1, k_2).

These corollaries suggest that to evaluate an r-round cipher, using all round keys is not necessary; considering only $r - 2$ round keys is sufficient.

Moreover, since the evaluations consider all keys, it is sufficient to evaluate f or g if $\forall x[f(x) = g(x)+k]$ holds. We introduce the following equivalence relation for achieving this purpose.

Lemma 10.

$$f \sim g \stackrel{\text{def}}{\Leftrightarrow} \exists k \in \mathrm{GF}(2)^n, \forall x \in \mathrm{GF}(2)^n[f(x) = g(x) + k]$$

is an equivalence relation over $S(\mathrm{GF}(2)^n)$.

It is trivial using this equivalence relation to show that it is sufficient for considering a complete set of the representatives of $S(\mathrm{GF}(2)^n)/\sim$.

Lemma 11. *For any* x_0, y_0,

$$\{f \in S(\mathrm{GF}(2)^n)|f(x_0) = y_0\}$$

is a complete set of representatives in $S(\mathrm{GF}(2)^n)/\sim$.

Using this lemma, it is sufficient for us to consider only the elements of, for example, $f(0) = 0$ in $S(\mathrm{GF}(2)^n)$.

5.3   Experimental Results

We calculated the maximum differential probability of F-function $\delta_f^{\max}$ and the maximum differential probability of cipher $\delta_E^{\max}$ for at least 3-round ciphers using the lemmas of the previous sections.

We calculated all f for the case that number of bits of F-function is equal to 3,[3] and randomly generated f in the case that number of bits of F-function is greater than 3. We show the results in Tables 1 and 2.[4] *Ratio* here means the ratio of the number of F-functions which derives pairs $(\delta_f^{\max}, \delta_E^{\max})$ to all bijective functions (or all randomly generated bijective functions in the case of Table 2).

In the tables, ∗ denotes the items that do not satisfy the evaluation inequality of Lemma 2, the maximum average of differential probability is replaced with maximum differential probability, $\dfrac{\delta_E^{\max}}{2^{2n}} \leq (\dfrac{\delta_f^{\max}}{2^n})^2$. These tables show that the inequality is 4.5 times looser for some F-functions. However, these tables also show that the maximum differential probability is smaller than $(\dfrac{\delta_f^{\max}}{2^n})^2$ for some F-functions.

We obtained the following interesting examples.

Example 1. The following example shows that the statement of [3, Theorem 1] does not hold.

$$\begin{cases} n = 4,\ r = 3 \\ f = \begin{pmatrix} 0\ 1\ 2\ 3\ 4\ 5\ \ 6\ \ 7\ 8\ \ 9\ 10\ 11\ 12\ 13\ 14\ 15 \\ 0\ 1\ 2\ 3\ 4\ 8\ 15\ 11\ 7\ 12\ \ 6\ 13\ \ 5\ 10\ 14\ \ 9 \end{pmatrix} \end{cases} \Rightarrow \begin{cases} \dfrac{\delta_f^{\max}}{2^4} = \dfrac{1}{4} \\ \dfrac{\delta_E^{\max}}{2^8} = 1.5 \times (\dfrac{\delta_f^{\max}}{2^4})^2 \end{cases}$$

Example 2. The following example shows that the maximum differential probability of cipher is less than the square of maximum differential probability of F-function.

$$\begin{cases} n = 3,\ r = 5 \\ f = \begin{pmatrix} 0\ 1\ 2\ 3\ 4\ 5\ 6\ 7 \\ 0\ 1\ 3\ 7\ 6\ 2\ 5\ 4 \end{pmatrix} \end{cases} \Rightarrow \begin{cases} \dfrac{\delta_f^{\max}}{2^3} = 1 \\ \dfrac{\delta_E^{\max}}{2^6} = \dfrac{1}{4} \times (\dfrac{\delta_f^{\max}}{2^3})^2 \end{cases}$$

Example 3. The following example shows that the maximum differential probability of cipher is 4.5 times greater than the square of maximum differential probability of F-function.

$$\begin{cases} n = 3,\ r = 5 \\ f = \begin{pmatrix} 0\ 1\ 2\ 3\ 4\ 5\ 6\ 7 \\ 0\ 1\ 2\ 6\ 7\ 5\ 4\ 3 \end{pmatrix} \end{cases} \Rightarrow \begin{cases} \dfrac{\delta_f^{\max}}{2^3} = \dfrac{1}{4} \\ \dfrac{\delta_E^{\max}}{2^6} = 4.5 \times (\dfrac{\delta_f^{\max}}{2^3})^2 \end{cases}$$

---

[3] We omit less than 3-bit F-function since its cases are trivial.

[4] We calculated cases of $n > 5$, however, we could not find an interesting case.

Table 1. Differential probability of an F-function and differential probability of a cipher ($n = 3$)

n	r	$\delta_f^{\max}$	$\delta_E^{\max}$	Ratio
3	3	2	4	4/15
		4	16	7/15
		8	64	4/15
4	2	*8	12/45	
		4	8	6/45
			16	14/45
			*32	1/45
		8	32	9/45
			64	3/45
5	2	*10	2/45	
		*12	4/45	
		*18	6/45	
		4	8	6/45
			16	12/45
			*32	2/45
			*48	1/45
		8	16	6/45
			32	3/45
			64	3/45
6	2	*10	2/45	
		*12	4/45	
		*18	6/45	
		4	12	6/45
			*18	12/45
			*32	2/45
			*48	1/45
		8	16	6/45
			32	3/45
			64	3/45
7	2	*12	2/45	
		*16	4/45	
		*18	6/45	
		4	12	6/45
			*18	12/45
			*32	2/45
			*48	1/45
		8	16	6/45
			32	3/45
			64	3/45

Table 2. Differential probability of an F-function and differential probability of a cipher ($n > 3$)

n	r	$\delta_f^{\max}$	$\delta_E^{\max}$	Ratio(%)
4	3	4	16	1.4
			*18	1.0
			*20	0.1
			*24	0.0
		6	36	49.2
		8	64	37.8
		10	100	8.7
		12	144	1.7
		16	256	0.1
5	3	4	*18	0.0
			*20	0.0
		6	36	21.0
		8	64	62.6
		10	100	14.7
		12	144	1.5
		14	196	0.1
		16	256	0.0
		18	324	0.0

There exists 16 differentials which have differentially weak keys in 1/32 key space, and all weak keys for each differential are different. Thus, a key which is in one half of the key space of the cipher is differentially weak.

6 Conclusion

This paper has extended the evaluation of maximum differential probability and maximum linear probability with keys for DES-like ciphers in the case that the F-function is bijective. As a result, strict evaluations we derived for 2-round and some 3-round ciphers. These results are the same as those gained by evaluating the maximum average of differential probability and the maximum average of linear probability, parameters were used as security measures against differential cryptanalysis and linear cryptanalysis, respectively.

Moreover, we have evaluated the maximum differential probability with keys over 3-round using computers. As a result, it is proved that there are cases in which the maximum differential probability is 4.5 times greater than the maximum average of differential probability.

There are three open problems.

1. obtaining general case evaluation
2. characterizing the F-functions whose maximum differential probability with keys is small
3. constructing design procedures of key scheduling which does not produce weak keys against differential cryptanalysis

References

1. K. Aoki and K. Ohta. Strict Evaluation of the Maximum Average of Differential Probability and the Maximum Average of Linear Probability. *IEICE Transactions Fundamentals of Electronics, Communications and Computer Sciences (Japan)*, Vol. E80-A, No. 1, pp. 2–8, 1997. (A preliminary version written in Japanese was presented at SCIS96-4A).
2. E. Biham and A. Shamir. Differential Cryptanalysis of DES-like Cryptosystems. *Journal of Cryptology*, Vol. 4, No. 1, pp. 3–72, 1991. (The extended abstract was presented at CRYPTO'90).
3. A. Canteaut. Differential cryptanalysis of Feistel ciphers and differentially δ-uniform mappings. In *Workshop on Selected Areas in Cryptography (SAC'97)*, pp. 172–184, Ottawa, Ontario, Canada, 1997. School of Computer Science, Carleton University, Entrust Technologies, and the Interac Association.
4. Y. Kaneko, S. Moriai, and K. Ohta. On Strict Estimation Method of Provable Security against Differential and Linear Cryptanalysis. In Y. Han, T. Okamoto, and S. Qing, editors, *Information and Communications Security — First International Conference ICICS'97*, Volume 1334 of *Lecture Notes in Computer Science*, pp. 258–268. Springer-Verlag, Berlin, Heidelberg, New York, 1997.
5. X. Lai, J. L. Massey, and S. Murphy. Markov Ciphers and Differential Cryptanalysis. In D. W. Davies, editor, *Advances in Cryptology — EUROCRYPT'91*, Volume 547 of *Lecture Notes in Computer Science*, pp. 17–38. Springer-Verlag, Berlin, Heidelberg, New York, 1991.

6. M. Matsui. Linear Cryptanalysis Method for DES Cipher. In T. Helleseth, editor, *Advances in Cryptology — EUROCRYPT'93*, Volume 765 of *Lecture Notes in Computer Science*, pp. 386–397. Springer-Verlag, Berlin, Heidelberg, New York, 1994. (A preliminary version written in Japanese was presented at SCIS93-3C).

7. M. Matsui. New Structure of Block Ciphers with Provable Security against Differential and Linear Cryptanalysis. In D. Gollmann, editor, *Fast Software Encryption, Third International Workshop, Cambridge, UK, February 1996, Proceedings*, Volume 1039 of *Lecture Notes in Computer Science*, pp. 205–218. Springer-Verlag, Berlin, Heidelberg, New York, 1996. (Japanese version was presented at SCIS96-4C).

8. K. Nyberg and L. R. Knudsen. Provable Security Against a Differential Attack. *Journal of Cryptology*, Vol. 8, No. 1, pp. 27–37, 1995. (A preliminary version was presented at CRYPTO'92 rump session).

A   Proof of Lemma 6

We prove this lemma using Fig. 1.

$$2 \Prob_{Y(0)}[Y(0) \bullet \alpha = Y(2) \bullet \beta] - 1 \tag{2}$$

$$= \frac{1}{2^{2n}} \sum_{L(0),R(0)} \exp_{-1}(L(0) \bullet \alpha_L + R(0) \bullet \alpha_R + (L(0) + f(R(0) + k_1)) \bullet \beta_L$$
$$+ (R(0) + f(L(0) + f(R(0) + k_1) + k_2)) \bullet \beta_R) \tag{3}$$

$$= \frac{1}{2^{2n}} \sum_{L(0),R(0)} \exp_{-1}(R(0) \bullet (\alpha_R + \beta_R) + f(R(0) + k_1) \bullet \beta_L$$
$$+ L(0) \bullet (\alpha_L + \beta_L) + f(L(0) + f(R(0) + k_1) + k_2) \bullet \beta_R) \tag{4}$$

$$= \frac{1}{2^{2n}} \sum_{L(0),R(0)} \exp_{-1}(R(0) \bullet (\alpha_R + \beta_R) + f(R(0) + k_1) \bullet \beta_L$$
$$+ (f(R(0) + k_1) + k_2) \bullet (\alpha_L + \beta_L)$$
$$+ (L(0) + f(R(0) + k_1) + k_2) \bullet (\alpha_L + \beta_L)$$
$$+ f(L(0) + f(R(0) + k_1) + k_2) \bullet \beta_R) \tag{5}$$

$$= \frac{1}{2^{2n}} \sum_{R(0)} \exp_{-1}(R(0) \bullet (\alpha_R + \beta_R) + f(R(0) + k_1) \bullet \alpha_L + k_2 \bullet (\alpha_L + \beta_L)) \tag{6}$$
$$\times \sum_{L(0)} \exp_{-1}((L(0) + f(R(0) + k_1) + k_2) \bullet (\alpha_L + \beta_L)$$
$$+ f(L(0) + f(R(0) + k_1) + k_2) \bullet \beta_R)$$

$$= \frac{1}{2^{2n}} \sum_{R(0)} \exp_{-1}((R(0) + k_1) \bullet (\alpha_R + \beta_R) + f(R(0) + k_1) \bullet \alpha_L$$
$$+ k_1 \bullet (\alpha_R + \beta_R) + k_2 \bullet (\alpha_L + \beta_L))$$
$$\times \lambda_f(\alpha_L + \beta_L, \beta_R) \tag{7}$$

$$= \exp_{-1}(k_1 \bullet (\alpha_R + \beta_R) + k_2 \bullet (\alpha_L + \beta_L))$$
$$\times \frac{\lambda_f(\alpha_R + \beta_R, \alpha_L)\lambda_f(\alpha_L + \beta_L, \beta_R)}{2^{2n}} \tag{8}$$

B   Proof of Lemma 7

We prove this lemma using Fig. 2.

$$2 \operatorname*{Prob}_{Y(0)}[Y(0) \bullet \alpha = Y(3) \bullet (\beta_L, 0)] - 1 \tag{9}$$

$$= \frac{1}{2^{2n}} \sum_{L(0),R(0)} \exp_{-1}(L(0) \bullet \alpha_L + R(0) \bullet \alpha_R$$
$$+ (R(0) + f(L(0) + f(R(0) + k_1) + k_2)) \bullet \beta_L) \tag{10}$$

$$= \frac{1}{2^{2n}} \sum_{L(0),R(0)} \exp_{-1}(L(0) \bullet \alpha_L + R(0) \bullet (\alpha_R + \beta_L)$$
$$+ f(L(0) + f(R(0) + k_1) + k_2) \bullet \beta_L) \tag{11}$$

$$= \frac{1}{2^{2n}} \sum_{R(0)} \exp_{-1}(R(0) \bullet (\alpha_R + \beta_L) + (f(R(0) + k_1) + k_2) \bullet \alpha_L)$$
$$\times \sum_{L(0)} \exp_{-1}((L(0) + f(R(0) + k_1) + k_2) \bullet \alpha_L$$
$$+ f(L(0) + f(R(0) + k_1) + k_2) \bullet \beta_L) \tag{12}$$

$$= \frac{1}{2^{2n}} \sum_{R(0)} \exp_{-1}((R(0) + k_1) \bullet (\alpha_R + \beta_L) + f(R(0) + k_1) \bullet \alpha_L$$
$$+ k_1 \bullet (\alpha_R + \beta_L) + k_2 \bullet \alpha_L) \times \lambda_f(\alpha_L, \beta_L) \tag{13}$$

$$= \exp_{-1}(k_1 \bullet (\alpha_R + \beta_L) + k_2 \bullet \alpha_L) \times \frac{\lambda_f(\alpha_R + \beta_L, \alpha_L)\lambda_f(\alpha_L, \beta_L)}{2^{2n}} \tag{14}$$

C   Proof of Lemma 8

We prove this lemma using Fig. 3.

$$2 \operatorname*{Prob}_{Y(0)}[Y(0) \bullet \alpha = Y(3) \bullet (\beta_L, \alpha_L)] - 1 \tag{15}$$

$$= \frac{1}{2^{2n}} \sum_{L(0),R(0)} \exp_{-1}(L(0) \bullet \alpha_L + R(0) \bullet \alpha_R$$
$$+ (R(0) + f(L(0) + f(R(0) + k_1) + k_2)) \bullet \beta_L$$
$$+ (L(0) + f(R(0) + k_1)$$
$$+ f(R(0) + f(L(0) + f(R(0) + k_1) + k_2) + k_3)) \bullet \alpha_L) \tag{16}$$

$$= \frac{1}{2^{2n}} \sum_{L(0),R(0)} \exp_{-1}(R(0) \bullet (\alpha_R + \beta_L) + f(L(0) + f(R(0) + k_1) + k_2) \bullet \beta_L$$

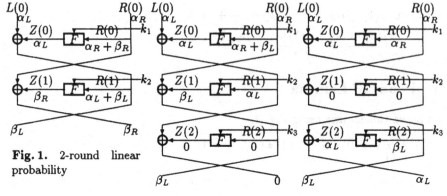

Fig. 1. 2-round linear probability

Fig. 2. 3-round linear probability with trivial 3-rd round

Fig. 3. 3-round linear probability with trivial 2-nd round

$$+ f(R(0) + k_1) \bullet \alpha_L$$
$$+ f(R(0) + f(L(0) + f(R(0) + k_1) + k_2) + k_3) \bullet \alpha_L \tag{17}$$
$$= \frac{1}{2^{2n}} \sum_{R(0)} \exp_{-1}(R(0) \bullet (\alpha_R + \beta_L) + f(R(0) + k_1) \bullet \alpha_L + (R(0) + k_3) \bullet \beta_L)$$
$$\times \sum_{L(0)} \exp_{-1}((R(0) + f(L(0) + f(R(0) + k_1) + k_2) + k_3) \bullet \beta_L$$
$$+ f(R(0) + f(L(0) + f(R(0) + k_1) + k_2) + k_3) \bullet \alpha_L \tag{18}$$
$$= \frac{1}{2^{2n}} \sum_{R(0)} \exp_{-1}((R(0) + k_1) \bullet \alpha_R + f(R(0) + k_1) \bullet \alpha_L + k_1 \bullet \alpha_R + k_3 \bullet \beta_L)$$
$$\times \lambda_f(\beta_L, \alpha_L) \quad \text{(since } f: \text{bijective)} \tag{19}$$
$$= \exp_{-1}(k_1 \bullet \alpha_R + k_3 \bullet \beta_L) \times \frac{\lambda_f(\alpha_R, \alpha_L)\lambda_f(\beta_L, \alpha_L)}{2^{2n}} \tag{20}$$

D   Proof of Lemma 9

We define transformed key $K^* = (k_1^*, k_2^*, \ldots, k_r^*)$ corresponding to original key $K = (k_1, k_2, \ldots, k_r)$.

We assume changing $(k_a, k_b) \mapsto (k_a^*, k_b^*)$ (a: odd, b: even). If we define

$$k_i^* = \begin{cases} k_i + \Delta k_a & i: \text{odd} \\ k_i + \Delta k_b & i: \text{even} \end{cases}, \quad \mu = (\Delta k_b, \Delta k_a), \quad \text{and} \quad \nu = \begin{cases} (\Delta k_a, \Delta k_b) & r: \text{odd} \\ (\Delta k_b, \Delta k_a) & r: \text{even} \end{cases},$$

$E_{K^*}(Y(0) + \mu) = E_K(Y(0)) + \nu$ holds.

E   Proof of Corollary 1

It is sufficient to prove that differential probability satisfies (1).

$$\forall \mu, \nu \quad \underset{Y(0), Y^*(0)}{\text{Prob}} [E(Y(0) + \mu) + \nu + E(Y^*(0) + \mu) + \nu = \beta | \Delta Y(0) = \alpha] \quad (21)$$

$$= \underset{Y(0), Y^*(0)}{\text{Prob}} [E(Y(0) + \mu) + E(Y^*(0) + \mu) = \beta$$

$$|(Y(0) + \mu) + (Y^*(0) + \mu) = \alpha] \quad (22)$$

$$= \underset{Y(0)+\mu, Y^*(0)+\mu}{\text{Prob}} [E(Y(0)) + E(Y^*(0)) = \beta | Y(0) + Y^*(0) = \alpha] \quad (23)$$

$$= \underset{Y(0), Y^*(0)}{\text{Prob}} [E(Y(0)) + E(Y^*(0)) = \beta | \Delta Y(0) = \alpha] \quad (24)$$

F   Proof of Corollary 2

It is sufficient to prove that linear probability satisfies (1).

$$\forall \mu, \nu \quad (2 \underset{Y(0)}{\text{Prob}} [(Y(0) + \mu) \bullet \alpha = (E(Y(0)) + \nu) \bullet \beta] - 1)^2 \quad (25)$$

$$= (2 \underset{Y(0)}{\text{Prob}} [Y(0) \bullet \alpha = E(Y(0)) \bullet \beta + (\mu \bullet \alpha + \nu \bullet \beta)] - 1)^2 \quad (26)$$

$$= (2 \underset{Y(0)}{\text{Prob}} [Y(0) \bullet \alpha = E(Y(0)) \bullet \beta] - 1)^2 \quad (27)$$

G   Proof of Lemma 10

Completeness $\forall k \in GF(2)^n [x+k \in S(GF(2)^n)]$ holds, and composite function
of bijective functions is bijective.
Reflexive law If $k = 0 \in GF(2)^n$, $\forall f \in GF(2)^n [f(x) = f(x) + k]$ holds. So,
$f \sim f$.
Symmetric law If $f \sim g$, $\exists k \in GF(2)^n [f(x) = g(x) + k]$ holds. Thus, $g(x) = f(x) + k$ holds, i.e. $g \sim f$.
Transitive law If $f \sim g$ and $g \sim h$ holds, since $\exists k \in GF(2)^n [f(x) = g(x) + k]$
and $\exists l \in GF(2)^n [g(x) = h(x) + l]$ holds, then $f(x) = g(x) + k = (h(x) + l) + k = h(x) + (l + k)$ holds. So, since $l + k \in GF(2)^n$ holds, $f \sim h$ holds.

H   Proof of Lemma 11

We define $R = \{f \in S(GF(2)^n) | f(x_0) = y_0\}$. We prove that R is a set of representatives. Let $f, g \in R$ and assume $C(f) = C(g)$. In this case, $f \sim g$ holds, and given the definition of R, $f(x_0) = g(x_0) = y_0$ holds. Thus, $f = g$ holds since $f(x) = g(x) + 0$.
  We prove $C(f) = \{f(x) + k | k \in GF(2)^n\}$. $g \in \{f(x) + k | k \in GF(2)^n\} \Leftrightarrow \exists k \in GF(2)^n [g(x) = f(x) + k] \Leftrightarrow g \sim f \Leftrightarrow g \in C(f)$.

$S(\mathrm{GF}(2^n)) \supseteq \bigcup_{f \in R} C(f)$ holds. On the other hand, since $^{\#}S(\mathrm{GF}(2^n)) = 2^n!$

holds and $^{\#}(\bigcup_{f \in R} C(f)) = {}^{\#}R \times {}^{\#}C(f) = (2^n - 1)! \times 2^n = 2^n!$ holds, so

$S(\mathrm{GF}(2^n)) = \bigcup_{f \in R} C(f)$ holds. That is, R is a complete set of representatives

in $S(\mathrm{GF}(2)^n)/\sim$.

Cryptanalysis of RC4-like Ciphers

S. Mister[1] and S. E. Tavares[2]

[1] Security Technology Group, Entrust Technologies Limited
750 Heron Road, Ottawa, Ontario, Canada, K1V 1A7
serge.mister@entrust.com
[2] Department of Electrical and Computer Engineering
Queen's University, Kingston, Ontario, Canada, K7L 3N6
tavares@ee.queensu.ca

Abstract. RC4, a stream cipher designed by Rivest for RSA Data Security Inc., has found several commercial applications, but little public analysis has been done to date. In this paper, alleged RC4 (hereafter called RC4) is described and existing analysis outlined. The properties of RC4, and in particular its cycle structure, are discussed. Several variants of a basic "tracking" attack are described, and we provide experimental results on their success for scaled-down versions of RC4. This analysis shows that, although the full-size RC4 remains secure against known attacks, keystreams are distinguishable from randomly generated bit streams, and the RC4 key can be recovered if a significant fraction of the full cycle of keystream bits is generated (while recognizing that for a full-size system, the cycle length is too large for this to be practical). The tracking attacks discussed provide a significant improvement over the exhaustive search of the full RC4 keyspace. For example, the state of a 5 bit RC4-like cipher can be obtained from a portion of the keystream using 2^{42} steps, while the nominal keyspace of the system is 2^{160}. More work is necessary to improve these attacks in the case where a reduced keyspace is used.

1 Introduction

Stream ciphers are often used in applications where high speed and low delay are a requirement. Although many stream ciphers are based on linear feedback shift registers, the need for software-oriented stream ciphers has lead to several alternative proposals. One of the more promising algorithms, RC4[9], designed by R. Rivest for RSA Data Security Inc., has been incorporated into many commercial products including BSAFE and Lotus Notes, and is being considered in upcoming standards such as TLS[1].

In this paper, the RC4[1] algorithm is described and known attacks reviewed. A detailed discussion of "tracking" attacks is provided and estimates of the complexity of cryptanalysis for simplified versions of RC4 are given.

---

[1] While RC4 remains a trade secret of RSA Data Security Inc., the algorithm described in [11] is believed to be output-compatible with RC4. This paper discusses the algorithm given in [11], and is referred to as RC4 for convenience.

S. Tavares and H. Meijer (Eds.): SAC'98, LNCS 1556, pp. 131–143, 1999.
© Springer-Verlag Berlin Heidelberg 1999

2   Background

The following description of RC4 is based on that given in [11]. It generalizes
RC4 to use n-bit words, but $n = 8$ is the most commonly used value. To use
RC4, a key is first used to initialize the 2^n word s-box S and counters i and
j through Algorithm 1. The keystream $\mathcal{K}$ is then generated using Algorithm 2.
The s-box entries and the counters i and j are n-bit words.

Algorithm 1 (RC4 Initialization). *Let $k_0 \ldots k_{l-1}$ denote the user's key, a
set of l n-bit words.*

1. For z from 0 to $2^n - 1$
   (a) Set $K_z = k_{z \bmod l}$.
2. For z from 0 to $2^n - 1$
   (a) Set $S_z = z$.
3. Set $j = 0$.
4. For i from 0 to $2^n - 1$
   (a) Set $j = j + S_i + K_i \bmod 2^n$.
   (b) Swap S_i and S_j.
5. Set $i = 0$ and $j = 0$.

Algorithm 2 (Keystream Generation).

1. Set $i = i + 1 \bmod 2^n$.
2. Set $j = j + S_i \bmod 2^n$.
3. Swap S_i and S_j.
4. Output $S_{S_i + S_j \bmod 2^n}$ as the next word in the keystream.

The RC4 keystream generation algorithm is depicted in Fig. 1.

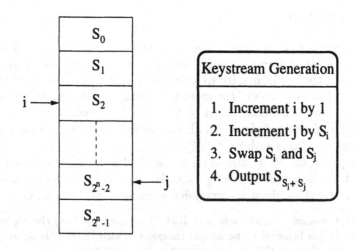

Fig. 1. RC4 Keystream Generation

The initialization algorithm is a key-dependent variant of the keystream generation algorithm, and is used to initialize the s-box S to a "randomly chosen" permutation. The nominal key length could be up to $n \cdot 2^n$ bits, but since it is used to generate only a permutation of 2^n values, the entropy provided by the key can be at most $\log_2(2^n!)$ bits, which will be referred to as the effective key length. Table 1 shows the nominal and effective key lengths for different values of n. In the remainder of this paper, the mod 2^n is sometimes omitted for brevity.

Table 1. Nominal and Effective Key Sizes for RC4-n

RC4 Word Size	Nominal Key Length (bits)	Effective Key Length (bits)
2	8	4.58
3	24	15.30
4	64	44.25
5	160	117.66
6	384	296.00
7	896	716.16
8	2048	1684.00
9	4608	3875.17

3   Published Results

This section is based on [7].

3.1   A Class of Weak Keys

In 1995, Andrew Roos posted a paper to the sci.crypt newsgroup[10] describing a class of weak keys, for which the initial byte of the keystream is highly correlated with the first few key bytes. The weak keys are those satisfying

$$k_0 + k_1 \equiv 0 \bmod 2^n \ .$$

The weak keys occur because the keystream initialization algorithm swaps a given entry of the s-box exactly once (corresponding to when the pointer i points to the entry) with probability $1/e$. In addition, for low values of i, it is likely that $S_j = j$ during the initialization. The reduction in search effort from this attack is $2^{5.1}$, but if linearly related session keys are used, the reduction in effort increases to 2^{18}.

3.2   Linear Statistical Weaknesses in RC4

In [3], the author derives a linear model of RC4 using the linear sequential circuit approximation (LSCA) method. The model has correlation coefficient $15 \cdot 2^{-3n}$, and requires $64^n/225$ keystream words. The model is successful in part because the s-box evolves slowly.

3.3   A Set of Short Cycles

Suppose that $i = a$, $j = a+1$, and $S_{a+1} = 1$ for some a. Then, after one iteration, $i = a+1$, $j = a+2$, and $S_{a+2} = 1$. Thus, the original relationship is preserved. Each such cycle has length $2^n \cdot (2^n - 1)$, and $(2^n - 2)!$ such cycles exist. Note however that, because RC4 is initialized to $i = j = 0$, these cycles never occur in practice. These observations were first made in [2] and outlined in [5].

4   Cycle Structures in RC4

4.1   Comparison with Randomly Chosen Invertible Mappings

The state of RC4 is fully determined by the two n-bit counters i and j and the s-box S. Since the number of states is finite, it must ultimately be periodic as the keystream generation function is iterated. Because the keystream generation function is invertible, the sequence of states is periodic. The length of the period depends on the word size n and the particular starting state, as illustrated in Table 2 for $n = 2$ and $n = 3$. For each period, the number of distinct cycles of that period is listed, followed by the number of initial states in each cycle, expressed as a formal sum. The last three columns will be explained in the next section. For comparison, Fig. 2 plots the expected cycle lengths for a randomly chosen permutation and those observed for RC4. For the randomly chosen permutation, the minimum and maximum lengths observed for the kth longest cycle in a set of 1500 arbitrarily chosen permutations is plotted.

It has also been observed that RC4 keystream sequences are slightly biased[4]. Define the gap at i for a sequence s to be the smallest integer $t \geq 0$ such that $s_i = s_{i-t-1}$. For a random sequence in which each element takes on one of 2^n values, the probability that $t = k$ is given by

$$\left(\frac{2^n - 1}{2^n}\right)^k \cdot \frac{1}{2^n} \ .$$

Table 3 shows the ratio of the actual to the expected gap probability, based on a sample of approximately 2^{30} elements of an arbitrarily chosen RC4 keystream. For all values of n, gaps of length 0 are more likely than expected, and gaps of length 1 are less likely than expected. In support of this, it has also been observed that the probability that $S_i = 0$ is lower than expected and that the probability that $S_i = 2^n - 1$ is higher than expected after a gap of length 0.

Table 2. Possible Periods for RC4 with Word Length 2 and 3

n	Period	# of Cycles	Number of Initial States	Shift Generator	Offset	Shifts Found (indexes are right shifts)
2	196	1	12	1	49	{0,1,2,3}
	164	1	12	1	41	{0,1,2,3}
3	955496	2	$15010 + 15274 = 30284$	2	238874	{0,2,4,6}, {0,2,4,6}
	322120	1	5144	1	40265	{0,1,2,3,4,5,6,7}
	53000	1	816	1	6625	{0,1,2,3,4,5,6,7}
	44264	1	688	5	5533	{0,1,2,3,4,5,6,7}
	29032	4	$482 + 505 + 488 + 457 = 1932$	4	14516	{0,4}, {0,4}, {0,4}, {0,4}
	9624	2	$153 + 149 = 302$	6	2406	{0,2,4,6}, {0,2,4,6}
	9432	1	140	3	3537	{0,1,2,3,4,5,6,7}
	4696	8	$80 + 77 + 70 + 93$ $+61 + 77 + 81 + 83 = 622$	0	4696	{0}, {0}, {0}, {0}, {0}, {0}, {0}, {0}
	3008	8	$50 + 41 + 32 + 35$ $+46 + 55 + 43 + 38 = 340$	0	3008	{0}, {0}, {0}, {0}, {0}, {0}, {0}, {0}
	648	1	8	1	81	{0,1,2,3,4,5,6,7}
	472	2	$7 + 7 = 14$	6	118	{0,2,4,6}, {0,2,4,6}
	456	1	12	1	57	{0,1,2,3,4,5,6,7}
	264	2	$5 + 7 = 12$	2	66	{0,2,4,6}, {0,2,4,6}
	120	2	$2 + 2 = 4$	7	15	{0,1,2,3,4,5,6,7}, {0,1,2,3,4,5,6,7}
	24	2	$1 + 1 = 2$	4	12	{0,4}, {0,4}

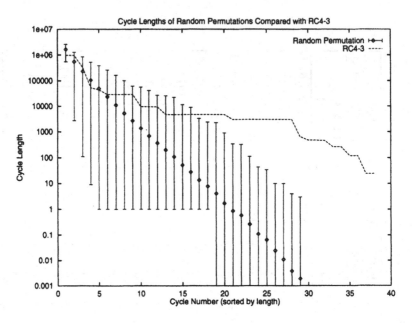

Fig. 2. Comparison of Cycle Lengths for RC4 and Random Permutations

Table 3. Deviation of RC4 Gap Lengths from those of Random Keystream

n	Gap						
	0	1	2	3	4	5	6
2	1.04082	0.952381	0.834467	0.870748	0.902998	1.72	1.26133
3	1.01828	0.956577	1.01042	0.994535	1.02179	1.00909	1.00284
4	1.00365	0.993622	1.0009	1.00126	1.00276	1.00039	1.00059
5	1.00099	0.99859	0.999946	1.0009	1.00081	1.00024	1.00046
6	0.999762	0.999901	1.00024	1.00039	1.00036	1.00014	0.999714

4.2   Partitioning of RC4 Cycles

As observed independently in [5], individual RC4 cycles can be partitioned into pieces of "equivalent" subsets, as follows. Define by $S^{\gg d}$ the s-box obtained by rotating the s-box entries to the right (or down) by d (formally, $S' = S^{\gg d}$ if $S'_t = S_{t-d \bmod 2^n}$). Let the right shift by d of an RC4 state (i, j, S) be defined by $(i+d, j+d, S^{\gg d})$. The following theorem holds:

Theorem 1. *Suppose that, for a given key, an RC4-n system goes through the state (i', j', S') and that the cycle length for this key is T. Then any cycle going through one or more states of the form $(i'+d, j'+d, S'^{\gg d})$ (where d is an integer) has period T and the shift relationship between the states is maintained as the two systems evolve. In addition, if the output sequences are compared word for word as the systems evolve beyond those states, the outputs will always differ if $d \not\equiv 0 \bmod 2^n$ and will always agree otherwise.*

Proof. Compare the state evolutions of the two systems (i', j', S') and $(i'' = i' + d, j'' = j' + d, S'' = S'^{\gg d})$. The steps for one round of keystream generation are:

   1. Set $i'' = i'' + 1 \bmod 2^n$.
   2. Set $j'' = j'' + S''_{i''} \bmod 2^n$.
   3. Swap $S''_{i''}$ and $S''_{j''}$.
   4. Output $S''_{S''_{i''}+S''_{j''} \bmod 2^n}$ as the next word in the keystream.

or

   1. Set $i'' = i'' + 1 \bmod 2^n$.
   2. Set $j'' = j'' + S'_{i''-d} \bmod 2^n$.
   3. Swap $S''_{i''}$ and $S''_{j''}$.
   4. Output $S'_{S'_{i''-d}+S'_{j''-d}-d \bmod 2^n}$ as the next word in the keystream.

which becomes:

   1. Set $i'' = i' + d + 1 \bmod 2^n$.
   2. Set $j'' = j' + d + S'_{i'} \bmod 2^n$.
   3. Swap $S''_{i'+d}$ and $S''_{j'+d}$.
   4. Output $S'_{S'_{i'}+S'_{j'}-d \bmod 2^n}$ as the next word in the keystream.

Thus, the shift relationship between the two systems is preserved, and only the output is different provided $d \not\equiv 0 \bmod 2^n$. Because the systems are identical except for the output, the periods of the two systems must also be the same. □

Consider a cycle of period T, and an arbitrary RC4 state (i', j', S') in that cycle. Then all shifts of this state belong to a cycle of length T. If there are only a few cycles of length T (as will be the case if T is large), then more than one may appear in the same cycle. The following theorem holds:

Theorem 2 (Cycle Partitioning). *Let γ be a cycle of period T and let $D = (i', j', S')$ be any state in the cycle. Let $d_1, \ldots, d_{k-1}$ $(k < 2^n)$ be the right shifts of D in the order they appear as RC4 evolves from state D ($d_0 = 0$ is understood). Then the distance (expressed as the number of encryptions) between successive shifts is given by T/k, and for any other state, D', in the same cycle, the right shifts of D' $d_1, \ldots, d_{k-1}$ are the only right shifts of D' appearing in the cycle, and appear in that order. Figure 3 illustrates this partitioning, with $\alpha = T/k$ and $k = 4$.*

Proof. Denote by D_t the right shift by d_t of D, and by $D(s)$ the RC4 state obtained by performing s encryptions starting at state D. Let l be the greatest distance between two consecutive shifts and denote the corresponding shifts d_a and $d_{a+1 \bmod k}$. Suppose that for some b, the distance s between d_b and $d_{b+1 \bmod k}$ was smaller than l. By Theorem 1, D_a remains a right shift of D_b as the systems evolve. Since $D_b(s)$ is a right shift of D_b, $D_a(s)$ must be a right shift of D_a. Thus, the distance between D_a and $D_{a+1 \bmod k}$ is less than or equal to s. But that distance is l, contradicting the assumption that $s < l$. Therefore, no smaller distance exists, and the shifts are at equal distances from each other. The second part of the theorem follows by Theorem 1 and the fact that any state in the cycle can be obtained by repeated encryption starting at any state D. □

In fact, only certain orderings of the shifts present in a given cycle are possible because Theorem 2 implies that $d_{i+1} - d_i$ is constant in a cycle. d_1 must then be a generator for the shifts in the cycle, and $d_i = i \cdot d_1 \bmod 2^n$. The last three columns of Table 2 confirm this statement for RC4-2 and RC4-3. In this table, the "Shift Generator" entry is the value of d_1, and the entry "Offset" is the distance between successive shifts. Finally, the "Shifts Found" table enumerates the right shifts of the initial state found in each cycle. All of these results were obtained experimentally. Note that in all cases, $T/k =$ Offset as required by the theorem. For example, for the cycles of length 472, a distance of 118 exists between shifts, the shifts appear in the order $\{0, 6, 4, 2\}$, and $472/4 = 118$. The entry $\{0\}$ in the "Shifts Found" column indicates that no shifts of the initial state appear in the cycle.

5   Tracking Analysis

Algorithm 3 below outlines a basic attack which can be mounted against RC4. In essence, the algorithm keeps track of all states which RC4 could be in, given that a particular keystream has been generated.

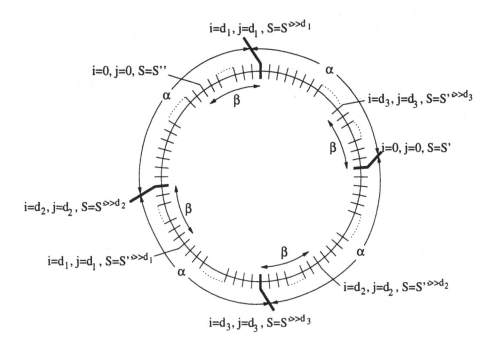

Fig. 3. Partitioning of an RC4 Cycle

Algorithm 3 (Forward Tracking).

1. *Mark all entries S_t as unassigned.*
2. *Set $i = 0$, $j = 0$, and $z = 0$.*
3. *Repeat:*
   (a) *Set $i = i + 1 \bmod 2^n$.*
   (b) *If S_i is unassigned, continue with the remainder of the algorithm for each possible assignment of S_i.*
   (c) *Set $j = j + S_i \bmod 2^n$.*
   (d) *If S_j is unassigned, continue with the remainder of the algorithm for each possible assignment of S_j.*
   (e) *Swap S_i and S_j.*
   (f) *Set $t = S_i + S_j \bmod 2^n$.*
   (g) *If S_t is unassigned and K_z does not yet appear in the s-box, set $S_t = K_z$.*
   (h) *If $S_t \neq K_z$, the state information is incorrect. Terminate this round.*
   (i) *Increment z.*
   (j) *If z is equal to the length of the keystream, output the current state as a solution and terminate the run.*

The forward tracking algorithm is illustrated in Fig. 4. In this diagram, the observed keystream is the all zero sequence, and the system is RC4-2. Figure 5 shows the number of nodes visited by the tracking algorithm as a function of depth for various word sizes n. Two cases are considered for the observed keystream; an arbitrarily chosen nonzero keystream, and a zero keystream. The zero keystream can be analysed more quickly than a more general keystream. To obtain approximate data for the $n = 5$ case in Fig. 5, at a given depth the depth-first search was only carried out for selected nodes. The total number of nodes visited was then calculated assuming that the number of nodes in each subtree would be the same for all nodes. Similar work has been done by Luke O'Connor[8].

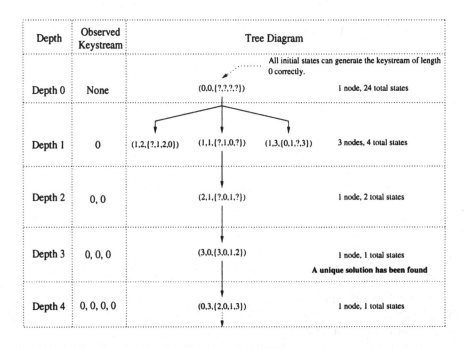

Fig. 4. Forward Tracking Algorithm for $n = 2$, and a 0 Keystream of Length 4

Several variations of this algorithm are possible. Backtracking, in which the keystream is processed in reverse, appears to be easier to implement efficiently because s-box entries are fixed sooner. Probabilistic variants, which use truncated tracking analysis or other information to determine the "best" node to follow in the tracking analysis, may be able to analyse more keystream, avoiding keystreams which result in a more difficult search. These variants are discussed in [6].

The performance of these algorithms can be used to provide an upper bound on the complexity of RC4 cryptanalysis. Table 4 shows the attack complexity observed to date. All of the results are based on backtracking, except the nonzero keystream $n = 5$ entry, based on an estimate from a probabilistic backtracking attack.

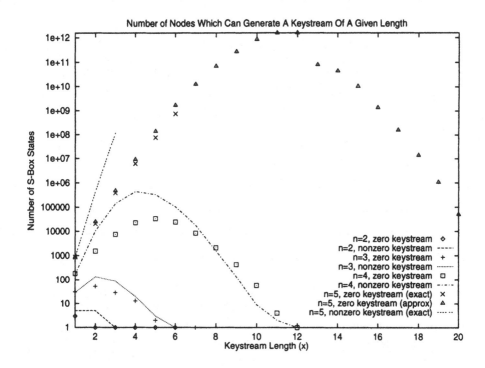

Fig. 5. Number of Nodes During Forward Tracking

6   An Attack on a Weakened Version of RC4

Suppose that RC4 was modified by replacing its initialization function with the following:

Algorithm 4 (Weak RC4 Initialization). *Let $k_0 \ldots k_{l-1}$ denote the user's key.*

1. *Calculate γ such that $\log_2 (\gamma!) = 8 \cdot l$.*
2. *For z from 0 to $2^n - 1$*
   (a) Set $K_z = k_{z \bmod l}$.
3. *For z from 0 to $2^n - 1$*
   (a) Set $S_z = z$.
4. *Set $j = 0$.*
5. *For i from 0 to $2^n - 1$*
   (a) Set $j = j + S_i + K_i \bmod 2^n$.
   (b) Swap $S_{i \bmod \gamma}$ and $S_{j \bmod \gamma}$.
6. *Set $i = 0$ and $j = 0$.*

The system still has $8 \cdot l$ bits of entropy. However, because the tracking analysis can easily be confined to searching the reduced keyspace, it is likely to succeed

Table 4. Estimated Upper Bound on the Complexity of Cryptanalysis of RC4-n

RC4 Word Size	Nominal Key Space	Effective Keyspace	Attack Complexity (arbitrary keystream)	Attack Complexity (zero keystream)
2	2^8	$2^{4.58}$	2^4	2^3
3	2^{24}	$2^{15.30}$	2^8	2^7
4	2^{64}	$2^{44.25}$	2^{20}	2^{17}
5	2^{160}	$2^{117.66}$	2^{69}	2^{42}
6	2^{384}	$2^{296.00}$	?	?
7	2^{896}	$2^{716.16}$	?	?
8	2^{2048}	$2^{1684.00}$	?	?

very quickly even for the full-size cipher. Table 5 summarizes the performance of the tracking attack for this weakened variant of RC4. Table 5 was obtained by performing a tracking attack on 20 keystreams generated with randomly chosen keys for each value of γ. The maximum observed complexity is reported. This result shows that RC4 depends heavily on its key schedule for its security. The attack complexity indicated in Table 5 is not monotonically increasing because it is often the case that the attack succeeds in substantially less than the maximum number of steps.

Table 5. Tracking Attack Complexity for RC4 with a Weakened Key Schedule ($n = 8$)

γ	Effective Keyspace	Attack Complexity
15	2^{40}	2^{14}
20	2^{61}	2^{19}
25	2^{83}	2^{23}
30	2^{107}	2^{17}
31	2^{112}	2^{28}
32	2^{117}	2^{26}
33	2^{122}	2^{26}

7   Conclusion

RC4 remains a secure cipher for practical applications. Several theoretical attacks exist but none have been successful against commonly used key lengths. Nonetheless, tracking analysis does substantially reduce the complexity of cryptanalysis compared to the maximum key length which could be specified. Tracking analysis would show promise if it were possible to use knowledge of the actual key length to limit the state space to be searched. In this regard, RC4's resilience is mainly due to the fact that the key schedule effectively prevents partial knowledge of the s-box state from providing information about the key. If this were not the case, tracking analysis would be successful even for the full-size cipher.

References

1. T. Dierks and C. Allen. The TLS protocol version 1.0. Internet Draft, ftp://ftp.ietf.org/internet-drafts/draft-ietf-tls-protocol-05.txt, November 1997.
2. H. Finney. An RC4 cycle that can't happen. Posting to sci.crypt, Sept. 1994.
3. J. D. Golić. Linear statistical weakness of alleged RC4 keystream generator. In Walter Fumy, editor, *LNCS 1233, Advances in Cryptology - EUROCRYPT '97*, pages 226–238, Germany, 1997. Springer.
4. R. J. Jenkins Jr. Re: RC4? Posting to sci.crypt, Sept 1994.
5. R. J. Jenkins Jr. ISAAC and RC4. Internet document at http://ourworld.compuserve.com/homepages/bob_jenkins/isaac.htm, 1996.
6. S. Mister. Cryptanalysis of RC4-like stream ciphers. Master's thesis, Queen's University, Kingston, Ontario, 1998.
7. S. Mister and S. E. Tavares. Some results on the cryptanalysis of RC4. In *Proceedings of the 19th Biennial Symposium on Communications*, pages 393–397, Kingston, Ontario, June 1-3, 1998.
8. L. O'Connor. Private communication, August 1998.
9. R. L. Rivest. The RC4 encryption algorithm. RSA Data Security Inc., March 1992.
10. A. Roos. A class of weak keys in the RC4 stream cipher. Posting to sci.crypt, Sept. 1995.
11. B. Schneier. *Applied Cryptography*. John Wiley & Sons, Inc., Toronto, Canada, 2nd edition, 1996.

Key Preassigned Traceability Schemes for Broadcast Encryption

D. R. Stinson and R. Wei

Department of Combinatorics and Optimization
University of Waterloo
Waterloo Ontario, N2L 3G1, Canada

Abstract. Traceability schemes for broadcast encryption are defined by Chor, Fiat and Naor in [6] to protect against a possible coalition of users producing an illegal decryption key. Their scheme was then generalized by Stinson and Wei in [17]. These schemes assume that every user can decrypt the secret value. In this paper we discuss key preassigned traceability schemes, in which only the users in a specified privileged subset can decrypt. A new scheme is presented in this paper, which has better traceability than previous schemes. We also present a new threshold traceability scheme by using ramp scheme. All the constructions are explicit and could be implemented easily.
Keywords: key preassigned scheme, broadcast encryption, traceability, secret sharing schemes, combinatorial designs.

1 Introduction

Most networks can be thought of as broadcast networks, in that any one connected to the network can access to all the information that flows through it. In many situations, such as a pay-per-view television broadcast, the data is only available to authorized users. To prevent an unauthorized user from accessing the data, the trusted authority (TA) will encrypt the data and give the authorized users keys to decrypt it. Some unauthorized users might obtain some decryption keys from a group of one or more authorized users (called *traitors*). Then the unauthorized users can decrypt data that they are not entitled to. To prevent this, Chor, Fiat and Naor [6] devised a traitor tracing scheme, called a *traceability scheme*, which will reveal at least one traitor on the confiscation of a pirate decoder. This scheme was then generalized by Stinson and Wei in [17]. There are some other recent papers discussing this topic (see [10, 12, 13]).

The basic idea of a traceability scheme is as follows. Suppose there are a total of b users. The TA generates a set T of v *base keys* and assigns ℓ keys chosen from T to each user. These ℓ keys comprise a user's *personal key*, and we will denote the personal key for user i by U_i. A *broadcast message*, M, consists of an enabling block, B, and a cipher block, Y. The *cipher block* is the encryption of the actual plaintext data X using a *secret key*, S. That is, $Y = e_S(X)$, where $e(\cdot)$ is the encryption function for some cryptosystem. The *enabling block* consists of data which is encrypted by some method, using some or all of the v keys in

S. Tavares and H. Meijer (Eds.): SAC'98, LNCS 1556, pp. 144–156, 1999.

the base set, the decryption of which will allow the recovery of the secret key S. Every authorized user should be able to recover S using his or her personal key, and then decrypt the cipher block using S to obtain the plaintext data, i.e., $X = d_S(Y)$, where $d(\cdot)$ is the decryption function for the cryptosystem.

Some traitors may conspire and give an unauthorized user a *pirate decoder*, E. E will consist of a subset of base keys such that $E \subseteq \cup_{i \in C} U_i$, where C is the coalition of traitors. An unauthorized user may be able to decrypt the enabling block using a pirate decoder. The goal of the TA is to assign keys to the users in such a way that when a pirate decoder is captured and the keys it possesses are examined, it should be possible to detect at least one traitor in the coalition C, provided that $|C| \leq c$ (where c is a predetermined threshold).

In all the traceability schemes discussed in [6, 17, 10, 12] it is assumed that every user can decrypt the enabling block. This means that the data supplier should assign the keys after he or she has determined who the authorized users are. In practice, however, this restriction may be inconvenient, as changes between authorized and unauthorized users may be frequent.

In this paper, we investigate traceability schemes in which the personal keys can be assigned before the authorized users are determined. We will call these schemes *key preassigned schemes*. Key preassigned schemes (for broadcast encryption) have been discussed by several researchers. The first scheme was introduced by Berkovits in [1]. Several recent papers have studied broadcast encryption schemes (see [3, 4, 8, 13, 15, 16], for example). Broadcast schemes enable a TA to broadcast a message to the users in a network so that a certain specified subset of authorized users can decrypt it. However, most of these broadcast schemes have not considered the question of traceability. We will briefly review the traceability of these schemes and then give some key preassigned schemes which have better traceability than the previous schemes. We will also discuss threshold tracing schemes which are more efficient but less secure in some respect. We will use combinatorial methods to describe the schemes and give some explicit constructions. The efficiency of the schemes is measured by considering the information rate and broadcast information rate.

There are two aspects of security in our schemes. One property of the scheme is to prevent unauthorized users from decrypting the enabling block; this is the usual question investigated in broadcast encryption. The second property is the ability of tracing a pirate decoder which is made by a coalition of users (which of course could be authorized users). Although these two properties both protect against coalitions, they have different effects. The first property can prevent the coalition of unauthorized users from decrypting the enabling block, but it does not protect against construction of a pirate decoder. The second property cannot prevent a coalition from decrypting the enabling block, but it enables the TA to trace at least one traitor if the decoder is found.

We will discuss unconditionally secure (in an information theoretic sense) schemes. These schemes do not depend on any computational assumption.

2   Definitions and notations

In this section, we give basic definitions and the notations used in this paper.

2.1   Broadcast encryption schemes

The definition of a broadcast encryption scheme we use in this paper will be the same as the one given in [15]. As in a traceability scheme, there is a trusted authority (TA) and a set of users $\mathcal{U} = \{1, 2, \cdots, b\}$, and the TA generates a set of v base keys and assigns a subset of the base keys to each user as his or her personal key. At a later time, a *privileged subset*, P, of authorized users is determined. The TA chooses a secret key S and broadcasts an enabling block B_P (which is an encryption of S) that can be decrypted by every authorized user, but which cannot be decrypted by certain *forbidden subsets* disjoint from P.

Let $\mathcal{P}$ denote the collection of possible privileged subsets and let $\mathcal{F}$ denote the collection of possible forbidden subsets. In this paper, we will consider the case when $\mathcal{P} = 2^{\mathcal{U}}$, so $\mathcal{P}$ contains all subsets of users, and $\mathcal{F}$ contains all f-subsets of users, where f is a fixed integer. To make things simpler (and since we want to focus on the traceability first), we will mainly consider the situation when $f = 1$. In the case $\mathcal{P} = 2^{\mathcal{U}}$ and $f = 1$, the privileged subset can be chosen to be any subset of users, and the enabling block cannot be decrypted by an individual unauthorized user. (It may be possible for subsets of unauthorized users to jointly decrypt the message, however.)

For $1 \leq i \leq b$, let $\mathbf{U}_i$ denote the set of all possible subsets of base keys that might be distributed to user i by the TA. Thus the personal key $U_i \in \mathbf{U}_i$. Let $\mathbf{S}$ denote the set of possible secret keys, so $S \in \mathbf{S}$. Let $\mathbf{B}_P$ be the set of possible enabling blocks for privileged subset P; thus $B_P \in \mathbf{B}_P$. Usually, U_i, S and B_P consist of tuples from a finite field $\mathbb{F}_q$. We define the *information rate* to be

$$\rho = \min\left\{ \frac{\log \mathbf{S}}{\log \mathbf{U}_i} : 1 \leq i \leq b \right\}.$$

and the *broadcast information rate* to be

$$\rho_B = \min\left\{ \frac{\log \mathbf{S}}{\log \mathbf{B}_P} : P \in \mathcal{P} \right\}.$$

In general, to decrease the size of the broadcast, i.e., to increase ρ_B, it is necessary to decrease ρ, and vice versa. Since it is trivial to construct a broadcast encryption scheme with $\rho = 1$ and $\rho_B = 1/b$, we are mainly interested in schemes with $\rho_B > 1/b$.

2.2   Traceability

Suppose a "pirate decoder" E is found. (We assume that the pirate decoder can be used to decrypt some enabling blocks.) If there exists a user i such that $|E \cap U_i| \geq |E \cap U_j|$ for all users $j \neq i$, then i is defined to be an *exposed user*. A c-traceability scheme is defined as follows.

Definition 21 Suppose any exposed user i is a member of the coalition C whenever a pirate decoder E is produced by C (so $E \subseteq \cup_{i \in C} U_i$) and $|C| \leq c$. Then the scheme is called a *c-traceability scheme*.

When a scheme is c-traceable, $\mathcal{P} = 2^{\mathcal{U}}$, and the forbidden subsets consist of all f-subsets of users, we call it a (c, f)-*key preassigned traceability scheme* and denote it as a (c, f)-KPTS. For the case $f = 1$, we denote the scheme as a c-KPTS.

Remark. The difference between Definition 21 and the one in [13] is that the size of the pirate decoder is not specified here. For example, the pirate decoder might be smaller or larger than a legitimate decoder. The only requirement is that a pirate decoder should be able to decode some enabling blocks.

A set system is a pair $(X, \mathcal{A})$, where X is a set of *points* and $\mathcal{A}$ is a collection of subsets of X called *blocks*. We will use set systems with the following property, which is modified from [17, Theorem 2.2].

Definition 22 A *traceability scheme system* is a set system $(X, \mathcal{A})$, where every block has size k for some integer k, with the property that for every choice of $c' \leq c$ blocks $A_1, A_2, \cdots, A_{c'} \in \mathcal{A}$, and for any t-subset $E \subseteq \cup_{j=1}^{c'} A_j$, where $t \geq k$, there does not exist a block $A \in \mathcal{A} \backslash \{A_1, A_2, \cdots, A_{c'}\}$ such that $|E \cap A_j| \leq |E \cap A|$ for $1 \leq j \leq c'$. Such a system will be denoted by (c, k)-TSS.

In this definition, the blocks correspond to legitimate decoders and E corresponds to a pirate decoder. We will be able to assume that $|E| \geq k$ due to the encryption scheme we use.

2.3   Secret sharing schemes

Let $\mathcal{U}$ be the set of b users, $\Gamma \subseteq 2^{\mathcal{U}}$ be a set of subsets called *authorized subsets*, and let $\Delta \subseteq 2^{\mathcal{U}}$ be a set of subsets called *unauthorized subsets*. In a (Γ, Δ)-*secret sharing scheme*, the TA has a secret value K. The TA will distribute secret information called *shares* to each user of $\mathcal{U}$ in such a way that any authorized subset can compute K from the shares they jointly hold, but no unauthorized subset has any information about K. The paper [14] contains an introduction to secret sharing schemes.

Let $r < t \leq b$. An (r, t, b)-*ramp scheme* is a secret sharing scheme in which the authorized subsets are all the subsets of $\mathcal{U}$ with cardinality at least t and the unauthorized subsets are all the subsets of $\mathcal{U}$ with cardinality at most r. When $r = t - 1$, the ramp scheme becomes a *threshold scheme* which is denoted by (t, b)-threshold scheme. The Shamir scheme provides a construction of a (t, b)-threshold scheme in which each share is an element of $\mathbb{F}_q$ and the secret is also an element of $\mathbb{F}_q$, for any prime power $q \geq b + 1$.

2.4   Key predistribution schemes

Fiat-Naor *key predistribution schemes* (or *KPS*) (see [8]) are used in the KIO construction for broadcast encryption schemes given in [15]. Let $f \leq b$ be an integer. The forbidden subsets consist of all subsets of size at most f. In a Fiat-Naor scheme, the TA chooses a secret value x_F for each possible forbidden subset F, and gives that value to each user in $\mathcal{U} \backslash F$. Let $P \subseteq \mathcal{U}$. The value

$$K_P = \sum_{F \cap P = \emptyset} x_F$$

is the *key* for the privileged subset P. K_P can be computed by any member of P, but K_P cannot be computed by any forbidden subset F disjoint from P (where $|F| \leq f$).

3   Traceability of previous broadcast schemes

Since key preassigned broadcast encryption schemes were proposed in [1], several constructions have been given. A summary of these results can be found in Stinson [15]. In [15], the KIO construction is described, and which is further discussed in [16]. We will not review these schemes here — we only wish to indicate that these schemes usually do not have any traceability, or have, at most, 1-traceability. (However, note that if in a scheme, every user has disjoint keys, then the scheme is "totally traceable". Thus the trivial scheme in [15] has b-traceability.)

Staddon first discussed the traceability of key preassigned broadcast schemes in her PhD thesis [13]. She constructed some schemes called "OR protocols" that have higher traceability. We briefly review the OR protocols now. In OR protocols, the size of a forbidden subset is f and the size of the privileged subset is $w = b - f$. These values are fixed ahead of time. The TA produces a key K_t for each subset P_t of $\mathcal{U}$, where $|P_t| = \lceil \frac{w}{n} \rceil$, and gives that key to every user in P_t, where n is a given positive integer. When the TA wants to broadcast an enabling block for a privileged subset P, he uses the n keys in the set

$$\mathcal{L}_P = \{K_t : P_t \subseteq P\}$$

to encrypt it, in such a way that any user who has at least one of these n keys is able to decrypt it.

It is shown in [13] that the OR protocol construction has $\Theta(\sqrt{n})$-traceability for $n > 2$ and b sufficiently large relative to n and f. However, the proof is based on the assumption that the pirate decoder always is the same size as a personal key, i.e., that it always contains

$$\binom{b-1}{\lceil \frac{w}{n} \rceil - 1}$$

keys. This assumption may not be practical. In fact, unauthorized users who possess even one key might be able to decrypt the enabling block if the key happened to belong to the set $\mathcal{L}_P$. Thus the OR protocol has no traceability if we consider the traceability under Definition 21, where we allow a pirate decoder to have fewer keys than a personal key.

The traceability schemes in [6,17] have the desirable property that any possible decoder must consist of the keys from the base key set, otherwise they will be useless for decoding. In some other proposed schemes, an enabling block can be decrypted using keys not in the base set. In such a scheme, the traceability property is defeated. We describe the traceability scheme proposed in [10] to illustrate this point.

In the scheme of [10] (which is not key preassigned), the TA chooses a random polynomial

$$f(x) = a_0 + a_1 x + a_2 x^2 + \cdots + a_c x^c.$$

The TA then computes $f(i)$ and gives it to user i secretly, so that the personal key of user i will be $(i, f(i))$. When TA wants to encrypt the secret key S, he broadcasts the enabling block $(S + a_0, a_1, a_2, \cdots, a_c)$. If a pirate decoder contains a pair $(u, f(u))$, then u will be the exposed user. However, two users i and j can construct a pirate decoder as follows. They choose two random non-zero numbers α and β and compute the following:

$$b_0 = \frac{\alpha f(i) + \beta f(j)}{\alpha + \beta}, b_1 = \frac{\alpha i + \beta j}{\alpha + \beta}, \cdots, b_c = \frac{\alpha i^c + \beta j^c}{\alpha + \beta}.$$

Since

$$a_0 = b_0 - a_1 b_1 - \cdots - a_c b_c,$$

the $(c + 1)$-tuple $(b_0, \ldots, b_c)$ can be used as a decoder. In this scenario, the traitors i and j cannot be exposed by the usual traitor tracing method.

4   The new scheme

In this section, we present our traceability schemes which will use a KIO type construction. The basic idea of the KIO construction is that the secret key is split into shares, using a threshold scheme (or a ramp scheme), and then the shares are encrypted, thus forming the enabling block. Our scheme is a key preassigned broadcast encryption scheme where $\mathcal{U} = \{1, \ldots, b\}$, $\mathcal{P} = 2^{\mathcal{U}}$ and $\mathcal{F}$ consists of all f-subsets of $\mathcal{U}$. We consider the case $f = 1$ first.

Suppose $(X, \mathcal{A})$ is a (c, k)-TSS, where $X = \{1, 2, \cdots, v\}$ and $\mathcal{A} = \{A_1, A_2, \cdots, A_b\}$. The block A_j determines the personal key given to user j, for $1 \leq j \leq b$. For each $u \in X$, let

$$R_u = \{j \in \mathcal{U} : u \in A_j\}.$$

The main steps in the protocol are as follows:

1. For every set R_u as defined above, the TA constructs a Fiat-Naor key pre-distribution scheme on user set R_u, with $\mathcal{F}_u = \{\{j\} : j \in R_u\} \cup \{\emptyset\}$ and $\mathcal{P}_u = 2^{R_u}$. Thus, for each u, $1 \leq u \leq v$, the TA chooses $|R_u| + 1$ secret

values, denoted x_{R_u} and $x_{R_u,j}$ ($j \in R_u$). These values are chosen at random from a finite field $\mathbb{F}_q$. The value x_{R_u} is given to each $i \in R_u$ and $x_{R_u,j}$ is given to each $i \in R_u \backslash \{j\}$. These keys form the personal key for user i.

We will assume the existence of a function Index on the set of base keys such that $\text{Index}(x) = j$ if x is a key from the jth Fiat-Naor scheme. These keys might be stored as pairs, e.g., (x_{R_u}, u) and $(x_{R_u,j}, u)$, so that the users know which keys are from which Fiat-Naor scheme.

2. Suppose the TA wants to encrypt the secret key $S \in \mathbb{F}_q$ for a privileged subset P. For the purposes of illustration, suppose $P = \{1, 2, \cdots, w\}$. The TA first uses a (k, n)-threshold scheme to split S into n shares $y_1, y_2, \cdots, y_n$, where $A_P = \cup_{i=1}^{w} A_i$ and $n = |A_P|$ (note that $n \leq v$, so a (k, v)-threshold scheme can be used here, if desired).

3. For each $j \in A_P$, the TA computes the secret key K_j of Fiat-Naor scheme on R_j for the privileged subset $R_j \cap P$, i.e.,

$$K_j = x_{R_j} + \sum_{i \in R_j \backslash P} x_{R_j,i}.$$

4. Each share y_j is encrypted using an encryption function $e(\cdot)$ with key K_j. The enabling block consists of the list of encrypted values

$$(e_{K_j}(y_j) : j \in A_P).$$

Since each user in P has k values in A_P, he can compute k keys $K_{i_1}, K_{i_2}, \cdots, K_{i_k}$ and then obtain k shares, $y_{i_1}, y_{i_2}, \cdots, y_{i_k}$. Using the reconstruction function of the threshold scheme, the user is able to recover the value of the secret key, S.

A user not in P cannot compute any of the keys K_i, since the Fiat-Naor scheme is secure against individual unauthorized users. Thus, the user cannot get any information about the n shares.

Now we consider traceability. Suppose a pirate decoder E is found. The TA can compute the Index of the decoder as

$$\text{Index}(E) = \{\text{Index}(x) : x \in E\}.$$

Note that the cardinality of the set $\text{Index}(E)$ is at least k, otherwise the decoder will be useless. The TA can then use this Index to find an exposed user, since the set system $(X, \mathcal{A})$ is a (c, k)-TSS.

The information rate of this scheme is

$$\rho = \frac{1}{kr}$$

where r_x is the number of blocks containing x, i.e., $r_x = |R_x|$, and $r = \max\{r_x : x \in X\}$. The broadcast information rate is

$$\rho_B = \frac{1}{n} \geq \frac{1}{v}.$$

The following theorem summarizes the properties of the scheme.

Theorem 41 *Suppose $(X, \mathcal{A})$ is a (c, k)-TSS in which $|X| = v$ and $|\mathcal{A}| = b$. Then there is a c-KPTS for a set of b users, having information rate $\rho \geq 1/(k\,r)$ and broadcast information rate $\rho_B \geq 1/v$.*

Remark. For the case $f > 1$, we need only change the construction of the Fiat-Naor scheme on each R_u so that the possible forbidden subsets are all subsets of R_u having size at most f. This will cause the information rate of the scheme to decrease, while the broadcast information rate remains the same.

The following small example will illustrate the scheme.

Example 41 *A 2-KPTS with 82 users.*

Let $X = \{0, 1, \cdots, 40\}$ and suppose $\mathcal{A}$ contains the following 82 blocks, where the calculations are in $\mathbb{Z}_{41}$, for $i = 0, 1, 2, \cdots, 40$:

$$A_i = \{1 + i, 10 + i, 18 + i, 16 + i, 37 + i\}$$
$$A_{41+i} = \{36 + i, 32 + i, 33 + i, 2 + i, 20 + i\}$$

The set system $(X, \mathcal{A})$ is a $(41, 5, 1)$-balanced incomplete block design (see [7]). This set system has the property that each pair of points appears in exactly one block, and every point appears in exactly 10 blocks. It is in fact a $(2, 5)$-TSS (see Theorem 63).

The block A_i is associated with user i. For each $u \in X$, the TA constructs a Fiat-Naor scheme on R_u. For example, for $u = 1$, it can be seen that

$$R_1 = \{0, 32, 24, 26, 5, 47, 51, 50, 81, 63\},$$

so $|R_1| = 11$. The TA will choose 11 secret values in $\mathbb{F}_q$ for some prime power q, and every user in R_1 will receive 10 of the 11 values. A Fiat-Naor scheme is implemented in this way on each R_u, and thus every user has 50 values in his or her personal key.

Now, suppose the TA wants to encrypt a secret key $S \in \mathbb{F}_q$, where the privileged subset is $P = \{0, 1, 2, \cdots, 59\}$, so $w = 60$. The TA uses a $(5, 41)$-threshold scheme to split S into 41 shares, $y_0, \ldots, y_{40}$. For example,

$$K_1 = x_{R_1} + x_{R_1, 63} + x_{R_1, 81}.$$

The enabling block will be the list of encrypted values

$$(e_{K_0}(y_0), \cdots, e_{K_{40}}(y_{40})).$$

Any user in P can decrypt the enabling block. For example, consider user 5. The block $B_5 = \{6, 15, 23, 21, 42\}$. Then user 5 obtains five of the 41 secret keys, namely, K_6, K_{15}, K_{23}, K_{21} and K_{42}, and recovers the five shares y_6, y_{15}, y_{23}, y_{21} and y_{42}. From these five shares S can be obtained.

Any user not in P cannot decrypt the enabling block. For example, let us consider user 63. If $j \notin B_{63}$, then user 63 does not have K_{R_j} and cannot compute K_j. On the other hand, if $j \in B_{63}$, then user 63 does not have $K_{R_j,63}$ and cannot compute K_j either. Thus user 63 cannot compute any of the shares in the threshold scheme.

Finally, let's show that the scheme is 2-traceable. If a pirate decoder E is found, then the TA can compute $\mathsf{Index}(E)$ as described above. $\mathsf{Index}(E)$ must contain at least 5 numbers, otherwise it cannot decode anything. Suppose that the decoder was made by two users, say i and j. Since $\mathsf{Index}(E) \subseteq (B_i \cup B_j)$ it must be the case that $|\mathsf{Index}(E) \cap B_i| \geq 3$ or $|\mathsf{Index}(E) \cap B_j| \geq 3$. Since any two blocks intersect in at most one point, $|\mathsf{Index}(E) \cap B_h| \leq 2$ if $h \neq i,j$. Thus user i or user j (or both) will be exposed users.

5   Threshold tracing

In the schemes of Section 4, the Index of any pirate decoder should contain at least k values, otherwise the decoder cannot get any information from the broadcast. However, as indicated in [11] (the final version of [6]), such security is not needed in many applications. For example, in pay-TV applications pirate decoders which decrypt only part of the content are probably useless. Thus [11] defined the concept of a *threshold traceability scheme*. In a threshold traceability scheme, the tracing algorithm only can trace the decoders which decrypt with probability greater than some threshold p. In this section, we discuss some *key preassigned threshold traceability schemes*, denoted by *KPTTS*. Our approach is quite different from the methods used in [11]. We will use ramp schemes to construct KPTTS.

We can obtain a ramp scheme from an orthogonal array.

Definition 51 An *orthogonal array* $\mathrm{OA}(t, k, s)$ is an $s^t \times k$ array, with entries from a set Y of $s \geq 2$ symbols, such that in any t columns, every $t \times 1$ row vector appears exactly once.

The following lemma ([7, Chapter VI.7]) provides infinite classes of orthogonal arrays, for any integer t.

Lemma 51 *If q is a prime power and $t < q$, then there exists an $OA(t, q+1, q)$.*

Suppose there is an $\mathrm{OA}(t, v + t - r, q)$ which is public knowledge. The secret information K is a $(t - r)$-tuple from $\mathbb{F}_q$. The TA chooses secretly a row in the OA such that the last $t - r$ columns of that row contains the tuple K. It is easy to see that there are q^r such rows. The TA then gives each of the v users one value from the first v columns of that row. Since any t of these values determine a row of the OA uniquely, t users can get K by combining their shares. However, from any r values, the users cannot obtain any information about K, since these r values together with last $t - r$ columns of any row in the OA determine that row. (For more detailed description of this construction, the reader can consult [9].

Our KPTTS is similar to the KPTS constructed in Section 4. The only difference is that we use a $(0, k, n)$-ramp scheme to split the message into shares in a KPTTS, instead of the (k, n)-threshold scheme used in the KPTS.

In the KPTTS, the base key set and preassigned keys are the same as in the KPTS. However, when the TA wants to send a secret message $M \in (\mathbb{F}_q)^k$ to a privileged subset, the TA uses a $(0, k, n)$-ramp scheme to split M into n shares. The TA uses the same method of KPTS to encrypt the n values, and broadcasts the resulting list of n values. Similar to the KPTS, any user in the privileged subset can compute k keys, so he or she can recover the n values from the ramp scheme, but the users not in the privileged subset cannot get any information from the encryption.

Now suppose that a pirate decoder E is found. If the size of $\mathsf{Index}(E)$ is not less than k, then the TA can find an exposed user as he did in the KPTS. When the size of $\mathsf{Index}(E)$ is less than k, the TA may not be able to trace the users in the coalition. So let us see what a decoder E could do, if the $\mathsf{Index}(E)$ contains $k - 1$ values. Note that the ramp scheme is constructed from an $OA(k, k + v, q)$. For any $k - 1$ values, there are q rows which contain these $k - 1$ values. Among these q rows, only one row carries the secret message M. Hence the decoding threshold of the KPTTS is

$$p = \frac{1}{q}.$$

The information rate of the KPTTS is the same as that of the KPTS, but the broadcast information rate of the KPTTS is much better. In the KPTTS, we have

$$\rho_B = \frac{k}{n} \geq \frac{k}{v}.$$

Similar to the KPTS, the KPTTS is also based on the set systems TSS. We will discuss the construction of TSS in the next section.

6   Constructions of traceability set systems

To construct our traceability schemes, we need to find traceability set systems. Some constructions for these types of set systems were given in [17]; they are based on certain types of combinatorial designs. (A comprehensive source for information on combinatorial designs is Colbourn and Dinitz [7].) We present a useful lemma for constructing TSS, and mention some applications of it.

Lemma 61 *Suppose there exists a set system $(X, \mathcal{A})$ satisfying the following conditions:*

1. $|A| = k \geq c^2 \mu + 1$ for any $A \in \mathcal{A}$;
2. $|A_i \cap A_j| \leq \mu$ for any $A_i, A_j \in \mathcal{A}$, $i \neq j$.

Then the set system is a (c, k)-TSS.

Proof. Let $E \subseteq \cup_{i=1}^c A_i$ with $|E| \geq k$. Since $k \geq c^2\mu + 1$, there is a block A_s, $1 \leq s \leq c$, such that $|E \cap A_s| \geq c\mu + 1$. For any $A \in \mathcal{A}\backslash\{A_1, A_2, \cdots A_c\}$, we have

$$|E \cap A| \leq |A \cap (\cup_{i=1}^c A_i)|$$
$$\leq c\mu$$
$$< c\mu + 1$$
$$\leq |E \cap A_s|.$$

Hence, the set system is a (c, k)-TSS.   □

As a first application of Lemma 61, we give a construction using t-designs.

Definition 61 *A* t-(v, k, λ) *design is a set system* $(X, \mathcal{A})$, *where* $|X| = v$ *and* $|A| = k$ *for all* $A \in \mathcal{A}$, *such that every* t-*subset of* X *appears in exactly* λ *blocks of* $\mathcal{A}$.

Theorem 62 *Suppose there exists a* t-$(v, k, 1)$ *design. Then there exists a* (c, k)-*TSS, where* $c = \lfloor \sqrt{(k-1)/(t-1)} \rfloor$.

Proof. Any two blocks of a t-$(v, k, 1)$ design intersect in at most $t - 1$ points. Apply Lemma 61 with $\mu = t - 1$.   □

There are many results on t-$(v, k, 1)$ designs for small values of t, i.e., for $2 \leq t \leq 6$. See [7] for a summary of known results. We can construct interesting TSS using designs with $t = 2$. For example, it is known that there is a 2-$(v, 5, 1)$ design for all $v \geq 5$, $v \equiv 1, 5 \bmod 20$. These designs give rise to an infinite family of $(2, 5)$-TSS. Applying Theorem 41 we have the following KPTS.

Theorem 63 *There exists a* 2-*KPTS for all* $v \geq 5$, $v \equiv 1, 5 \bmod 20$, *for a set of* $b = v(v-1)/20$ *users, having* $\rho = \frac{4}{5(v-1)}$ *and* $\rho_B = \frac{1}{v}$.

Note that Example 41 is the case $v = 41$ of the above theorem.
Similarly, we have

Theorem 64 *There exists a* 2-*KPTTS for all* $v \geq 5$, $v \equiv 1, 5 \bmod 20$, *for a set of* $b = v(v-1)/20$ *users, having* $\rho = \frac{4}{5(v-1)}$ *and* $\rho_B = \frac{5}{v}$.

A 3-$(q^2 + 1, q + 1, 1)$ design, known as an *inversive plane*, exists for any prime power q. The following result concerns the KPTS and KPTTS that can be constructed from inversive planes.

Theorem 65 *For any prime power* q, *there exist a* c-*KPTS and a* c-*KPTTS, where* $c = \lfloor \sqrt{\frac{q}{2}} \rfloor$, *with information rate* $\rho \approx \frac{1}{q^3}$ *and broadcast information rates* $\rho_B \approx \frac{1}{q^2}$ *for KPTS and* $\rho_B \approx \frac{1}{q}$ *for KPTTS.*

In [11], it is proved that there exists a threshould traceability scheme with broadcast information rate $\rho_B = O(\frac{1}{4c})$. However, the proof of that is not explicit. Our construction is explicit and the threshold of our scheme is usually better than that of the scheme in [11]. Also our scheme is key preassigned.

Many other constructions of TSS can be given using combinatorial objects such as packing designs, orthogonal arrays, universal hash families, etc. The constructions are similar to those found in [16, 17].

7   Some remarks

We make a couple of final observations in this section.

- The (c, f)-KPTS scheme discussed in this paper is a generalization of the traceability schemes in [6, 17]. The schemes in [6, 17] are in fact the case of $f = 0$ of our main construction. When $f = 0$, there is no protection against an unauthorized user decrypting the enabling block.
- Most broadcast schemes and traceability schemes in the literature are described as unconditionally secure schemes. If the encryption function $e(\cdot)$ used in the scheme in this paper is addition in a finite field $\mathbb{F}_q$, then our scheme is also unconditionally secure. However, the drawback of using the above unconditionally secure encryption scheme is that the resulting KPTS and KPTTS will be a one-time scheme. On the other hand, if we desire only computational security, then we can replace $e(\cdot)$ by any cryptosystem that is computationally secure against a known plaintext attack, and we will obtain a KPTS that can be used for many broadcasts. This simple modification can be applied to other one-time schemes described in previously published papers.

Acknowledgment

The authors' research is supported by the Natural Sciences and Engineering Research Council of Canada. We would also like to acknowledge Tran van Trung for helpful discussions concerning this research.

References

1. S. Berkovits, How to broadcast a secret, *Advances in Cryptology: EURO-CRYPT'91, Lecture Notes in Computer Science*, **547** (1992), 536-541.
2. J. Bierbrauer, T. Johansson, G. Kabatianskii and B. Smeets, On families of hash functions via geometric codes and concatenation, *Advances in Cryptology - CRYPTO'93, Lecture Notes in Computer Science*, **773** (1994), 331-342.
3. C. Blundo, and A. Cresti, Space requirement for broadcast encryption, *Advances in Cryptology: EUROCRYPT'94, Lecture Notes in Computer Science*, **950** (1995), 287-298.
4. C. Blundo, L.A. Frota Mattos and D.R. Stinson, Trade-offs between communication and storage in unconditionally secure schemes for broadcast encryption and interactive key distribution, *Advances in Cryptology: CRYPTO'96, Lecture Notes in Computer Science*, **1109** (1996), 387-400.
5. J. L. Carter and M. N. Wegman, Universal classes of hash functions, J. Computer and System Sci., **18** (1979), 143-154.
6. B. Chor, A. Fiat and M. Naor, Tracing traitors, *Advances in Cryptology: CRYPTO'94, Lecture Notes in Computer Science* **839** (1994), 257-270.
7. C.J. Colbourn and J.H. Dinitz, eds., *CRC Handbook of Combinatorial Designs*, CRC Press, Inc., 1996.

8. A. Fiat and M. Naor, Broadcast encryption, *Advances in Cryptology: CRYPTO'93, Lecture Notes in Computer Science*, **773** (1994), 480-491.
9. W-A. Jackson and K. M. Martin, A combinatorial interpretation of ramp schemes, *Austral. J. Combinatorics* **14** (1996), 51-60.
10. K. Kurosawa and Y. Desmedt, Optimum traitor tracing and asymmetric schemes, *Advances in Cryptology: EUROCRYPT'98, Lecture Notes in Computer Science*, **1403** (1998), 145-157.
11. M. Naor and B. Pinkas, Threshold traitor tracing, *Advances in Cryptology: CRYPTO'98, Lecture Notes in Computer Science* **1462** (1998), 502-517.
12. B. Pfitzmann, Trials of traced traitors, *Information Hiding, Lecture Notes in Computer Science*, **1174** (1996), 49-64.
    collusions, In
13. J.N. Staddon, *A combinatorial study of communication, storage and traceability in broadcast encryption systems*, PhD thesis, University of California at Berkeley, 1997.
14. D.R. Stinson, An explication of secret sharing schemes, *Designs, Codes and Cryptography*, **2** (1992), 357-390.
15. D.R. Stinson, On some methods for unconditionally secure key distribution and broadcast encryption, *Designs, Codes and Cryptography*, **12** (1997), 215-243
16. D.R. Stinson and Tran van Trung, Some new results on key distribution patterns and broadcast encryption, *Designs, Codes and Cryptography*, **14** (1998), 261-279.
17. D.R. Stinson and R. Wei, Combinatorial properties and constructions of traceability schemes and frameproof codes, *SIAM J. Discrete Math*, **11** (1998), 41-53.

Mix-Based Electronic Payments

Markus Jakobsson[1] and David M'Raïhi[2]

[1] Information Sciences Research Center, Bell Laboratories
Murray Hill, NJ 07974
markusj@research.bell-labs.com
[2] Gemplus Corporation
3 Lagoon Drive, Suite 300
Redwood City, CA 94065-1566
david.mraihi@gemplus.com

Abstract. We introduce a new payment architecture that limits the power of an attacker while providing the honest user with privacy. Our proposed method defends against all known attacks on the bank, implements revocable privacy, and results in an efficient scheme which is well-suited for smartcard-based payment schemes over the Internet.

1 Introduction

Since the conception of anonymous payment schemes by Chaum [7] in the early eighties, a lot of attention has been given to new schemes for transfer of funds. Anonymity and its drawbacks, in particular, has been a busy area of research lately, giving rise to many solutions for balancing the need for privacy against the protection against abuse. However, all of these schemes have been based on the same basic architecture as the pioneering scheme, which we suggest may be an unnecessary limitation. By rethinking the underlying privacy and fund-transfer architecture, we show that a stronger attack model can be adopted, and costs curbed.

It is well known that it never pays to make any link of a chain stronger than the weakest link. What has not been given much – if any – attention is that, in fact, it can be *damaging* to do so. The reason is that an attacker may selectively avoid using the weakest link components, thereby potentially enjoying protocol properties corresponding to the *strongest* link of the chain. More specifically, an attacker is implicitly given the option to use alternative tools in lieu of the given privacy-limiting portions of the scheme. For example, as appears to be the case for most Internet based payment schemes, the *connection* is the weakest link in terms of privacy, as a user reveals his IP address to some entity when establishing a connection. An attacker could physically mail in disks and ask to have the encrypted responses posted to newsgroups – and therefore successfully hide his IP address. An honest user would not be granted this same level of privacy: his address would only be concealed by means of an anonymizer, such as a mix-network (e.g., [6, 1, 21, 26, 33]) or a crowd [27], both of which are based on the cooperation of some trusted participants. Therefore, if one link of the

S. Tavares and H. Meijer (Eds.): SAC'98, LNCS 1556, pp. 157–173, 1999.
© Springer-Verlag Berlin Heidelberg 1999

chain, corresponding to the privacy of one system component, is stronger than others, this potentially benefits an attacker without directly adding any value to the honest users.

We argue that it is unnecessary to employ a scheme with properties that can only be used by attackers. Moreover, by degrading the privacy of components until all components deliver the degree of privacy of the weakest link, efficiency improvements can be found. Thus, we can strengthen the defense against attacks (some of which might not be known to date) at the same time as we make the system less costly, without making the system less attractive to its honest users.

An example from the literature of this type of "double improvement" is the work of Jakobsson and Yung [17–20], in which the *bank robbery attack* (in which the attacker obtains the secret keys of the banks) was introduced and prevented against, and the mechanisms employed to achieve this goal were used to improve the versatility and efficiency of the resulting scheme. Their scheme has two modes – one efficient off-line mode, for the common case, and one rather costly on-line mode, to be used only after a successful bank robbery. By making sure that an attacker can never enjoy a higher degree of privacy than an honest user, and implementing mechanisms for a quorum of servers to selectively revoke privacy, we obtain the same degree of security as the *on-line mode* of the scheme by Jakobsson and Yung, at a cost that is comparable to the *off-line mode* of the same scheme.

Our scheme is, unlike other payment schemes with privacy, not based on blind signatures (or variations thereof), but its privacy is derived solely from the use of a mix-network. It does not allow an attacker any higher degree of privacy than honest users, but all users enjoy the following degree of privacy: As long as no quorum of bank servers is corrupted or decides to perform tracings, and no payer or payee reveals how they paid or were paid, the bank servers can only learn the number of payments to and from each account during each time period (e.g., one month.) Moreover, as is appropriate, the transactions are independent in the sense that knowledge about one transaction gives no knowledge of another transaction (other than reducing the number of remaining possibilities.)

We note that if one bank server and one merchant collude, then they can together remove the privacy for transactions that this merchant was involved in[1]. Whereas this is a lower degree of protection than what many other schemes offer, it makes sense in a setting where users are not concerned with the bank potentially learning about *a few* of their transactions, as long as the majority of the transactions are secret. We believe that this is an adequate degree of protection to avoid profiling of users performing every-day transactions, which is a degree of privacy that society seems to agree on should be granted.

The communication and computational costs for the different transactions are very similar to those of other schemes, but the amount of data stored by payers and merchants is significantly lower, making our suggested scheme particularly suitable for smartcard implementations.

---

[1] This limitation can be avoided at the cost of a lower propagation rate, as will be discussed onwards.

2   Using Smartcards for Payment Schemes

Let us for a moment take a step back from what features are *desirable* in a payment scheme, and focus on what is *absolutely necessary*. In order not to have to rely on any physical assumption, we need some form of authentication and encryption for security and privacy. Also, in order for the user not to have to put excessive trust in other entities, public key solutions have to be used. Therefore, let us call signature generation and encryption the minimal requirements (where the user performs all or some of the computation involved in performing the transactions.)

The most suitable and least expensive way of meeting these requirements in an environment with portable payment devices would be to develop and use a very simple smartcard with a hardware accelerator for modular multiplications, and with a minimum of memory (since what is expensive in the design of smartcards is EEPROM and RAM). Such a product would be perfect for generating and verifying signatures (e.g. DSA [25] or RSA [28] signatures), and the use of a co-processor would allow very fast public-key certificate verification. It would offer the necessary operations required by all e-commerce applications. However, the use of such a smart card desperately clashes with existing solutions for anonymous e-commerce, since these require high amounts of storage, complicated protocols, and in many cases rely to some extent on tamper-resistance. (Note, however, that there are numerous payment schemes that are smart-card based, e.g., [10, 23]. This is, in fact, a case in point, since these schemes either require expensive special-purpose smart cards, or limit the safe usage of the payment scheme due to its reduced security.)

The main advantage of our scheme is that we overcome the major drawbacks related to the original e-coin paradigm: using a mix-decryption scheme to provide controlled anonymity, we build a counter-based scheme where the participants do not need to store anything but their secret keys and a small amount of user-related information. In our setting, all participants could therefore be represented as owners of an inexpensive and simple smart-card of the type discussed above. We demonstrate how to build a payment scheme offering privacy, using only features of such a minimalistic cryptographic smartcard. Our solution offers:

- flexibility and plug-and-play ability to users: the same device enables home-banking, e-commerce and potentially extensions to applications such as privacy-enhanced email,
- strong protection to users, issuers and governments by the use of controlled and balanced privacy.

3   Intuitive Approach

Instead of using the common token approach, we will use an account-based approach, where a payer requests to have a payment made from his account to another account. The payer's identity will be known to the bank, who will deduct the appropriate amount from his account after a valid, authenticated payment request has been received. However, the merchant's identity will not be obvious from this request. More specifically, it will be encrypted using a public key, whose corresponding secret key is distributively held by a set of servers, which are controlled by banks and government entities. At given time intervals, or after a certain number of payment orders have been received, these will be *collectively* decrypted by a set of servers holding secret key shares. This is done in a manner that does not reveal the relationship between any given encrypted payment order and its corresponding cleartext payment order, the latter which contains the account number of the merchant to be paid. After the collective decryption of all encrypted payment orders, all the accounts indicated by these will be credited appropriately.

In order to increase the propagation speed between initiation of payment (the submission of the encrypted payment order) and the acceptance of a payment by the merchant, we can let the payer prove to the merchant that the payment order has been accepted by the bank, and that the eventual decryption of the payment order will result in the crediting of the merchant's account being made.

We note that in either case, there is no need for the merchant to be in direct contact with the bank (there is no deposit), and neither the payer nor the merchant needs to store a large amount of information, but just a sufficient amount to generate vs. verify payment orders. The main computation will be done by the much more powerful bank servers, and without any strong online or efficiency requirements being imposed.

4   Related Work

When the concept of electronic payments was introduced (see [7,9]) there was a strong focus on *perfect* privacy. Lately, as possible government policies have started to be considered (e.g., [34]), and attacks exploiting user privacy have been discovered (e.g., [31,17]), the attention has shifted towards schemes with *revocable* privacy.

In [2], Brickell, Gemmell and Kravitz introduced the notion of trustee-based tracing, and demonstrated a payment scheme that implemented computational privacy that could be revoked by cooperation between the bank and a trustee. The concept of *fair blind signatures* was independently introduced by Stadler, Piveteau and Camenisch [32]; this is a blinded signature for which the blinding can be removed by the cooperation between the signer (the bank) and a trustee. This type of signatures were employed in payment schemes in [5,4], and a smart-card based variation was suggested in [24]. In [12,11], methods were introduced allowing the trustee not to have to be on-line during the signature generation, by employing so called *indirect discourse proofs*, in which the withdrawer proves to the bank that the trustee will later be able to trace.

In the above schemes, a cooperating bank and trustee were able to trace all properly withdrawn coins, but not coins obtained in other ways. The underlying attack model was strengthened by Jakobsson and Yung [17], by the introduction of the *bank robbery attack*, in which an attacker compromises the secret keys used for signing and tracing, or forces the signers to produce signatures using an alternate generation protocol; their proposed solution protects against this attack and permits *all* coins to be traced, and any coin to be successfully blacklisted after a short propagation delay. Also, the degree of trust needed was decreased, by the use of methods to assure that the tracing and signing parties do not cheat each other. This work was improved on in [18–20] by the distribution of the parties and the introduction of a minimalistic method for privacy revocation (i.e., verifying if a given payment corresponds to a given withdrawal.)

We achieve the same protection against attacks on the bank as the above constructions, and additionally, by shifting the model from a coin-based to an account-based model, make bank robberies futile: If the secret keys of the banks should be compromised, this only allows the attacker to *trace* payments, and *not* to mint money. Additionally, we meet the level of functionality introduced in [18,19] in terms of methods for tracing, and distribution of trust and functionality. Finally, by the shift in architecture, we remove the propagation delay for blacklisting - in fact, we avoid having to send out blacklists to merchants altogether.

However, our scheme requires the payer to be able to communicate with a bank or clearing center for each payment. Also, it does not offer the same granularity of payments that the use of *challenge semantics* in [17,18] could, but potentially requires multiple payments to match a particular amount (much like for many coin based schemes).

On the other hand, we are able to reduce the amount of information stored by the payer (who only has to store his secret key.) Moreover, the merchants neither have to store any payment-related information (after having verified its correctness), nor do they ever have to connect to the bank. (In some sense, one might say that the deposit is performed by the payer.) This allows for a very efficient implementation in a smartcard based Internet setting (where the smartcards are used by the payers to guarantee that only authorized users can obtain access to funds.)

The anonymity in our scheme is based solely on the use of a mix-network, a primitive introduced by Chaum [6]. We demonstrate an implementation based on a general mix-network (e.g., [6,33]), and state and prove protocol properties. We then consider the use of a particular, recently proposed type of mix-network [1,21,26], and discuss how its added features can further strengthen the payment scheme. The result is an efficient and practical payment scheme, well suited for Internet implementation, that prevents by all known attacks by ensuring that an attacker *never* can obtain any more privacy than an honest user is offered. The privacy is controlled by a conglomerate of banks and ombudsmen, a quorum of which have to cooperate in order to revoke it.

Another scheme that uses a mix-network (or some other primitive implementing anonymous communication) is the payment scheme of Simon [30]. That scheme is similar to ours also in that it takes the "minimalist approach" as well, building an anonymous payment scheme from a small set of simple components, and in that the account information to some extent is kept by the bank. More specifically, Simon's scheme uses a mix-network to communicate preimages to function values that are registered as having a value. When the merchant receives such a payment, he deposits it by sending (again, using the mix-network) the preimage along with a new function value, to which only the merchant knows the preimage. The bank cancels the old function value and associates value with the new function value. Whereas this allows linking of payments through a chain, it still offers a high degree of privacy to its users. Although Simon's scheme is not directly concerned with limiting the privacy an attacker can enjoy, and there are several privacy-related attacks that can be performed (since there is no direct notion of revocation of privacy), it still appears to be the case that the scheme is secure against some of the strongest attacks (e.g., bank robbery.) This is due to the fact that the representation of value is effectively controlled by the bank by the storage of data by the bank, and not solely linked to having a bank signature.

An interesting observation is the close relationship between our proposed payment scheme and election schemes. More specifically, it can be seen that our payment scheme can be used as an election scheme with only minor modifications; the resulting election scheme[2] implements most recently proposed features, and allows for multi-bit votes to be cast.

5   Model

Participants: There are five types of participants, all modeled by polynomial-time Turing machines: *payers, merchants, banks*[3], a *transaction center*, and the *certification authority*. The payers and merchants have accounts with banks (but not necessarily the same banks); the transaction center processes transfers between accounts of payers and merchants, and is controlled by a conglomerate of banks; the certification authority issues certificates on all other participants' public keys, and may be controlled by banks and government organizations. The transaction center knows for each payer's public key the identity of the payer's bank. (If a user has several banks, he correspondingly has several public keys, one per bank.)

Trust: The payers and merchants trust that the banks will not steal their money. The payers trust for the privacy of a particular transaction that the merchant of the transaction does not collude with bank servers or otherwise attempt to reveal their identities. The payers trust for their general privacy that there is not a quorum of dishonest, cooperating banks constituting a corrupt transaction center.

Computation: We base our general scheme on the existence of a mix-network. Such a scheme can be produced based on any one-way function. We also

---

[2] Instead of encrypting the account number of the payee, the payer/voter would encrypt his vote.

[3] It is possible to substitute some bank servers for ombudsman servers, whose aim is to limit illegal tracing transactions by bank servers. These will not have to be trusted with funds, and only hold tracing keys. Due to a shortage of space, we do not elaborate on how exactly to implement these.

present a scheme using a recently proposed type of ElGamal based mix-network [1, 21, 26].

6   Requirements

We will require the following from our scheme:

Unforgeability: It is not possible for a coalition of participants not including a quorum of bank servers to perform valid payments for a value v exceeding the value charged to these participants.

Impersonation safety: It is not possible for a coalition of cheating participants to perform a transaction resulting in an honest participant being charged more than what he spent.

Overspending blocking: It is not possible for a coalition of participants to perform payments that are accepted by merchants as valid, for an amount exceeding their combined limits, or exceeding what they will be charged for.

Payment blocking: It is always possible for a user to block all payments from his account, going into effect immediately.

Revocability: The bank of any user can restrict the rights of the user to perform payments. If a user has no bank agreeing to a payment, the payment will not be performed.

Framing-freeness: It is not possible for a coalition of participants not including an honest user to produce a transcript corresponding to a payment order from the honest user.

Uniform anonymity: The probability for any coalition of participants, not including a quorum of bank (and ombudsman) servers, to determine from whom a payment was made and to whom (assuming that neither of these parties collaborate with the same servers) is non-negligible better than a guess, uniformly at random from all possible pairs, given all pairs of payers and merchants corresponding to the input and output of the mix-network. It is not possible for a coalition of participants to obtain a higher degree of anonymity by forcing alternative protocols to be used.

Traceability: Any quorum of bank servers[4] are able to perform the following actions:

1. Given identifying information about a payer (and possibly a particular payment transaction of the payer), establish the identity of the receiver of the corresponding payment(s).
2. Given identifying information about a merchant (and possibly a particular payment transaction to the merchant), establish the identities of the corresponding payer(s).
3. Given identifying information about a payer; a particular payment transaction from the same; a merchant; and a particular payment transaction to the same, establish whether these correspond to each other.

---

[4] As noted before, some of these may be controlled by an ombudsman.

7   Quick review: Mix-networks

Before presenting our scheme, let us briefly review what a mix-network is: In general terms, it is a set of servers that serially decrypt and permute lists of incoming encrypted messages. Here, the messages are either encrypted using all the individual public keys of the servers, or using one public key, where the corresponding secret key is shared by the mix servers.

The scheme implements privacy as long as at least one of the active mix servers does not reveal what random permutation he applied, and the encryption scheme is probabilistic, so that it is not possible to compute the same encrypted messages that constituted the input given the decrypted messages that constitute the output.

Some of the schemes introduced are robust, meaning that it is not possible for some subset of cheating servers participating in the mix-decryption to make the final output incorrect (without this being detected by the honest participating servers.)

We present a more detailed explanation of one particularly useful class of mix-networks in section 9.

8   Paying and Tracing (General Version)

8.1   Setup

For each time interval between accounting sessions (in which the payment orders get decrypted and the merchants credited) a new public key may used (we will elaborate on when this is necessary onwards.) These public keys can be broadcast beforehand. The corresponding secret keys are known only to the transaction center, and are kept in a distributed manner so that any quorum of mix-servers can calculate it.

Each party is associated with a public key, which is registered with the transaction center if the party in question is authorized to perform payments. Only this party knows the corresponding secret key, which is used to sign payment orders and possibly other data which needs to be authenticated. The signature scheme employed for this is assumed to be existentially unforgeable.

8.2   Paying

There are three phases of the payment scheme: *negotiation, initiation of payment,* and *completion of payment.* In the first, the payer and the merchant produce a description of the transaction; in the second, the transaction is being committed to and the payer's account debited; and in the third, the merchant's account is credited accordingly. Note that the transaction becomes binding in the second phase already, so the transfer of the merchandise bought can be initiated right after the second phase has finished, and the third phase may be performed at a later point. This is a scenario rather similar to the credit card scenario, in which the payer commits to the transfer much before the merchant receives the funds.

1. **Negotiation**

   In the negotiation phase, the payer and the merchant agree on an exchange, i.e., a description of the merchandise, how it will be delivered, the price of the merchandise, etc. We let m be a hashed-down description of this contract. Furthermore, the merchant gives the payer his account number a (which specifies the bank as well as identifies the merchants account in this bank) and a serial number s for this transaction. If the payment will constitute of several coins, a suite $s_1, \ldots, s_k$ of such serial numbers is given, each one representing one partial payment. A payment order o is of the form $m|a|s$. (If a suite of serial numbers is used, then a corresponding suite of payments orders $o_1, \ldots, o_k$ will be generated.)

2. **Initiation of Payment**

   (a) The payer encrypts a batch of payment orders $o_1, \ldots, o_n$ using the public key encryption scheme of the *transaction center* (which corresponds to the mix-network). These payment orders do not have to correspond to one transaction only, or have the same denomination, but may stem from multiple simultaneous transactions and be of different values. The result is a batch of encrypted payment orders, $\tilde{o}_1, \ldots, \tilde{o}_n$. If different denominations are supported in the system, a description d is appended, describing what the values of the different transactions are. If only one denomination is supported, d is the empty string. The payer signs $\tilde{o}_1, \ldots, \tilde{o}_n, d$ using his private key. The resulting signature is σ. The payer sends $\tilde{o}_1, \ldots, \tilde{o}_n, d, \sigma$ to the transaction center.

   (b) The transaction center verifies the signature σ. One of two approaches is taken in order to avoid replay attacks: either, the payer needs to prove knowledge of some part of the posted message in a way that depends on time, or the transaction center keeps the posted messages in a sorted list and ignores any repeated posting. If the latter approach is taken, the payment orders are encrypted using a new public key for each payment period (the time in between two accounting phases.)

   The transaction center then may verify that the payer has sufficient funds. This may be required of certain banks, or for amounts above some threshold; alternatively, verifications may be performed randomly or in batches to lower the processing times and costs.

   Each valid encrypted payment order $\tilde{o}_i$ is added to an internal list of payments to be performed; there is one such list for each denomination. The payers' accounts are debited accordingly, either immediately or in batches. The transaction center signs each[5] individual valid encrypted payment order $\tilde{o}_i$, resulting in a signature σ_i. The transaction center uses different public keys for different denominations, or appends a description of the denomination before signing. It sends $\sigma_1, \ldots, \sigma_n$ to the payer.

---

[5] Suites of transactions orders corresponding to the same transaction may be signed together to improve efficiency.

(c) The payer sends a suite of signed, encrypted payment orders (of the format $\tilde{o}_i, \sigma_i$) to the corresponding merchant. He also sends a proof of what the encrypted messages are (in some cases, this amounts plainly to show the plaintext.)

(d) The merchant verifies that σ_i is the transaction center's signature on the payment order $\tilde{o}_i$, and that the decrypted value o_i is of the form $m|a|s$ for valid a valid merchandise description m, the merchant's account number a, and a sequence number s not yet received.

If the above verification goes through, then the merchant stores the sequence number, and delivers the merchandise purchased.

3. **Completion of Payment**

At given intervals, the transaction center decrypts all the payment orders using the mix-decryption scheme employed. The result is a plaintext list of payment orders for each denomination. The items of these lists indicate what accounts are to be credited, and by how much. The banks corresponding to the accounts indicated credit these accounts accordingly.

Remark 1: If steps 2c and 2d are excluded, we can avoid the privacy assumption that the merchant is not colluding with a bank server. The cost for this is a lower propagation speed, i.e., the purchase will not be possible to complete until the bank servers has decrypted the payment order batch.

Remark 2: Note that the payment orders can be generated by the payers without any real interaction between the payer and the merchant. This is possible if m is a publicly available description (e.g., the hash of an advertisement sent out by the merchant,) a is publicly available as well, and $s_1, \ldots, s_k$ is selected uniformly at random from a space that is large enough to avoid collisions.

Remark 3: In order to limit the amount of storage needed for the merchant, the sequence numbers may contain a date and time stamp, and will only be accepted - and stored - within an interval associated with this timestamp.

8.3   Tracing

If a general mix-network is employed, only the two basic tracing operations can be performed efficiently, tracing from a payer to a payment order, and from a payment order to a payer. We will later look at how this can be extended (and simplified) for an ElGamal based mix-network.

1. **Payer → Payment Order**

The trace is performed simply by decrypting the encrypted payment order $\tilde{o}$ in question, arriving at the plaintext payment order o, which specifies the account to which the payment is performed.

2. **Payment Order → Payer**

The trace is performed as follows: the mix-servers reverse the computation of o from $\tilde{o}$ step by step, proving to each other that each step is valid. Depending on what type of mix-network is employed, it may be necessary to keep intermediary values, such as partial decryptions and the permutations employed; if these are not available, then they can be re-generated by decrypting the entire bulletin board again.

9 Paying and Tracing (ElGamal Version)

If the recently proposed type of mix-decryption scheme (different methods independently proposed by Abe [1], by Jakobsson [21], and by Ogata, Kurosawa, Sako, and Takatani [26]) is used by the transaction center, this allows several improvements to be made. Let us first briefly describe this mix-decryption scheme, and then elaborate on the advantageous consequences of using it.

1. The input to the scheme is a list of ElGamal encrypted messages (encrypted with the public key of the transaction center, i.e., mix center.) The output is a permuted list of the corresponding decrypted values.
2. The decryption process can be performed by any k out of n servers, who share the secret key for decryption. The participants with whom the encryptions originated need not be involved.
3. No subset of less than k out of n mix servers can perform the decryption.
4. The decryption process is robust, meaning that if any server(s) should cheat, then the cheating will be detected, and their identities become known to the remaining servers.
5. No subset of cheating servers can correlate items of the input to items of the output (unless they already know the corresponding plaintext messages.)
6. The decryption process is efficient.

If this scheme is employed, the following holds:

1. As long as there exists a quorum of honest mix-servers (controlled by the banks), the decryption process will be possible[6].
2. As long as at least one of the participating mix-servers is honest, the correctness of the output is guaranteed[7].
3. As long as there is no dishonest quorum of mix-servers, the privacy of users will be guaranteed.

The tracing scheme can be simplified as well, and makes it unnecessary for the banks to store the contents of the bulletin board, or intermediary computation, in order to perform traces. The following three types of tracing can be performed:

1. **Payer → Payment Order**
   The trace is performed simply by decrypting the encrypted payment order $\tilde{o}$ in question, arriving at the plaintext payment order o, which specifies the account to which the payment is performed.

---

[6] This is an important issue, since otherwise it would be possible for one corrupt bank or ombudsman (controlling one mix-server) to stop all payments in one payment period (unless the payers volunteer to repeat their payments!)

[7] This is important, or a subset of mix-servers could manipulate the payments without being detected.

2. **Payment Order → Payer**

   Here, the given decrypted payment order is to be compared to the encrypted input items. When a match is found, the corresponding payer identifier is output. This can be done by first blinding both the payment order and all the input items[8] using the same distributively held blinding factor, and then decrypt all the blinded input items, after which a match is found. If the mix-decryption scheme by Jakobsson [21] is used, then the trace can be performed by computing a tag, corresponding to the input to the mix-network, and comparing this tag to the partially decrypted (but still blinded) encrypted payment orders. We refer to [21] for a more detailed description.

3. **(Payment Order, Payer) → (yes/no)**

   This trace is performed by computing the above tag from the encrypted payment order in question, and verify whether this tag corresponds to the decrypted payment order (without revealing any other information than this one bit.) This can be done using the verification protocol for undeniable signatures [8], or using a small modification of the first method described above.

10 Claims

The basic scheme satisfies the following requirements (for a quorum size of all participating bank servers, who are assumed to follow the protocol, although they may be curious to learn additional information): *unforgeability, impersonation safety, overspending blocking, payment blocking / revocability, uniform anonymity*, and *framing-freeness*. Additionally, the two first modes of tracing detailed in *traceability* are possible.

The ElGamal based scheme satisfies (without any condition on quorum size or behavior of non-cooperating servers) the above requirements, and full *traceability*. We prove these claims in the Appendix.

11 Performance Analysis

This section demonstrates that our proposed scheme is sufficiently efficient to be practical in a mobile setting, in which the user's device is a portable token, such as a smart-card with restricted capacities in terms of computational power and memory storage.

Although our solution is also competitive in a PC-based setting, it is the mobile setting that is the most restrictive. We note also that whereas a good performance w.r.t. the mix-decryption is important, it is not vital for the efficiency of the scheme, since the mix-decryption is performed off-line and in batches, and does not involve the payer or merchant, nor requires these to wait for the result of

---

[8] An ElGamal encryption (a, b) can be blinded using a blinding factor δ by computing (a^δ, b^δ). This results in a blinding of the plaintext message m to m^δ, but in encrypted form.

the operation. In fact, the employment of the mix-network will be transparent to the users in the case where a threshold-based scheme (such as ElGamal) is used. (For a treatment of the efficiency of the mix-decryption schemes in question, we refer to the corresponding papers where these are treated in detail.)

Table 1 summarizes computational performance of two smartcards with cryptographic accelerators: Smartcard the new smartcard product from Siemens [29], used for instance in the Gemplus GPK range, and the Hitachi[16][9] enhanced chip for smartcards. Performance of the Siemens SLE66X160S has been carefully evaluated, running routines on emulator at a 3.68 MHz clock frequency and with the worst-case set of parameters (all the exponent bits set).

Data Length l (bits)	512	768	1024
H8/3113	90 ms	350 ms	700 ms
SLE66CX160S	186 ms	593 ms	1045 ms

Table 1. - Exponentiation Timings for exponent=1_l

Let us now consider a full implementation of the scheme. The smartcard used is based on the Siemens chip (GPK card) giving us timings around 300 ms for the 1024-bit RSA signature (with CRT) and 200 ms for the El-Gamal encryption. The total time expected for processing a payment order is approximately the time required for encrypting and signing data: The communication time for processing such transactions is around 100 ms when transmitted according to 7816-4 standard at 115,200 bd. Since the user does not need to update a transaction log, writing to EEPROM (which is quite time consuming) is avoided.

We performed various practical tests confirming that the total time for issuing a payment order should be around 1 second, as detailed in table 2. We assume that:

1. the time spent for checking at merchant and center is negligible (we consider a protocol overhead of 100 ms)
2. communication is performed at 115,200 bds
3. 512-bit modulus for ElGamal and 1024-bit modulus for RSA are practical and secure parameter sizes

12   Acknowledgments

We want to thank Moti Yung and David Kravitz for interesting discussions before resp. during the conception and writing of the article.

---

[9] Timings are estimated since the chip has not yet been introduced, but is expected to be released during 1998.

Operation	Processing Time
512-bit El-Gamal Encryption	227 ms
Hashing Ciphertext (SHA)	32 ms
1024-bit RSA Signature	289 ms
Sending to center	110 ms
Sending to merchant	110 ms
Checking Overhead	100 ms
Total	868 ms

Table 2. Transaction Time Evaluation

References

1. M. Abe, "Universally Verifiable Mix-net with Verification Work Independent of the Number of Mix-centers," Eurocrypt '98.
2. E. Brickell, P. Gemmell, D. Kravitz, "Trustee-based Tracing Extensions to Anonymous Cash and the Making of Anonymous Change," Proc. 6th Annual ACM-SIAM Symposium on Discrete Algorithms (SODA), 1995, pp. 457-466.
3. H. Bürk, A. Pfitzmann, "Digital Payment Systems Enabling Security and Unobservability," Computers and Security 8/5 (1989), pp. 399-416
4. J. Camenisch, U. Maurer, M. Stadler, "Digital Payment Systems with Passive Anonymity-Revoking Trustees," Computer Security - ESORICS 96, volume 1146, pp. 33-43.
5. J. Camenisch, J-M. Piveteau, M. Stadler, "An Efficient Fair Payment System," 3rd ACM Conf. on Comp. and Comm. Security, 1996, pp. 88-94.
6. D. Chaum, "Untraceable electronic mail, return addresses, and digital pseudonyms," Communications of the ACM, ACM 1981, pp. 84-88
7. D. Chaum, "Blind Signatures for Untraceable Payments," Advances in Cryptology - Proceedings of Crypto '82, 1983, pp. 199-203.
8. D. Chaum, H. Van Antwerpen, "Undeniable Signatures," Crypto '89, pp. 212-216
9. D. Chaum, A. Fiat and M. Naor, "Untraceable Electronic Cash," Advances in Cryptology - Proceedings of Crypto '88, pp. 319-327.
10. Digicash, see www.digicash.com
11. G.I. Davida, Y. Frankel, Y. Tsiounis, and M. Yung, "Anonymity Control in E-Cash Systems," Financial Cryptography 97.
12. Y. Frankel, Y. Tsiounis, and M. Yung, "Indirect Discourse Proofs: Achieving Efficient Fair Off-Line E-Cash," Advances in Cryptology - Proceedings of Asiacrypt 96, pp. 286-300.
13. M. Franklin and M. Yung, "Towards Provably Secure Efficient Electronic Cash," Columbia Univ. Dept of C.S. TR CUCS-018-92, April 24, 1992. (Also in Icalp-93, July 93, Lund Sweden, LNCS Springer Verlag).
14. M. Franklin and M. Yung, "Blind Weak Signatures and its Applications: Putting Non-Cryptographic Secure Computation to Work," Advances in Cryptology - Proceedings of Eurocrypt '94
15. Gemplus, "Gemplus Public Key (GPK) Cards", October 1996.
16. Hitachi, "Product directory: H8 Range", 1997.
17. M. Jakobsson and M. Yung, "Revocable and Versatile Electronic Money," 3rd ACM Conference on Comp. and Comm. Security, 1996, pp. 76-87.

18. M. Jakobsson and M. Yung, "Applying Anti-Trust Policies to Increase Trust in a Versatile E-Money System," Financial Cryptography '97.

19. M. Jakobsson and M. Yung, "Distributed 'Magic Ink' Signatures," Eurocrypt '97, pp. 450-464

20. M. Jakobsson, "Privacy vs. Authenticity," PhD Thesis, University of California, San Diego, Department of Computer Science and Engineering, 1997. Available at http://www-cse.ucsd.edu/users/markus/.

21. M. Jakobsson, "A Practical Mix," Eurocrypt '98, available at www.bell-labs.com/user/markusj/

22. S. Low, N. Maxemchuk, "Anonymous Credit Cards," 2nd ACM Conference on Computer and Communications Security, 1994, pp. 108-117.

23. Mondex, see www.mondex.com

24. D. M'Raïhi, "Cost-Effective Payment Schemes with Privacy Regulation," Advances in Cryptology - Proceedings of Asiacrypt '96.

25. National Institute for Standards and Technology, "Digital Signature Standard (DSS)," Federal Register Vol 56(169), Aug 30, 1991.

26. W. Ogata, K. Kurosawa, K. Sako, K. Takatani, "Fault Tolerant Anonymous Channel," Information and Communications Security, '97, pp. 440 - 444

27. M. Reiter, A. Rubin, "Crowds: Anonymity for Web Transactions," ACM Transactions on Information and System Security, April, 1998.

28. R. Rivest, A. Shamir and L. Adleman, "A method for Obtaining Digital Signatures and Public-Key Cryptosystems", Communications of the ACM, v. 21, n. 2, Feb 1978, pp. 120-126.

29. Siemens, "ICs for Chip Cards", Customer Information, June 1997.

30. D. Simon, "Anonymous Communication and Anonymous Cash," Crypto '96, pp. 61-73.

31. S. von Solms and D. Naccache, "On Blind Signatures and Perfect Crimes," Computers and Security, 11 (1992) pp. 581-583.

32. M. Stadler, J-M. Piveteau, J. Camenisch, "Fair Blind Signatures," Advances in Cryptology - Proceedings of Eurocrypt '95, 1995.

33. P. Syverson, D. Goldschlag, M. Reed, "Anonymous connections and onion routing," IEEE Symposium on Security and Privacy, 1997.

34. B. Witter, "The Dark Side of Digital Cash," Legal Times, January 30, 1995

Appendix: Proofs of Claims

We outline the proofs of the claims stated in section 10; where applicable, the proofs have one part relating to the general scheme (assuming that all mix servers of the mix-network used cooperate) and one to the ElGamal scheme (assuming a quorum of specified size to cooperate). The final version will include full proofs, which are kept short here due to space shortage.

Theorem 1: The schemes implement unforgeability.

Proof of Theorem 1: *(Sketch)*
First, in order to post a payment order that will be accepted, a valid signature of the payer has to be produced; by the soundness of the underlying existentially unforgeable signature scheme, this can only be done by the account owner in question. Second, in order for a party to be credited, it is necessary that one

of the the output items of the mix-network specifies their account number. If a general mix-network is used, the mix-servers are assumed to perform the valid decryption (if they do not, they can only alter who gets credited, and not the number of payments performed); if the suggested ElGamal mix-network is used, then by the robustness of this, the outputs are guaranteed to be valid if at least one participating mix server is honest. Therefore, only the participating payers will be charged, and only the intended payees debited. The amount of the credits cannot exceed the amount of debits. $\Box$

Theorem 2: The schemes implement impersonation safety.

Proof of Theorem 2:
By the soundness requirements of the identification schemes used to obtain access to accounts, only authorized parties are going to get access to their accounts, as long as they keep their secret keys secret. Also, it is not possible for an adversary to produce a signature by a payer not cooperating with him. This follows from the assumption that the signature scheme used by the payer to sign an encrypted payment order is existentially unforgeable: a new signature cannot be produced without knowledge of the secret key. If the same signed message is reposted in the same payment period, then it will be removed, either after the payer has failed to prove knowledge of some part of the post, or since duplicates are ignored (depending on the approach taken.) Since we, for the approach where duplicates are removed, require the payment orders to be encrypted using a public key specific to the time period between two accounting sessions (corresponding to the mix-decryptions), and the relationship between the secret keys of intervals is unknown, it is not possible to force an old encryption to be accepted in a new time interval.

Theorem 3: The schemes implement overspending blocking.

This follows trivially from Theorem 1, and the fact that for each payment one signature has to be generated, and for each such signature, one account is billed.

Theorem 4: The schemes implement payment blocking / revocability.

Proof of Theorem 4: *(Sketch)*
Since a payment can only be made from an account by producing a signature, and sending this to the transaction center, and the signatures used for this will identify the account owners, it is possible for an account owner to stop all payments from his account by requesting that no signature of his is accepted by the transaction center. This corresponds to either putting his public key on a blacklist, or by removing it from the list of valid public keys. Similarly, the same can be done by the bank of an account holder to block the account holder access to his funds. $\Box$

Theorem 5: The schemes implement framing-freeness.

Proof of Theorem 5: *(Sketch)*
This follows from Theorem 2, and the fact that nobody but a user (including the bank servers) knows the secret key of this user. Therefore, it is not possible to

produce a set of transcripts indicating that a party performed a given payment, without the cooperation from this party. □

Theorem 6: The schemes implement uniform anonymity.

Proof of Theorem 6: *(Sketch)*
The only way a payment can be initiated is by posting an encrypted and signed payment order on the bulletin board of the transaction center. The signature identifies the payer; if the identity of the signer cannot be established (or he is not authorized to make payments,) then the posted message will be ignored. Then, the only way that merchants can be credited is by decrypting all the payment orders. The link between encrypted an decrypted items can only be established by a quorum of mix-servers. If fewer than this could correlate the input and output items, this contradicts the assumption that the ElGamal encryption scheme is probabilistic. By the traceability option (see next theorem), the anonymity of any valid encrypted or decrypted payment order can be removed, and therefore, all participants enjoy the same degree of anonymity. □

Theorem 7: The general scheme implements the two first methods for traceability; the ElGamal based scheme implements full traceability.

Proof of Theorem 7: *(Sketch)*
We have established above that for each account that is credited, the transaction center has a signature of the party whose account will be debited: the link between the two can always be established by either decrypting a single posted encrypted and signed payment order (tracing from payer to payment order), or by (potentially partially) re-encrypting a decrypted payment order (arriving at the payer information from the payment order.) In addition, the third tracing option (comparison) can be performed in the ElGamal based scheme, by the use of the verification protocol for undeniable signatures. □

Over the Air Service Provisioning

Sarvar Patel

Bell Labs, Lucent Technologies
67 Whippany Rd, Whippany, NJ 07981, USA
sarvar@bell-labs.com

Abstract. Mobile users should be able to buy their handsets and then get service from any service provider without physically taking the handset to the provider's location or manually entering long keys and parameters into the handset. This capability to activate and provision the handset remotely is part of the current North American wireless standards and is referred to as 'over the air service provisioning' (OTASP). We examine current proposals and point out some of their limitations. Often the knowledge shared between the mobile user and the network is not fully specified and hence not exploited. We depart from this norm by first providing a classification of various sharing of secrets and secondly we make explicit the assumed shared knowledge and use it to construct various schemes for OTASP. We present a different OTASP scheme for each of the following assumptions: 1) availability of a land line, 2) public key of a CA in the handset, 3) weak secret shared by the mobile user and the network, and 4) secret of the mobile user which can only be verified by the network.

1 Introduction

In the future, more users will continue to migrate to mobile phones and the existing users of analog phones will also migrate to authenticated digital mobile phones. In order to have the mobile handset activated, first, authorizing information from the user is needed by the service provider, and then parameters like long lived keys, telephone numbers, and other information need to be securely distributed to the handset. This entire process will be referred to as service provisioning. It is important that the mobile users have their handsets provisioned in the most convenient way possible. Ideally, the best methods for provisioning of handsets should allow for:

1. Interchangeability of handsets: Users should be able to buy handsets from any vendor as done currently with wired phones.
2. Interchangeability of service providers: Users should be able to get service from any service providers in their region.
3. Spontaneity: Users should be able to activate service when they want without having to wait to receive special codes, PINs, or even provisioned handsets.
4. Remote provisioning: Users should be able to provision their handsets remotely, over the air, without having to take the phone to a special service location.

S. Tavares and H. Meijer (Eds.): SAC'98, LNCS 1556, pp. 174–189, 1999.
© Springer-Verlag Berlin Heidelberg 1999

5. Convenience: The service provisioning process should not require the users to enter long parameters manually into the phones.

This does not mean that other arrangements are not possible, for example where the handset is bought with service already activated. And furthermore, service will only be provided by the service providers if authorizing information (e.g. credit card information) is given by the mobile user and verified.

1.1 Example OTASP Scenario

A mobile user buys a handset and now wants to get service. The user first places a wireless call to the regional service providers (one of the calls allowed by the unprovisioned handset) using its handset id number. The service operator comes on-line and requests authorizing information (e.g. credit card information) which the user provides. Then the network downloads to the phone its long-lived key, telephone number, and other parameters. Now the mobile user has service and can immediately make calls if so desired. This entire exchange happens securely, so that others cannot eavesdrop on the authorizing information nor can they successfully impersonate as the network to the user or impersonate as the user to the network.

1.2 North American OTASP Proposal

Currently, the North American cellular standard specifies an OTASP protocol [13] using Diffie-Hellman key agreement. First the network transfers a 512 bit prime p and a generator g. Then a Diffie Hellman key exchange occurs as the mobile and the network pick their respective random numbers R_M and R_N and calculate and exchange the exponentials $g^{R_M} \bmod p$ and $g^{R_N} \bmod p$. Each raise the exponential by their secret random exponent to form $g^{R_M R_N} \bmod p$. The 64 least significant bits of $g^{R_M R_N} \bmod p$ serve as the long-lived key, called the A-key.

The Diffie Hellman key exchange is secure against passive attacks. Furthermore, active attacks may be difficult to mount on the radio interface if transmission of messages need to be blocked or substituted, thus offering some "practical" security against active attacks. For our discussion, we sidestep the radio issues by assuming the existence of active attackers. Below we list some limitations of the proposal.

Limitations

1. The first problem with this use of the Diffie-Hellman key exchange is that it is unauthenticated and susceptible to a man-in-the-middle attack. An attacker can impersonate the network and then in turn impersonate the user to the network. This way the attacker can select and know the A-key as it relays messages between the user and the network to satisfy the authorization requirements. The man in the middle attack has to be carried on the voice channel also.

2. The parameters g and p are unverified which means that an attacker can replace them, for example, with a prime p' such that $p' - 1$ has small factors which will allow the attacker to solve the discrete log and recover R_M from the exponential $g^{R_M} \bmod p'$ sent by the mobile. This is another way for the attacker to discover the A-key on the mobile side while performing a man in the middle attack. A more powerful attack, which we describe next, is thwarted by a good choice made by the protocol. Lets say the attacker sends some $g^x \bmod p$ to the network which forms $g^{x R_N} \bmod p$. The attacker now has to send some y to the mobile such that $y^{R_M} \bmod p' = g^{x R_N} \bmod p$. Since R_M has been recovered by the attacker from $g^{R_M} \bmod p'$, this equation may be solvable for y. If so, the attacker now knows the A-key shared by the handset and the network. This attack is more powerful than the previous one because the same A-key, known to the attacker, would be residing in the mobile and the network. Fortunately, this attack is not possible because the protocol requires the mobile to receive the networks exponential, $g^{R_N} \bmod p$, along with g and p before it will send its exponential $g^{R_M} \bmod p$.

3. There are no provisions to check for the equivalent of weak keys for Diffie Hellman. For example if the attacker substitutes 1 or -1 as the exponentials exchanged on both sides then the Diffie Hellman key will be 1 or -1 and known or easily guessable to the attacker. Note that this attack does not require the attacker to be in the middle encrypting and decrypting the transmission under different keys. Once 1 or -1 has been substituted then the attacker only has to eavesdrop. Also, importantly the same Diffie Hellman key, 1 or -1, is agreed upon by both the handset and the network with probability $\frac{1}{2}$. If the prime p is not a safe prime then other values are also possible; as a precaution all of them should be checked, and if the Diffie Hellman key matches any of them then the session should be aborted by the network.

4. We present another attack which is undetectable by the network, unlike the previous attack, yet the same known A-key will reside in the network and the handset. The attacker performs a Diffie Hellman key exchange with the network and agrees on a 512 bit key, K. The attacker now wants the mobile to have the same 512 bit key, K. The attacker, pretending to be the network, sends to the mobile $(g, K + 1, K)$ instead of $(g, p, g^{R_N} \bmod p)$. The handset calculates $K^{R_M} \bmod (K + 1)$ as the mobile's key. This key will equal one if R_M is even and the key will equal K if R_M is odd. Thus with probability $\frac{1}{2}$ the attacker has forced the network and the handset to have the same 512 bit key, K and thus the same 64 bit A-key.

5. The protocol is a key agreement protocol which is normally fine, but a key distribution protocol where the network creates a well chosen and random (i.e. unpredictable) A-key and distributes it to the handset would have been preferable in this situation. This may allow the network to associate the A-key of a handset in some deterministic manner which is good for its key storage and management procedures. This can be done via the use of pseudo random functions (PRFs) which is indexed by a master key and maps handset/phone number into 64 bit A-keys [8]. Thus instead of storing thousands and perhaps millions of keys and protecting them, the network can store

just one master key and use the PRF to recover a particular A-key associated with the handset's number. This master key and the related algorithm can be performed in a protected device, thus the A-keys are never seen by anyone.

6. The size of the prime, 512 bits, may be too short.

The above attacks should not be surprising once it is decided that an unauthenticated Diffie Hellman key exchange will be used. In fact one can make a general statement that any unauthenticated key exchange will reveal the authorizing information (e.g. credit card number) to an active adversary. When a mobile requests an OTASP session, the active attacker blocks the session and, pretending to be the network, carries an OTASP conversation with the mobile user. Once the user reveals the credit card number, the attacker terminates the session and can use the credit card number to get service on other mobile phones or use it for other fraudulent purposes. Thus the security in such protocols lies in the practical difficulty of implementing active attacks. Despite its limitations, this is an interesting application of the Diffie Hellman key agreement protocol and makes service provisioning much more convenient.

1.3   Carroll-Frankel-Tsiounis Key Distribution Proposal

This is a key distribution proposal which has the benefit of allowing the network to randomly choose strong keys from the space of A-keys. The Carroll-Frankel-Tsiounis (CFT) proposal also uses Rabin to speed up computation as did the Beller-Chang-Yacobi [1] protocol and also assumes that each handset possesses the public key of a certificate authority (CA). However, it interestingly differs from other protocols in its extensive use of unforgeable signatures and encryptions which are semantically secure or plaintext aware. The protocol is as follows:

1. The network sends the mobile its public encryption key signed by the CA (unforgeable signatures are used here).
2. The mobile verifies the network's public encryption key and then generates a random session key SK and a random pad AP. It encrypts both the SK and AP using the network's public encryption key and sends it to the network (the encryption here is semantically secure).
3. The network recovers the SK and the AP and uses the SK to perform symmetric encryption of the A-key and the AP which it sends to the mobile (symmetric encryption here is plaintext aware).
4. The mobile verifies the AP in the decryption; the handset and the network now both possess the A-key which is used to derive the voice encryption key and set up an encrypted voice channel.
5. At this point the operator requests authorizing information (e.g. credit card information) from the user. If the user furnishes the information then the user has been authenticated to the network and service will be provided in the future.

We make some cautionary observations that the CFT protocol should not be viewed as a "solution to all problems," but the components that surround it should also be designed within the correct security model. We provide specific examples below:

The CFT protocol assumes that a unique A-key is generated for every OTASP attempt by the handset. If the A-key generation process does not guarantee this then the overall protocol will be insecure; as an example the same A-key may be generated for the same handset/phone number combination which would allow this attack: a handset uses its serial number and phone number (if assigned) to access the network for OTASP. At this point the attacker blocks the access. Instead the attacker picks a random session key SK and a random pad AP and sends them to the network using the blocked handset's serial/phone number. The network responds with the encrypted A-key which the attacker retrieves and aborts the connection. Now the attacker is in possession of the A-key for that handset. If the legitimate handset again accesses the network with its own session key and pad, the network will again transport the same A-key to the handset encrypting it with the session key. Now the handset will have the A-key and the user, on the encrypted voice channel, will give authorizing information thus successfully completing service provisioning. Unfortunately, the attacker already has the A-key and he also can use it later to make fraudulent calls. Thus in the CFT model, one should not directly use a PRF to associate an A-key to a handset/phone number without guaranteeing uniqueness at every OTASP attempt. Later, we will show that it is not necessary to have this restriction in a key distribution protocol.

Secondly, a mild form of denial of service attack is possible if the handset/phone numbers are not used in the cryptographic operations of the CFT protocol or the later verification stages. Actually, there are two forms of denial of service attacks. In the first one, the customer's credit card is not charged and OTASP is unsuccessful, while in the second one, the customer's credit card is charged and OTASP is unsuccessful. There is little that can be done to prevent the first form of the attack in any protocol, but we would like to assure the customers that if their credit cards are charged then OTASP will be successful and their service will be activated. An attacker can perform the second form of attack by substituting other numbers in place of the handset's true id number and user's phone number throughout the protocol. The protocol will be completed and credit cards charged, but the network will not have activated the true handset/phone number. Verification steps done after the CFT protocol can also be satisfied as long as they do not use handset/phone numbers as part of the cryptographic operations. Thus later attempt by the user to access the system will be rejected. This attack is possible because the handset id number used in communication is not part of the public key encryption of the SK and the AP sent by the mobile to the network. If it was then the network can know that no attacker could have substituted another false handset/phone number. Fortunately, the current North American does perform cryptographic operations using the handset/phone numbers in the verification steps following the A-key agreement. So the attack would be detectable at this time, and the credit cards should not be charged until after verification steps.

Finally the CFT protocol tries to minimize the impact due to lack of strong random sources and points out that if independent sources of randomness are used then a weak source for the SK is not problematic as long as the other AP is strongly random. We caution that this is not true if the CFT protocol is embedded in the current North American standard because candidates for the SK can be used to decrypt the symmetric encryption and give candidates for the A-key. These candidates can be further verified because the A-key is used to derive another session key called the SSD which is further used to answer challenges. An attacker can see if a candidate A-key results in the same response to the challenge as actually seen in a session. If not then the next candidate is tried until the true A-key is recovered. However, these off-line verifications of SK guesses would not be possible if the challenge/response protocol following the CFT protocol was replaced by a more secure authentication protocol (i.e. zero knowledge) which does not reveal information about the A-key. Nevertheless, the weak source for SK must have some minimum strength so that on-line verifications of A-key are not practical. Assume an attacker has good guesses for SK and hence, guesses for A-key. If these guesses are few then the attacker can use each A-key guess to establish a session (e.g. call origination). The attacker will be unsuccessful on all tries except the one with the true A-key.

2   Classification of Shared Knowledge

Different OTASP schemes are possible depending upon what shared knowledge is possessed by the mobile user/handset and the network. At one extreme, one can assume that the mobile handset and the network each have a public key and both keys are known to the handset and network. Similarly, for the symmetric cryptography case, we can assume that the mobile handset and the network both share a strong secret (64 bits or more). Next we can assume that the mobile handset possesses the public key of a CA and thus indirectly has knowledge of the network's public key. At the next level we can assume that the mobile user and the network only share a weak secret. Finally, the weakest assumption is that the network can only verify a secret relayed to it by the mobile user. That is the network does not even share a secret with the user, but can only verify it.

Orthogonal to this is the availability of a secure channel (land line phone) which can aid in OTASP. Although, we describe the various schemes with respect to the wireless environment, they can be used in other insecure environments (e.g. internet) where there is a need to remotely provision. For different situations, different assumptions may be valid. In some environment there may be an agreement on the certificate authority and hence on the root key to be pre-stored in every device. However, for other environments no easy agreement on a CA and its procedures may be possible and hence, the other non-CA based schemes need to be used.

3   Secure Password Scheme

We want to review in this section methods to use weak secrets (e.g. passwords) for authentication and key agreement. These methods will be used repeatedly. Standard protocols used in symmetric key authentication and key agreement are susceptible to off-line dictionary attacks. A typical protocol consists of a challenge R sent from user A to user B. User B responds with a function $f_P(R)$, using the challenge R and the password P. An eavesdropper hearing the pair R and $f_P(R)$ can use this information to verify guesses. Users do not pick passwords randomly from the entire 2^n possible space, but instead use memorable and meaningful passwords. Quite often their passwords are picked from names and words in dictionaries. So an attacker can perform off-line dictionary attack by picking each word, P', in the dictionary and forming $f_{P'}(R)$ and seeing if it matches $f_P(R)$. If it does not then the attacker tries the next guess until it finds a match which reveals the password P. No matter how complicated the protocol, if it only uses symmetric cryptography then it will be susceptible to a variation of the off-line dictionary attack.

3.1   Review of Secure Password Protocols

There has been some advance towards password protocols resistant to off-line dictionary attacks [10], [3], and [9]. Lomas et.al [10], were the first to propose a 3 party protocol using passwords which were protected against dictionary attacks. Bellovin and Merritt [3] made similar protocols called Encrypted Key Exchange (EKE) for two party authentication and key exchange using passwords and still had protection against dictionary attacks.

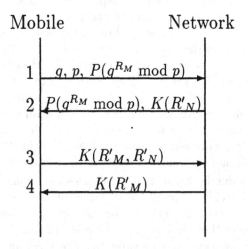

Fig. 1. The Diffie Hellman Encrypted Key Exchange (DH-EKE)

[11] showed that information leakage is possible in EKE which allows one to retrieve the passwords after hearing some exchanges. Furthermore [12] presented attacks against variations of EKE and Gong et.al protocols where varying or unverified parameters were used. The fixing or verifications of parameters seems necessary. RSA parameters are unverifiable and hence must be fixed. However, Elgamal and DH (Diffie Hellman) parameters like g and p are verifiable and hence can either be fixed or changing but verified during each exchange.

We now describe Bellovin and Merritt's secure password protocol, called Encrypted Key Exchange(EKE) based on the Diffie-Hellman key exchange (DH-EKE). The mobile and the network (see figure 1) have agreed upon a g and a p, here we assume that the g and the p are fixed. The mobile picks a random number R_M and calculates $g^{R_M} \bmod p$. The mobile further encrypts the value using the password P to get $P(g^{R_M} \bmod p)$ and sends it to the network. The network similarly calculates and sends $P(g^{R_N} \bmod p)$ to the mobile. Both decrypt the messages and exponentiate with their respective secret random exponents to form the Diffie Hellman key, $g^{R_M R_N} \bmod p$ which serves as the session key. The session key is then authenticated by random challenges to prevent replay and other attacks. Since the Diffie Hellman exponential g^{R_M} and g^{R_N} are encrypted using the password P, a man-in-the middle attack is not possible, furthermore, the passwords prove authentication because without knowing the passwords there would not be an agreement on the Diffie-Hellman keys. How does this stop the dictionary attacks? Well, because guessing the password P' allows one to recover the guesses for the exponentials $g^{R_M'}$ and $g^{R_N'}$, but the attacker cannot form the Diffie Hellman key $g^{R_M' R_N'}$ and verify the guess, hence off-line guesses from the dictionary cannot be verified.

Encrypting $g^R \bmod p$ with a symmetric cipher can leak information, and Bellovin and Merritt propose a random padding method, however, this also leaks information allowing an attacker to recover the password. Some countermeasures are available [3], [11].

4   Secure Channel (Land Line) Available

Assuming a land line connection is available we present two different schemes for OTASP which are secure against man in the middle attacks. The first method of performing a secure over the air key exchange, uses a secure password protocol (see Figure 2). First the user contacts the operator over a land phone line (First two lines in the figure refer to a land line connection and the third line refers to a wireless connection). The land phone line serves as our authenticated and private channel. The operator asks questions to verify that the user is authorized. Then the user is instructed to power on the handset and enter a 4 digit number provided by the operator. The network then uses this 4 digit number as the password to perform a secure password Diffie-Hellman key exchange as described in the previous section. Once a temporary DH key has been agreed then the channel can be encrypted and the messages authenticated using symmetric techniques as used for executing a secure session. This way the A-key and

Fig. 2. Over the Air Key Exchange - Using Land Lines with a Password Protocol

other parameters can be securely and privately downloaded to the handset. As an example, the telephone number can be privately downloaded and thus the anonymity of the user can be kept, including other parameters.

The second method of doing a secure over the air key exchange is to have the mobile handset display a string and have the user read it to the operator, this way the man in the middle is avoided. The protocol is described in figure 3 where the first and last passes occur with the operator on the land lines while the 2nd and 3rd occur on the wireless connection between the handset and the network. After the operator has authorizing information from the mobile user, the mobile

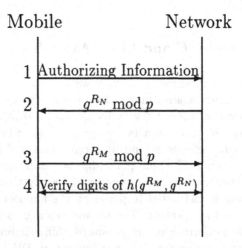

Fig. 3. An Over the Air Key Exchange Susceptible to Birthday Attacks

is turned on and in steps 2 and 3 the mobile and the network send g^{R_M}, and g^{R_N} to each other. The operator and the user verbally verify that some of the digits of $h(g^{R_M}, g^{R_N})$ are the same, where h is a cryptographic hash. If verified then the network and mobile use $h(g^{R_M R_N})$ as the A-key or else the network can initiate security violation procedures. This is good not just for wireless, but for any kind of remote update of the parameter with human verification.

There is a birthday attack which one has to be careful about. Lets say that we decide that 64 bits will be read back by the mobile user to the operator. It would seem that the attacker would have to try 2^{64} different values so that the hash of those values $h(g^{R_M}, g^{R_N'})$ and $h(g^{R_M'}, g^{R_N})$ are the same, a necessary condition to carry the man-in-the-middle attack. So the attacker tries all the values for R_N' and R_M' such that their hashes are the same. Although naive counting would suggest that about 2^{64} values need to be tried to get a significant chance of collision, we can by birthday arguments show that about 2^{32} values need to be tried to find a collision with probability near $1/2$. To slow things down one can put $g^{R_M R_N}$ as part of the hash: $h(g^{R_M}, g^{R_N}, g^{R_M R_N})$. This means the attacker has to do exponentiation along with hashes which makes things slower. If the user is asked to read back alphabetic characters then about 14 letters are needed to cover 64 bits. If alphanumerics are used then about 12 are enough. However 2^{32} complexity for an attack even in real time may not be enough security and if 2^{64} complexity is required then 128 bits must be read back by the user which is about 24 alphanumerics. There is a simpler method of making the protocol in Figure 3 resistant to the birthday attack by introducing a restriction on the sequence of the exchange of the Diffie Hellman exponentials. In particular, if we insist that the mobile will not send its g^{R_M} until it receives g^{R_N} from the network then the birthday attack is foiled. The man in the middle attacker was previously able to see both the exponentials g^{R_M} and g^{R_N} and thus it was able to exploit the birthday attack. Now the attacker has to commit an exponential to the mobile before it will see the mobile's exponential g^{R_M} thus reducing one degree of freedom for the attacker.

We present another version of the protocol with one more round which is not susceptible to the birthday attack, and furthermore is resistant to searches for consistent exponentials by a man in the middle attacker. Thus verification of a smaller string is sufficient. The protocol is described in figure Figure 4 where the first and the final step are over the land voice link while the middle three steps occur over the air link. On the wireless link, first the network sends the hash of its exponential, the user then sends its exponential, g^{R_M}, and finally the network sends its exponential g^{R_N}. The user first verifies the hash of the exponential sent by the network. Then both the user and the network calculate $h(g^{R_M}, h(g^{R_N}))$ and verbally verify its first 4 digits. A man-in-the-middle attacker cannot use birthday type attacks or do searches for consistent exponentials because as a network he has to commit to the exponential he is using (via the hash) before he sees the users exponential. Similarly, the attacker as a user has to commit to the exponential before the value of the networks exponential, associated with the hash, is revealed. Thus we need to verify a much smaller string (e.g. 4 digits).

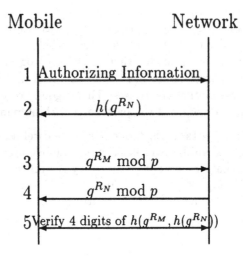

Fig. 4. An Over the Air Key Exchange Resistant to Birthday Attacks

There is an existing protocol, called VP1 [4] which is related to this protocol, but it is used for agreeing on a session key whereas we use our protocol to update parameters via an authenticated link. Secondly, VP1 is a 4 round protocol whereas we are a 3 round protocol on the wireless link. VP1 first has both the user and the network hash their exponentials and exchange them. Then the actual exponentials are exchanged. Finally, a hash of the function of the exponentials is calculated and 6 digits are verified by voice. Our protocol is more efficient in terms of the number of rounds and the messages exchanged. Since VP1 does not use an authenticated link, it ultimately does not protect against man-in-middle attacks, unless the end parties know each other's voice. Our method can be used for other environments. Also other variations can be built using different public key schemes.

The safe prime p and the generator g were assumed to be fixed and prestored in the handset. However, if that is not the case then the attacker can replace g and p with g' and p' which may allow it to calculate the discrete logarithm efficiently. If g and p are also sent over the air then they could also be used as part of the hash calculation, $h(g, p, g^{R_M}, g^{R_N})$ in order to detect the substitution of g and p by the attacker.

4.1 Man-in-the-Middle on Voice Channel

Although we have described the last two protocols as requiring a land line voice link one can execute the protocols without it. Thus a handset can be used to make the voice call itself to the network operator and a Diffie Hellman performed on the control channel. No man in the middle attack is possible on the control channel as described above. However, a man-in-the-middle attack is possible if it is also performed on the voice channel. First the attacker pretends to be the network to the user and gets the authorizing information. Secondly the attacker pretends to be the user to the network. If this is happening in real time then all the information from one conversation is relayed to the other (assume there are two attackers working in concert). Then on the control channel the man-in-the-middle attack can proceed because when it comes time to verify the digits on the voice link, the two attackers will say the different digits to the network and the user respectively. Although we have described this as happening in real time, it could happen that the attacker gets authorizing information from the user and then at a later time calls the network and gives it the information to get the A-key and other parameters.

5 Mobile and Network Share Strong Keys

If the handset and the network already share strong secrets either in form of each other's public keys or strong secret keys, then well known protocols can be used to exchange or update parameters. [1] is an example of a protocol using public keys to establish session keys. [2] is an example of secure protocols used with symmetric cryptography to establish session keys. The session keys can be used to encrypt a session and authorizing information can be provided to start service and the parameters can be provisioned into the handset. We are trying to get strong secrets agreed upon by both sides, but since we don't possess them initially we will try to bootstrap from weak secrets to strong secrets.

6 Mobile Has CA's Public Key

Manufactures can install a CA's public key in all the phones. This is much easier than installing a unique public key for each handset. The assumption is the one used in the CFT protocol and also used in SSL. Unlike the CFT protocol, we will use the handset id as part of our encryption thus blocking the denial of service attack. Furthermore, we want our protocol to allow a network to be able to pick a well chosen A-key for a specific handset/phone number and try to distribute that A-key to the handset repeatedly. The CFT protocol was not able to do this because it revealed its hand before it was necessary. That is, it revealed the A-key before authorization was given on the voice channel. In fact our protocol does not require a voice connection to transfer the authorizing (e.g. credit card) information from the mobile user to the network. The user can be prompted on the handset and the user can enter the information on the handset. We reveal the A-key only if the authorization step has been successfully completed. Here are the steps of our protocol:

1. The handset requests OTASP and identifies itself via a handset id number or a telephone number if already assigned.
2. The network sends its public key, signed by the CA, to the handset (unforgeable signatures).

3. The mobile generates a random session key SK and encrypts the handset id and the session key SK with the network's public key and sends it to the network (using probabilistic encryption [7] or semantically secure encryption).
4. The network decrypts and then uses session key SK to initiate a authenticated and encrypted session. Voice and messages are both encrypted and message authentication is provided using a message authentication key and a encryption key derived from the session key, SK.
5. The user provides the credit card information to the network.
6. If the credit card information is verified then the network distributes the A-key to the handset.

Note that unlike the CFT protocol, the entire OTASP has been encrypted and authenticated using the session key derived from the SK rather than the session key derived from the A-key.

7   Mobile User and Network Share Weak Secret

It may be that the user and the service provider have had a previous contact and now share some personal information about the user. The user might have already interacted with the service provider or the user may be an existing land line customer of the service provider and now wants to get mobile service. It is not clear, what the personal information is that the network and the user share. Is it the mothers maiden name? The last 4 digits of the social security number (SS#) or zipcode? Obviously there must be some information that the operator has which it uses to authenticate the user on the voice link. Assume its the last 4 digits of the SS#. If so then we can perform a secure password protocol using the last 4 digits of the SS#. First the user enters the last 4 digits of the SS# into the handset, and then using the handset itself a secure password protocol is executed and the A-key and other parameters are updated in the handset. Note all this happens without a voice call being placed either on the land line or on the handset. In a sense, a temporary password (last 4 digits of SS#) is used and then other parameters are updated. The protocol is the same as the one in Figure 3 except all the communications take place over the wireless phone. There is no need for a real time operator to be involved, although it is not precluded from the protocol.

Perhaps credit card information can serve as the shared secret. Since the network operators have access to credit card information about all users, this information can be used as a shared secret. If an operator using the name of a person and type of card (e.g. citibank visa), can know the credit card information, then that can serve as the shared secret. This may not be possible if the operator can only verify the credit card number, but does not know the credit card number from just the name of the user. When the user contacts the network operator, the operator will ask for the user's name and type of credit card. Then the user is prompted to enter the card number which will be used as the weak shared secret to initiate a secure password protocol. If successful then the A-key and other parameters can be distributed to the handset over an encrypted and authenticated session.

8   Network Can Verify Mobile User's Secret

In the worst case there is no shared secret, however, the mobile user has a secret (name + credit card number + expiration date) which the network can verify. The network does not know this information ahead of time, but can only verify it. Now we cannot use the previous techniques and hence come full circle to using something very similar to the current North American OTASP proposal. There does not seem to be any cryptographic techniques which will protect against active attackers in this situation.

We will also perform an unauthenticated Diffie-Hellman key exchange or another unauthenticated public key (e.g. Rabin) based key distribution. However, we will do it in such a way that performing a man in the middle attack is very difficult, involving service denial to much of the service region for extended periods of time. We do this by disguising an OTASP key exchange as a normal system access (e.g. call origination) followed by an encrypted and authenticated session. In order to do this, a random ID/telephone number for the handset should be used to make a call origination. The network, when it sees the random ID/telephone number will know that this is not a legitimate number. The network knows that this could happen either because there was error in transmission or some one is trying to initiate OTASP. The network then continues to pretend its a normal call, and sends random bits to the handset and the handset also sends random bits to the network. However the first 1024 bits of the disguised call can be exponentials $g^{R_M} \bmod p$ and $g^{R_N} \bmod p$. The key is derived and used to encrypt the rest of the session after some predetermined time, say after 10 seconds of data. Then the call should be placed to the operator in the network and the mobile user should relay the secret credit card information which the network will verify. If the information is verified then A-key can be transported to the handset along with other parameters.

If the exchange is disguised well then the only way for the attacker to act as man in the middle is to try to do so with most calls which are going on, hoping that it will find one that is truly an OTASP call. To have any significant probability of finding such calls it will have to be blocking most calls because an OTASP call is a rare call, once or few times in the lifetime of a phone. A call origination on the other hand is very frequent, thus the cost of the attack is expensive. So if such kind of blocking or denial of service occurs then it should become easier to find the attacker and becomes all the more important to find the source and put an end to the blocking. The security of such a protocol is not cryptographic but practical. We make the practical assumption that the adversary is not so powerful as to block most calls and still go undetected.

The security of this strengthened unauthenticated Diffie Hellman is analogous to the strength of passwords schemes in withstanding on-line attacks. An attacker guessing at the correct password in one session has a low probability of guessing it correctly, but guessing over thousands of sessions the attacker has a high probability of recovering a password for some user. Similarly, an attacker performing a man in the middle attack on a session has a low probability of success, but over thousands of sessions the probability of success is high.

The Diffie Hellman exchange is only strengthened if the OTASP call is indistinguishable from a normal call origination and this is very tricky to guarantee. For example, if g and p were sent during the OTASP call then the call is distinguishable because an attacker can check if p is a prime. If not then the attacker knows that this is a normal call origination and there is no need to perform a man in the middle; we had assumed that g and p are fixed and known. If we do want to send g and p as part of the session then [11] outlines some methods to do this safely. A general method of sending primes is to send the random string used in picking a prime rather than sending the actual prime. At the other end, the same random string is used to pick the same prime or a safe prime. To transfer g, a random string is sent to the other party who checks if the string is a generator of p. If so then it is used as the g else the increment is checked until a generator is found.

9   Conclusion

We started by examining the proposals for OTASP and their limitations. We then provided over the air service provisioning methods under various assumptions about knowledge shared between the mobile user/handset and the network. We started from strong assumptions like the knowledge of strong keys and moved to the weakest assumption, that the mobile and network do not share any secrets but if the mobile relays its secret the network can verify it. For this weak assumption we could only provide a practically secure scheme assuming a limited adversary. If agreement on public keys is desired, we can further have the handset and network generate their respective private and public keys and exchange them. Thus we are able to bootstrap from a small secret to a strong public key. The table (Figure 5) organizes the various protocols according to the different assumptions. The left column of the table lists the various assumptions about the secrets shared between the network and the handset. The top row of the table lists the various 4 assumptions about availability of a land line, or the availability of pre-stored and public constants like the CA's public encryption key or the g and p, or the possibility that nothing is pre-stored.

If a strong secret is shared then a standard two party session key agreement protocol can be used for all 4 assumptions. If a weak secret is shared between the handset and the network then the schemes of section 4 can be used with the availability of land lines; if the CA's public encryption key is pre-stored in the handset then the protocol of section 6 can be used; if g and p are pre-stored then the secure password scheme can be used and even when g and p are not pre-stored the secure password scheme can be used as long as the g and p are verified. Finally, the weakest assumptions on secrets is that the mobile user has a secret, but the network does not share it and can only verify it when presented by the

	Land line available	CA's public key in handset	g, p in handset	nothing pre-stored in handset
Strong secret	session key agreement	session key agreement	session key agreement	session key agreement
weak secret	secure password	CA scheme	secure password	secure password (g, p verify)
Verifiable secret, but unshared	secure password or verify hash	CA scheme	strengthened DH	strengthened DH with transfer g, p

Fig. 5. Summary Table

mobile user. In this case, the availability of land lines means that the protocols from section 4 can be used; if the CA's public encryption key is pre-stored then the protocol of section 6 can be used here also; if g and p are pre-stored then one has to resort to the strengthened version (section 8) of the unauthenticated Diffie Hellman key exchange which makes the OTASP call indistinguishable from a normal call; even if g and p are not pre-stored then one can use the protocol in section 8 as long as g and p are carefully transferred as outlined in section 8.

References

[1]   M. Beller, L. Chang, and Y. Yacobi, Privacy and authentication on a portable communication system, *IEEE J. Select. Areas Commun.*, 11: 821-829, 1993.

[2]   M. Bellare, and P. Rogaway, Entity authentication and key distribution, *Advances in Cryptology - Crypto*, 1993.

[3]   S. Bellovin, and M. Merritt, Encrypted key exchange: password-based protocols secure against dictionary attacks, *IEEE computer Society symposium on research in security and privacy*, 72–84 May 1992.

[4]   E. Blossom, The VP1 Protocol for Voice Privacy Devices, December 1996.

[5]   C. Carroll, Y. Frankel, and Y. Tsiounis, Efficient key distribution for slow computing devices: Achieving fast over the air activation for wireless systems,, *IEEE symposium on security and privacy*, May, 1998.

[6]   EIA/TIA, Cellular Radio Telecomm. Intersystem Operations IS-41C 1994.

[7]   S. Goldwasser, and A. Micali, Probabilistic encryption, *Journal of Computer and Systems Science*, 28: 270-299, 1984.

[8]   O. Goldreich, S. Goldwasser, and A. Micali, On the cryptographic applications of random functions, *Advances in Cryptology - Crypto*, 1984.

[9]   L. Gong, T. Lomas, R. Needham, J. Saltzer, Protecting poorly chosen secrets from guessing attacks, *IEEE J. Sel. Areas Commun.*, 11: 648-656, 1993.

[10]   T. Lomas, L. Gong, J. Saltzer, R. Needham, Reducing Risks from Poorly Chosen Keys, *ACM Operating Systems Review*, 23(5): 14-18, Dec. 1989.

[11]   S. Patel, Information Leakage in Encrypted Key Exchange, *Proceedings of DIMACS workshop on Network Threats*, 38: 33–40, December 1996.

[12]   S. Patel, Number theoretic attacks on secure password schemes, *IEEE symposium on security and privacy*, 236–247 May 1997.

[13]   TIA, IS-41-C enhancements for Over-The-Air-Service Provisioning, *TR45.2 Working Group*, December 1996.

Faster Attacks on Elliptic Curve Cryptosystems

Michael J. Wiener and Robert J. Zuccherato

Entrust Technologies, 750 Heron Road
Ottawa, Ontario, Canada K1V 1A7
{wiener, robert.zuccherato}@entrust.com

Abstract. The previously best attack known on elliptic curve cryptosystems used in practice was the parallel collision search based on Pollard's ρ-method. The complexity of this attack is the square root of the prime order of the generating point used. For arbitrary curves, typically defined over $GF(p)$ or $GF(2^m)$, the attack time can be reduced by a factor or $\sqrt{2}$, a small improvement. For subfield curves, those defined over $GF(2^{ed})$ with coefficients defining the curve restricted to $GF(2^e)$, the attack time can be reduced by a factor of $\sqrt{2d}$. In particular for curves over $GF(2^m)$ with coefficients in $GF(2)$, called anomalous binary curves or Koblitz curves, the attack time can be reduced by a factor of $\sqrt{2m}$. These curves have structure which allows faster cryptosystem computations. Unfortunately, this structure also helps the attacker. In an example, the time required to compute an elliptic curve logarithm on an anomalous binary curve over $GF(2^{163})$ is reduced from 2^{81} to 2^{77} elliptic curve operations.

1 Introduction

Public-key cryptography based on elliptic curves over finite fields was proposed by Miller [9] and Koblitz [6] in 1985. Elliptic curves over finite fields have been used to implement the Diffie-Hellman key passing scheme [2,4] and also the elliptic curve variant of the Digital Signature Algorithm [1,10]. The security of these cryptosystems relies on the difficulty of solving the *elliptic curve discrete logarithm problem*. If P is a point with order n on an elliptic curve, and Q is some other point on the same curve, then the elliptic curve discrete logarithm problem is to determine an l such that $Q = lP$ and $0 \le l \le n - 1$ if such an l exists. If this problem can be solved efficiently, then elliptic curve based cryptosystems can be broken efficiently.

There are known classes of elliptic curves in which solving the discrete logarithm problem is (relatively) easy. These classes include supersingular curves [8] and anomalous curves [12,14,15]. The elliptic curve discrete logarithm problem for supersingular curves can be reduced to the discrete logarithm problem in a small finite extension of the underlying finite field. The discrete logarithm problem in the finite field can then be solved in subexponential time. Anomalous curves are curves defined over the field $GF(p)$ and have exactly p points. The elliptic curve discrete logarithm problem for anomalous curves can be solved in

S. Tavares and H. Meijer (Eds.): SAC'98, LNCS 1556, pp. 190–200, 1999.

$O(\ln p)$ operations. Both supersingular and anomalous curves are easily identified and excluded from use in cryptographic operations.

The best attack known on the general elliptic curve discrete logarithm problem is parallel collision search [18] based on Pollard's ρ algorithm [11] which has running time proportional to the square root of the largest prime factor dividing the curve order. This method works for any cyclic group and does not make use of any additional structure present in elliptic curve groups. We show how this method can be improved for any elliptic curve logarithm computation by exploiting the fact that the negative of a point can be computed very rapidly.

Certain classes of elliptic curves have been proposed for use in cryptography because of their ability to provide efficiencies in implementation. Among these have been subfield curves and anomalous binary or Koblitz curves [7, 16]. Using the Frobenius endomorphism, we show that these curves also allow a further speed-up for the parallel collision search algorithm and therefore provide less security than was originally thought. This is the first time that the extra structure provided by these curves has actually been used to attack the cryptosystems upon which they are based. Independent work in this area has also been performed by Robert Gallant, Robert Lambert and Scott Vanstone [5] and by Robert Harley, who has used his results to solve the ECC2K-95 Certicom challenge problem.

2 Background

This section will provide the necessary background material on various properties of elliptic curves and will also describe the parallel collision search method for computing discrete logarithms.

2.1 Elliptic Curves Over $GF(p)$

Let $GF(p)$ be a finite field of characteristic $p \neq 2, 3$, and let $a, b \in GF(p)$ satisfy the inequality $4a^3 + 27b^2 \neq 0$. An *elliptic curve*, $E_{(a,b)}(GF(p))$, is defined as the set of points $(x, y) \in GF(p) \times GF(p)$ which satisfy the equation

$$y^2 = x^3 + ax + b,$$

together with a special point, $\mathcal{O}$, called the *point at infinity*. These points form an Abelian group under a well-defined addition operation which we now describe.

Let $E_{(a,b)}(GF(p))$ be an elliptic curve and let P and Q be two points on $E_{(a,b)}(GF(p))$. If $P = \mathcal{O}$, then $-P = \mathcal{O}$ and $P+Q = Q+P = Q$. Let $P = (x_1, y_1)$ and $Q = (x_2, y_2)$. Then $-P = (x_1, -y_1)$ and $P + (-P) = \mathcal{O}$. If $Q \neq -P$ then $P + Q = (x_3, y_3)$ where

$$x_3 = \mu^2 - x_1 - x_2$$
$$y_3 = \mu(x_1 - x_3) - y_1,$$

and

$$\mu = \begin{cases} \dfrac{y_2 - y_1}{x_2 - x_1} & \text{if } P \neq Q \\[2mm] \dfrac{3x_1^2 + a}{2y_1} & \text{if } P = Q. \end{cases}$$

2.2 Elliptic Curves Over $GF(2^m)$

We now consider non-supersingular elliptic curves defined over fields of characteristic 2. Let $GF(2^m)$ be such a field for some $m \geq 1$. Then a non-supersingular elliptic curve is defined to be the set of solutions $(x, y) \in GF(2^m) \times GF(2^m)$ to the equation

$$y^2 + xy = x^3 + ax^2 + b$$

where $a, b \in GF(2^m)$ and $b \neq 0$, together with the point on the curve at infinity, $\mathcal{O}$. We denote this elliptic curve by $E_{(a,b)}(GF(2^m))$ or (when the context is understood) E.

The points on an elliptic curve form an Abelian group under a well defined group operation. The identity of the group operation is the point $\mathcal{O}$. For $P = (x_1, y_1)$ a point on the curve, we define $-P$ to be $(x_1, y_1 + x_1)$, so $P + (-P) = (-P) + P = \mathcal{O}$. Now suppose P and Q are not $\mathcal{O}$, and $P \neq -Q$. Let P be as above and $Q = (x_2, y_2)$, then $P + Q = (x_3, y_3)$, where

$$x_3 = \mu^2 + \mu + x_1 + x_2 + a$$
$$y_3 = \mu(x_1 + x_3) + x_3 + y_1,$$

and

$$\mu = \begin{cases} \dfrac{y_2 + y_1}{x_2 + x_1} & \text{if } P \neq Q \\[2mm] \dfrac{x_1^2 + y}{x_1} & \text{if } P = Q. \end{cases}$$

2.3 Anomalous Binary and Subfield Curves

Anomalous binary curves (also known as Koblitz curves) are elliptic curves over $GF(2^n)$ that have coefficients a and b either 0 or 1. Since it is required that $b \neq 0$, they must be defined by either the equation

$$y^2 + xy = x^3 + 1$$

or the equation

$$y^2 + xy = x^3 + x^2 + 1.$$

Since these curves allow very efficient implementations of certain elliptic curve cryptosystems, they have been particularly attractive to implementors of these schemes [7, 16]. Anomalous binary curves are just a special case of subfield curves which have also been proposed for use in elliptic curve cryptography because they also give efficient implementations.

If $m = ed$ for $e, d \in \mathbb{Z}_{>0}$, then $GF(2^e) \subset GF(2^m)$. Using underlying fields of this type provide very efficient implementations [3, 13]. If a and b are actually elements of $GF(2^e)$, then we say that E is a *subfield curve*. Notice in this case that $E_{(a,b)}(GF(2^e)) \subset E_{(a,b)}(GF(2^m))$.

If e is small, so that the number of points in $E_{(a,b)}(GF(2^e))$ can be easily counted, there is an easy way to determine the number of points in $E_{(a,b)}$ $(GF(2^m))$. Denote by $\#E$ the number of points in E. Then it is well known that $\#E_{(a,b)}(GF(2^e)) = 2^e + 1 - t$ for some $t \leq 2\sqrt{2^e}$. The value t is known as the *trace* of the curve. If α and β are the two roots of the equation $X^2 - tX + 2^e = 0$, then $\#E_{(a,b)}(GF(2^m)) = 2^m + 1 - \alpha^d - \beta^d$. This is known as *Weil's Theorem*.

2.4   The Frobenius Endomorphism

An interesting property of anomalous binary curves is that if $P = (x, y)$ is a point on the curve, then so is (x^2, y^2). In fact $(x^2, y^2) = \lambda P$ for some constant λ. We can see this in the general case of subfield curves using the Frobenius endomorphism.

The Frobenius endomorphism is the function ψ that takes x to x^{2^e} for all $x \in GF(2^m)$. Since we are working in a field of characteristic 2, notice that $\psi(r(x)) = r(\psi(x))$ for all $x \in GF(2^m)$ and any rational function r with coefficients in $GF(2^e)$. If $P = (x, y)$ is a point on the subfield curve E, define $\psi(P) = (\psi(x), \psi(y))$. Also define $\psi(\mathcal{O}) = \mathcal{O}$. It can be shown from the curve's defining equation and the fact that $(a + b)^{2^e} = a^{2^e} + b^{2^e}$ for all $a, b \in GF(2^m)$ that if $P \in E$ then $\psi(P) \in E$. Thus if E is a subfield curve and $P, Q \in E$, then $\psi(P + Q) = \psi(P) + \psi(Q)$.

Now, consider a point $P \in E$ where E is a subfield curve and P has prime order p with p^2 not dividing $\#E$. By the above remarks we have $p\psi(P) = \psi(pP) = \psi(\mathcal{O}) = \mathcal{O}$. Hence $\psi(P)$ must also be a point of order P. Since $\psi(P) \in E$, we must have $\psi(P) = \lambda P$ for some $\lambda \in \mathbb{Z}$, $1 \leq \lambda \leq p - 1$. The value λ is constant among all points in the subgroup generated by P and is known as the *eigenvalue* of the Frobenius endomorphism.

It is known that for any point $P \in E$, the Frobenius endomorphism satisfies

$$\psi^2(P) - t\psi(P) + 2^e P = \mathcal{O}$$

where t is the trace as defined in Section 2.3. Therefore, it can also be shown that λ is one of the roots of the quadratic congruence

$$X^2 - tX + 2^e \equiv 0 \pmod{p}.$$

Hence, λ can be efficiently computed.

2.5   Parallel Collision Search

Given a point Q on an elliptic curve which is in a subgroup of order n generated by P, we seek l such that $Q = lP$. Pollard's ρ method [11] proceeds as follows.

Partition the points on the curve into three roughly equal size sets S_1, S_2, S_3 based on some simple rule. Define an iteration function on a point Z as follows

$$f(Z) = \begin{cases} 2Z & \text{if } Z \in S_1 \\ Z + P & \text{if } Z \in S_2. \\ Z + Q & \text{if } Z \in S_3. \end{cases}$$

Choose $A_0, B_0 \in [1, n-1]$ at random and compute the starting point $Z_0 = A_0 P + B_0 Q$. Compute the sequence $Z_1 = f(Z_0), Z_2 = f(Z_1), \ldots$ keeping track of A_i, B_i such that $Z_i = A_i P + B_i Q$. Thus,

$$(Z_{i+1}, A_{i+1}, B_{i+1}) = \begin{cases} (2Z_i, 2A_i, 2B_i) & \text{if } Z \in S_1 \\ (Z_i + P, A_i + 1, B_i) & \text{if } Z \in S_2. \\ (Z_i + Q, A_i, B_i + 1) & \text{if } Z \in S_3. \end{cases}$$

Note that A_i and B_i can be computed modulo n so that they do not grow out of control. Because the number of points on the curve is finite, the sequence of points must begin to repeat. Upon detection that $Z_i = Z_j$ we have $A_i P + B_i Q = A_j P + B_j Q$, which gives $l = \frac{A_i - A_j}{B_j - B_i} \bmod n$, unless we are very unlucky and $B_i \equiv B_j \pmod{n}$.

Actually, Pollard's function is not an optimal choice. In [17] it is recommended that the points be divided into about 20 sets of equal size $S_1, \ldots, S_{20}$ and that the iteration function be

$$f(Z) = \begin{cases} Z + c_1 P + d_1 Q & \text{if } Z \in S_1 \\ Z + c_2 P + d_2 Q & \text{if } Z \in S_2 \\ \quad \vdots & \quad \vdots \\ Z + c_{20} P + d_{20} Q & \text{if } Z \in S_{20} \end{cases} \tag{1}$$

where the c_i and d_i are random integers between 1 and $n-1$. The use of this iteration function gives a running time very close to that expected by theoretical estimates. In order to make computation of the values A_i and B_i more efficient, we suggest that constants $c_{11}, \ldots, c_{20}$ and $d_1, \ldots, d_{10}$ could be zero so that only one of the values A_i or B_i need to be updated at each stage.

Pollard's ρ method is inherently serial and cannot be directly parallelized over several processors efficiently. Parallel collision search [18] provides a method for efficient parallelization. Several processors each create their own starting points Z_0 and iterate until a "distinguished point" Z_d is reached. A point is considered distinguished if it satisfies some easily tested property such as having several leading zero bits. The triples (Z_d, A_d, B_d) are contributed to a memory common to all processors. When the memory holds two triples containing the same point Z_d, then the logarithm l can be computed as with Pollard's ρ method.

The expected number of iterations required to find the logarithm is $\sqrt{\frac{\pi n}{2}}$. The object of this paper is to reduce this number.

3   Faster Attacks for Arbitrary Curves

Notice that for elliptic curves over both $GF(p)$ and $GF(2^m)$, given a point $P = (x, y)$ on the curve it is trivial to determine its negative. Either $-P = (x, -y)$ (in the $GF(p)$ case) or $-P = (x, x + y)$ (in the $GF(2^m)$ case). Thus, at every stage of the parallel collision search algorithm, both Z_i and $-Z_i$ could be easily computed.

We would like to reduce the size of the space that is being searched by parallel collision search by a factor of 2. We can do this by replacing Z_i with $\pm Z_i$ at each step in a canonical way. A simple way to do this is to choose the one that has smallest y coordinate when its binary representation is interpreted as an integer.

When performing a parallel collision search, Z_i, A_i and B_i should be computed as normal. However, $-Z_i$ should also be computed, and whichever one of Z_i and $-Z_i$ has the smallest y coordinate should be taken to be Z_i. If Z_i has the smallest y coordinate, then everything progresses as normal. If $-Z_i$ has the smallest y coordinate then $-Z_i$ should replace Z_i, $-A_i$ should replace A_i and $-B_i$ should replace B_i. Notice that the equation $Z_i = A_i P + B_i Q$ is still maintained.

Thus, the search space for the parallel collision search is reduced to only those points which have a smaller y coordinate than their negative. Since exactly half of the points ($\neq \mathcal{O}$) have this property we have reduced the search space by a factor of 2. Because the extra computational effort in determining which of Z_i and $-Z_i$ to accept is negligible, the expected running time of the algorithm will be reduced by a factor of $\sqrt{2}$. This improvement in attack time is valid for any elliptic curve.

A technicality which affects the most obvious application of this technique is the appearance of trivial 2-cycles. Suppose that Z_i and Z_{i+1} both belong to the same S_j and that in both cases after f is applied, the negative of the resulting point is used. This is when $Z_{i+1} = -(Z_i + c_j P + d_j Q)$ (say) and $Z_{i+2} = -(Z_{i+1} + c_j P + d_j Q) = Z_i$. The occurrence of these 2-cycles is reduced by using the iteration function given in Equation (1) since it gives more choices for the multipliers. It does not reduce it enough so that efficient implementations are possible however. To reduce the occurrence of 2-cycles even further, we can use a look-ahead technique which proceeds as follows. Define $f_w(Z) \equiv Z + c_w P + d_w Q$. Suppose that $Z_i \in S_j$. Then $f(Z_i) = f_j(Z_i)$. Begin by computing $R = \pm f_j(Z_i)$, a candidate for Z_{i+1}. If $R \notin S_j$ then $Z_{i+1} = R$. If $R \in S_j$, then we treat Z_i as though it were in S_{j+1} (where $j + 1$ is reduced modulo 20), and compute a new candidate $R = \pm f_{j+1}(Z_i)$. If $R \notin S_{j+1}$, then $Z_{i+1} = R$, otherwise continue trying $j+2, j+3, \ldots$. If all 20 choices fail (a very low probability event), then just use $Z_{i+1} = \pm f_j(Z_i)$. The idea is to reduce the probability that two successive points will belong to the same set. Note that Z_{i+1} still depends solely on Z_i, a requirement for parallel collision search to work.

This modified iteration function causes the amount of computation to increase by an expected factor of approximately $\frac{20}{19}$, a small penalty which can be reduced by using more than 20 cases. The occurrence of 2-cycles is not completely eliminated, but is significantly reduced. If necessary, it can be reduced further by using more than 20 cases or by looking ahead two steps instead of just one. Another way to deal with 2-cycles is to consider them to be distinguished points.

4   Faster Attacks for Subfield Curves

We will now describe an attack on subfield curves that again uses parallel collision search and will reduce the running time by a factor of $\sqrt{d}$ when considering curves over $GF(2^{ed})$.

Let $E_{(a,b)}(GF(2^{ed}))$ be a subfield curve with $a, b \in GF(2^e)$ and let P be a point on the curve such that not both coordinates are from a proper subfield of $GF(2^{ed})$. In other words $P \in E_{(a,b)}GF(2^{ed})$, but $P \notin E_{(a,b)}(GF(2^{ef}))$ for any $f, 1 \le f \le d - 1$. Let P have prime order p such that p^2 does not divide the order of the curve and let d be odd. These conditions are not restrictive since most elliptic curve cryptosystems require the use of points P with prime order very close to the curve order, which usually implies the above conditions.

By these conditions we get that

$$\psi(P) = \lambda P \ne P,$$
$$\psi^2(P) = \lambda^2 P \ne P,$$
$$\vdots$$
$$\psi^{d-1}(P) = \lambda^{d-1} P \ne P,$$
$$\psi^d(P) = \lambda^d P = P$$

which implies that $d | p - 1$.

Remember that $\psi(x) = x^{2^e}$. Since we are working over a subfield of characteristic 2, squaring is always a very efficient operation. In particular when a normal basis representation is used, it is just a cyclic shift of the binary representation of the field element. Thus $\psi(P)$ can be computed very efficiently.

Similar to Section 3, we will use a parallel collision search and compute Z_i, A_i and B_i as usual. We can now also compute the $2d$ different points on the curve $\pm \psi^j(Z_i)$ for $0 \le j \le d - 1$. We would like to choose a "distinguished" or "canonical" representative from this set. We will first consider the d points $\psi^j(Z_i)$ and use the one whose x coordinate's binary representation has smallest value when interpreted as an integer. We can then choose either that point or its negative depending on which has smaller y coordinate when interpreted as an integer. This point will now replace Z_i. If we have chosen $\pm \psi^j(Z_i)$ to replace Z_i, we must then replace A_i with $\pm \lambda^j A_i$ and also replace B_i with $\pm \lambda^j B_i$ to maintain the relationship $Z_i = A_i P + B_i Q$. The powers of λ^j can be precomputed to obtain further efficiencies.

By performing the above operation at every step of the parallel collision search, we will be reducing the size of our search space by a factor of $2d$. Thus, the expected running time to compute the discrete logarithm will decrease by a factor of $\sqrt{2d}$.

The iteration function f used in the parallel collision search must be chosen carefully. In particular, notice that if the function is chosen to be a choice between just $2Z$, $Z+P$ and $Z+Q$ (as in the basic parallel collision search algorithm), then in some situations trivial cycles are likely to occur. Notice that for $i < j$, Z_j can be written as $Z_j = p_1(\lambda)Z_i + p_2(\lambda)P + p_3(\lambda)Q$ where p_1, p_2 and p_3 are polynomials in λ. Also notice that these polynomials will have small coefficients if $j - i$ is not too big. When using anomalous binary curves, the value λ satisfies $\lambda^2 + \lambda + 2$ or $\lambda^2 - \lambda + 2$. In either case, λ will be likely to be a root of the polynomials in the expression for Z_j, and hence a trivial cycle will be encountered. Experimentation shows that the modified iteration function described in Section 3 reduces the occurrences of these trivial cycles sufficiently for practical purposes.

4.1   Anomalous Binary Curves

Now consider the situation created by using anomalous binary curves. If $E_{(a,b)}$ $(GF(2^m))$ is such a curve, then $a, b \in \{0, 1\}$, so we are actually using subfield curves with $e = 1$ and $d = m$.

These curves are particularly well suited to this attack because the size of the space searched is reduced by a factor of $2m$, which reduces the expected running time by a factor of $\sqrt{2m}$. Thus the attacks on anomalous binary curves using this method are the most efficient among all subfield curves.

As an example, consider the anomalous binary curve $E_{(1,1)}(GF(2^{163}))$. This curve has been considered particularly attractive for implementing elliptic curve cryptosystems since its order is twice a prime close to 2^{162}. Many standards recommend that elliptic curve cryptosystems use curves divisible by a prime of at least 160 bits to obtain an expected attack time of at least 2^{80} operations [1, 2].

The conventional parallel collision search method for computing discrete logarithms on this curve is expected to take approximately 2^{81} operations. Using the improvements suggested above will reduce this expected running time by a factor of $\sqrt{2 \cdot 163}$ to approximately 2^{77} operations. This is below the required level of security imposed by the standards. Thus, this curve should not be used if a security level of 2^{80} is desired.

5   Efficiency Considerations

It has been shown that the number of group operations required to perform an elliptic curve logarithm can be reduced, but this is not much good if too much added computation is required in each step. In this section we show how to keep computation low. At each stage of the algorithm we know that the equation $Z_i = A_i P + B_i Q$ holds. We have at each stage that $A_{i+1} = \pm \lambda^j (A_i + c)$ (say) for some $0 \leq j \leq d - 1$ and some multiplier c. If we represent A_i as $A_i = (-1)^{u_i} \lambda^{v_i} w_i$, $u_i \in \{0, 1\}$, $0 \leq v_i \leq d - 1$, $0 \leq w_i \leq n - 1$, then we can compute w_{i+1} as $w_i + (-1)^{u_i} \lambda^{-v_i} c$, v_{i+1} as $v_i + j$, and u_i is negated if necessary. The coefficient B_i can be tracked similarly. If there is a precomputed table of

$\lambda^{-j}c$ for each $j = 0, \ldots, d - 1$ and each multiplier c, then the computation on each step consists of additions or subtractions modulo n, additions modulo d and sign changes. This is much cheaper than an elliptic curve addition and is not a significant part of the algorithm run-time.

We have implemented these ideas on the anomalous binary curve $E_{(0,1)}(GF(2^{41}))$. The iteration function used 20 multipliers and used the look-ahead scheme described in Section 3. Over 15 trials, the experimental run-times were consistent with the expected run-time of $\sqrt{\frac{(\pi/2)2^{41}}{2 \cdot 41}}$.

6   Other Attempts for Faster Attacks

Another way that one might try to take advantage of the Frobenius endomorphism is to use parallel collision search as usual, but to check whether any stored distinguished points are negatives of each other or can be mapped to each other with the Frobenius endomorphism. This is easiest when using a method for choosing distinguished points which leaves a point distinguished if the Frobenius map is applied.

Unfortunately, this approach will not work unless the iteration function is carefully chosen so that all members of one equivalence class map to the same new equivalence class. The principle behind parallel collision search is that each distinguished point stands for the entire set of points in the trail leading to the distinguished point. A collision occurs because one trail runs into another trail and is lead to the common distinguished point. When a collision occurs and is detected, the two distinguished points are identical. The probability of encountering two unequal distinguished points which have a Frobenius map and/or negation map is very low.

Another way to think of this is that the iteration function acts as a random mapping and not all distinguished points are equally likely to appear. In fact, distinguished points tend to have radically different sized tree structures leading into them. The conditional probabilities are such that if a distinguished point occurs, it is very likely to have a large tree structure leading into it, making it a likely candidate to appear again. However, the distinguished points which are Frobenius and/or negation maps of the one which has occurred are not likely to have large tree structures.

It should be noted that the methods presented in Section 4 may also apply to any elliptic curve that has an easily computed automorphism that has small order modulo the order of the point. For example, consider the curve

$$y^2 = x^3 - ax$$

over $GF(p)$, with point $P = (x_0, y_0)$ of prime order n. This curve has complex multiplication by $\mathbb{Z}[i]$. Let i_p be a solution to $x^2 \equiv -1 \pmod{p}$. Then, $\psi_i(P) = (-x_0, i_p y_0) = \lambda_i P$ where λ_i is a solution to $x^2 \equiv -1 \pmod{n}$. Since

$$\psi_i^0(P) = P$$

$$\psi_i^1(P) = \lambda_i P$$
$$\psi_i^2(P) = -P$$
$$\psi_i^3(P) = -\lambda_i P$$

are all distinct we can reduce the size of our search space by a factor of 4 to get a speed up of 2 over the general parallel collision search.

Also, consider the curve

$$y^2 = x^3 + b$$

over $GF(p)$, with point $P = (x_0, y_0)$ of prime order n. This curve has complex multiplication by $\mathbb{Z}[\omega]$ where ω is a cube root of unity. Let ω_p be a solution to $x^3 \equiv 1 \pmod{p}$. Then, $\psi_\omega(P) = (\omega_p x_0, y_0) = \lambda_\omega P$ where λ_ω is a solution to $x^3 \equiv 1 \pmod{n}$. Since

$$\pm\psi_\omega^0(P) = \pm P$$
$$\pm\psi_\omega^1(P) = \pm\lambda_\omega P$$
$$\pm\psi_\omega^2(P) = \pm\lambda_\omega^2 P$$

are all distinct we can reduce the size of our search space by a factor of 6 to get a speed up of $\sqrt{6}$ over the general parallel collision search.

7   Conclusion

Subfield and anomalous binary curves have been attractive to cryptographers for quite some time because of the efficiencies they provide both in curve generation and in the implementation of cryptographic algorithms. There have also been unsubstantiated warnings for quite some time that these curves may be more open to attack because of the greater structure that these curves have. The results of this paper show that this structure can in fact be used to obtain faster attacks. While the attack presented here still has a fully exponential running time, care should be exercised when choosing these curves regarding their expected security level. In certain circumstances these curves may still be attractive because of their efficiencies with respect to curves of similar security levels.

These results highlight the fact that more research must be done on the cryptanalysis of elliptic curve cryptosystems before we can be fully confident of the security level different curves offer. Two open questions remain:

- Can the ideas presented here be used, possibly in combination with other methods to reduce the attack time further?
- Can similar ideas be applied to other classes of curves or to curves whose coefficients do not lie in the subfield?

8   Acknowledgment

The authors would like to thank Niels Provos for pointing out the fact that the two curves mentioned in Section 6 also allow a speed up of the parallel collision search algorithm.

References

1. ANSI X9.62-199x: Public Key Cryptography for the Financial Services Industry: The Elliptic Curve Digital Signature Algorithm (ECDSA), January 13, 1998.
2. ANSI X9.63-199x: Public Key Cryptography for the Financial Services Industry: Elliptic Curve Key Agreement and Transport Protocols, October 5, 1997.
3. E. De Win, A. Bosselaers, S. Vandenberghe, P. De Gersem and J. Vandewalle, "A fast software implementation for arithmetic operations in $GF(2^n)$," *Advances in Cryptology, Proc. Asiacrypt96, LNCS 1163*, K. Kim and T. Matsumoto, Eds., Springer-Verlag, 1996, pp. 65-76.
4. W. Diffie and M. Hellman, New directions in cryptography, *IEEE Transactions on Information Theory* **22** (1976), pp. 644-654.
5. R. Gallant, R. Lambert and S. Vanstone, "Improving the Parallelized Pollard Lambda Search on Binary Anomalous Curves", Research Report No. CORR98-15, Department of Combinatorics and Optimization, University of Waterloo, Waterloo, Canada, (1998).
6. N. Koblitz, Elliptic curve cryptosystems, *Mathematics of Computation* **48** (1987), pp. 203-209.
7. N. Koblitz, "CM-curves with good cryptographic properties," *Advances in Cryptology, Proc. Crypto91, LNCS 576*, J. Feigenbaum, Ed., Springer-Verlag, 1997, pp. 279-287.
8. A. Menezes, T. Okamoto and S. Vanstone, Reducing elliptic curve logarithms to logarithms in a finite field, *IEEE Transactions on Information Theory*, **39** (1993), pp. 1639-1646.
9. V. Miller, "Uses of elliptic curves in cryptography," in *Advances in Cryptology - CRYPTO '85*, Lecture Notes in Computer Science, **218** (1986), Springer-Verlag, pp. 417-426.
10. National Institute for Standards and Technology, "Digital signature standard," *Federal information processing standard*, U.S. Department of Commerce, FIPS PUB 186, Washington, DC, 1994.
11. J.M. Pollard, Monte Carlo methods for index computation (mod p), *Mathematics of Computation*, **32** (1978), pp. 918-924.
12. T. Satoh and K. Araki, Fermat Quotients and the Polynomial Time Discrete Log Algorithm for Anomalous Elliptic Curves, *preprint*, 1997.
13. R. Schroeppel, H. Orman, S. OMalley and O. Spatscheck, "Fast key exchange with elliptic curve systems," *Advances in Cryptology, Proc. Crypto95, LNCS 963*, D. Coppersmith, Ed., Springer-Verlag, 1995, pp. 43-56.
14. I. Semaev, Evaluation of discrete logarithms in a group of p-torsion points of an elliptic curve in characteristic p, *Mathematics of Computation*, **67** (1998), pp. 353-356.
15. N. Smart, The discrete logarithm problem on elliptic curves of trace one, *preprint*, 1997.
16. J. Solinas, "An improved algorithm for arithmetic on a family of elliptic curves," *Advances in Cryptology, Proc. Crypto97, LNCS 1294*, B. Kaliski, Ed., Springer-Verlag, 1997, pp. 357-371.
17. E. Teske, "Speeding up Pollard's rho method for computing discrete logarithms," Technical Report No. TI-1/98, Technische Hochschule Darmstadt, Darmstadt, Germany, (1998).
18. P. van Oorschot and M. Wiener, Parallel collision search with cryptanalytic applications, *Journal of Cryptology*, to appear.

Improved Algorithms for Elliptic Curve Arithmetic in $GF(2^n)$

Julio López[1]* and Ricardo Dahab[2]**

[1] Dept. of Computer Science, University of Valle
A. A. 25130 Cali, Colombia
julioher@dcc.unicamp.br
[2] Institute of Computing, State University of Campinas
C.P. 6176, 13083-970, Campinas, SP, Brazil
rdahab@dcc.unicamp.br

Abstract. This paper describes three contributions for efficient implementation of elliptic curve cryptosystems in $GF(2^n)$. The first is a new method for doubling an elliptic curve point, which is simpler to implement than the fastest known method, due to Schroeppel, and which favors sparse elliptic curve coefficients. The second is a generalized and improved version of the Guajardo and Paar's formulas for computing repeated doubling points. The third contribution consists of a new kind of projective coordinates that provides the fastest known arithmetic on elliptic curves. The algorithms resulting from this new formulation lead to a running time improvement for computing a scalar multiplication of about 17% over previous projective coordinate methods.

1 Introduction

Elliptic curves defined over finite fields of characteristic two have been proposed for Diffie-Hellman type cryptosystems [1]. The calculation of $Q = mP$, for P a point on the elliptic curve and m an integer, is the core operation of elliptic curve public-key cryptosystems. Therefore, reducing the number of field operations required to perform the scalar multiplication mP is crucial for efficient implementation of these cryptosystems.

In this paper we discuss efficient methods for implementing elliptic curve arithmetic. We present better results than those reported in [8, 5, 2]; our basic technique is to rewrite the elliptic operations (doubling and addition) with less costly field operations (inversions and multiplications), and replace general field multiplications by multiplications by fixed elliptic coefficients.

The first method is a new formula for doubling a point, i.e., for calculating the sum of equal points. This method is simpler to implement than Schroeppel's method [8] since it does not require a quadratic solver. If the elliptic curve coefficient b is sparse, i.e., with few 1's in its representation, thus making the

---

* Research supported by a CAPES-Brazil scholarship
** Partially supported by a PRONEX-FINEP grant no. 107/97

S. Tavares and H. Meijer (Eds.): SAC'98, LNCS 1556, pp. 201–212, 1999.
© Springer-Verlag Berlin Heidelberg 1999

multiplication by the constant b more efficient than a general field multiplication, then our new formula should lead to an improvement of up to 12% compared to Schroeppel's method [8]. We also note that our formula can be applied to composite finite fields as well.

In [2], a new approach is introduced for accelerating the computation of repeated doubling points. This method can be viewed as computing consecutive doublings using fractional field arithmetic. We have generalized and improved the formulas presented in that paper. The new formulas can be used to speed-up variants of the sliding-window method. For field implementations where the cost-ratio of inversion to multiplication varies form 2.5 to 4 (typical values of practical software field implementations), we expect a speed-up of 7% to 22% in performing a scalar multiplication.

In [7], Schroeppel proposes an algorithm for computing repeated doubling points removing most of the general field multiplications, and favoring elliptic curves with sparse coefficients. Using his method, the computation of $2^i P, i \geq 2$ requires i field inversions, i multiplications by a fixed constant, one general field multiplication, and a quadratic solver. Since inversion is the most expensive field operation, this method is suitable for finite fields where field inversion is relatively fast. If the cost-ratio of inversion to multiplication is less than 3, this algorithm may be faster than our repeated doubling algorithm.

When field inversion is costly (e.g., for normal basis representation, the cost-ratio of inversion to multiplication is at least 7 [2,8]), projective coordinates offer an alternative method for efficiently implementing the elliptic curve arithmetic. Based on our doubling formula, we have developed a new kind of projective coordinates which should lead to an improvement of 38% over the traditional projective arithmetic coordinates [4] and 17% on the recent projective coordinates presented in [5], for calculating a multiple of a point.

The remainder of the paper is organized as follows. Section 2 presents a brief summary of elliptic curves defined over finite fields of characteristic two. In Section 3, we present our doubling point algorithm. Based on this method, we describe an algorithm for repeated doubling points in Section 4. In Section 5, we describe the new projective coordinates.An implementation of the doubling and adding projective algorithms is given in the appendix.

2   Elliptic Curves over $GF(2^n)$

A non-supersingular elliptic curve E over $GF(2^n)$ is defined to be the set of solutions $(x, y) \in GF(2^n) \times GF(2^n)$ to the equation,

$$y^2 + xy = x^3 + ax^2 + b \ ,$$

where a and $b \in GF(2^n), b \neq 0$, together with the point at infinity denoted by $\mathcal{O}$.

It is well known that E forms a commutative finite group, with $\mathcal{O}$ as the group identity, under the addition operation known as the "tangent and chord method". Explicit rational formulas for the addition rule involve several arithmetic operations (adding, squaring, multiplication and inversion) in the underlying finite field. In what follows, we will only be concerned with formulas for doubling a point P in affine coordinates; formulas for adding two different points in affine or projective coordinates can be found in [4,5].

Let $P = (x_1, y_1)$ be a point of E. The doubling point formula [4] to compute $2P = (x_2, y_2)$ is given by

$$
\begin{cases}
x_2 = x_1^2 + \dfrac{b}{x_1^2} , \\
y_2 = x_1^2 + (x_1 + \dfrac{y_1}{x_1}) \cdot x_2 + x_2 .
\end{cases}
\tag{1}
$$

Note that the x-coordinate of doubling point formula $2P$ depends only on the x-coordinate of P and the coefficient b, but doubling a point requires two general field multiplications, one multiplication by the constant b and one field inversion.

Schroeppel [6] improved the doubling point formula saving the multiplication by the constant b. His improved doubling point formula is :

$$
\begin{cases}
x_2 = M^2 + M + a , \\
y_2 = x_1^2 + M \cdot x_2 + x_2 , \\
M = x_1 + \dfrac{x_1}{y_1} .
\end{cases}
\tag{2}
$$

Observe that the x-coordinates of the previous doubling point formula lead to the quadratic equation for M:

$$
M^2 + M + a = x_1^2 + \frac{b}{x_1^2} .
\tag{3}
$$

If we assume that the cost of multiplying by a sparse fixed constant is comparable in speed to field addition, and that solving the previous quadratic equation is faster, then we obtain another method for doubling a point with an effective cost of one general multiplication and one field inversion. A description of this method, developed by Schroeppel, can be found in [8, pp. 370-371] and [5].

In the next section, we introduce a new doubling point formula which requires also a general field multiplication, one field inversion, but does not depend on a quadratic solver.

3   A New Doubling Point Formula

Given an elliptic curve point $P = (x_1, y_1)$, the coordinates of the doubling point $2P = (x_2, y_2)$ can be calculated by the following new doubling point formula:

$$
\begin{cases}
x_2 = x_1^2 + \dfrac{b}{x_1^2} , \\
y_2 = \dfrac{b}{x_1^2} + a x_2 + (y_1^2 + b) \cdot (1 + \dfrac{b}{x_1^4}) .
\end{cases}
\tag{4}
$$

To derive the above formula we transform the y-coordinate of the doubling point formula (2):

$$y_2 = x_1^2 + (x_1 + \frac{y_1}{x_1}) \cdot x_2 + x_2 = \frac{b}{x_1^2} + (\frac{y_1^2 + b + ax_1^2}{x_1^2}) \cdot x_2$$

$$= \frac{b}{x_1^2} + ax_2 + \frac{y_1^2 + b}{x_1^2} \cdot (\frac{x_1^4 + b}{x_1^2}) = \frac{b}{x_1^2} + ax_2 + (y_1^2 + b) \cdot (1 + \frac{b}{x_1^4}) .$$

3.1 Performance Analysis

We begin with the observation that our doubling formula eliminates the need for computing the field element M from formula (2), which requires either one general multiplication or a quadratic solver. The calculation of $2P$ requires one general field multiplication, two field multiplications by the fixed constant b, and one field multiplication by the constant a. This last multiplication can be avoided by choosing the coefficient a to be 0 or 1.[1] Thus, our formula favors elliptic curves with sparse coefficients, i.e., those having relatively few 1's in their representation.

In order to compare the running time of our formula with Schroeppel's method [8] for computing a scalar multiplication, we made the following assumptions:

- Adding and squaring field elements is fast.
- Multiplying a field element by a sparse constant is comparable to adding.
- The cost of solving the quadratic equation (3) and determining the right solution is about half of that of a field multiplication (this is true for the finite field implementation given in [6], but no efficient method is known for tower fields [7]).

The fastest methods for computing a scalar multiplication [6,3] perform five point doublings for every point-addition, on average. Table 1 compares our formula, in performing a scalar multiplication, for different values of the cost-ratio r of inversion to multiplication.

Table 1. The number of field multiplications for computing $2^5 P + Q$.

Cost-Ratio	New Formula #Mult.	Schroeppel [8] #Mult.	Improv. %
$r = 2$	19	21.5	12
$r = 2.5$	22	24.5	10
$r = 3$	25	27.5	9
$r = 4$	31	33.5	7

---

[1] E is isomorphic to E_1: $y^2 + xy = x^3 + \alpha x^2 + b$, where $Tr(\alpha) = Tr(a)$, $\alpha = 0$ or γ and $Tr(\gamma) = 1$ (if n is odd, we can take $\gamma = 1$), see [4, p. 39].

Therefore, for practical field implementations as those given in [6, 9, 2], our formula should lead to a running time improvement of up to 12% in computing a scalar multiplication. However, for elliptic curves selected at random (where the coefficient b is not necessarily sparse), both our and Schroeppel's method may not give a computational advantage. A better algorithm for computing $2^5 P$ is presented in the next section.

4   Repeated Doubling Algorithm

We present a method for computing repeated doublings, $2^i P, i \geq 2$, which is based on fractional field arithmetic and the doubling formula. The idea is to successively compute the elliptic points $2^j P = (x_j, y_j)$, $j = 1, 2, \ldots, i$, as triples $(\nu_j, \omega_j, \delta_j)$ of field elements, where $x_i = \frac{\nu_i}{\delta_i}$ and $y_i = \frac{\omega_i}{\delta_i^2}$. The exact formulation is given in

Theorem 1. *Let $P = (x, y)$ be a point on the elliptic curve E. Then the coordinates of the point $2^i P = (x_i, y_i), i \geq 2$, are given by*

$$x_i = \frac{\nu_i}{\delta_i} \ , \tag{5}$$

$$y_i = \frac{\omega_i}{\delta_i^2} \ . \tag{6}$$

where

$$\nu_{k+1} = \nu_k^4 + b\delta_k^4 \ , \quad \nu_0 = x$$
$$\delta_{k+1} = (\delta_k \cdot \nu_k)^2 \ , \quad \delta_0 = 1$$
$$\omega_{k+1} = b\delta_k^4 \cdot \delta_{k+1} + \nu_{k+1} \cdot (a\delta_{k+1} + \omega_k^2 + b\delta_k^4) \ , \quad \omega_0 = y, \ 0 \leq k < i.$$

Proof. We will prove by induction on i that $x_i = \frac{\nu_i}{\delta_i}$ and $y_i = \frac{\omega_i}{\delta_i^2}$. This is easily true for $i = 1$. Now assume that the statement is true for $i = n$; we prove it for $i = n + 1$:

$$x_{n+1} = \frac{b}{x_n^2} + x_n^2 = \frac{b\delta_n^2}{\nu_n^2} + \frac{\nu_n^2}{\delta_n^2}$$
$$= \frac{b\delta_n^4 + \nu_n^4}{\nu_n^2 \cdot \delta_n^2} = \frac{\nu_{n+1}}{\delta_{n+1}} \ ;$$

similarly, for y_{n+1} we obtain:

$$y_{n+1} = \frac{b}{x_n^2} + ax_{n+1} + (y_n^2 + b) \cdot (1 + \frac{b}{x_n^2})$$
$$= \frac{b\delta_n^2}{\nu_n^2} + a\frac{\nu_{n+1}}{\delta_{n+1}} + (\frac{\omega_n^2}{\delta_n^4} + b) \cdot (1 + \frac{b\delta_n^4}{\nu_n^4})$$

$$= \frac{b\delta_n^4}{\delta_{n+1}} + a\frac{\nu_{n+1}}{\delta_{n+1}} + \frac{(\omega_n^2 + b\delta_n^4)\cdot\nu_{n+1}}{\delta_{n+1}^2}$$

$$= \frac{\omega_{n+1}}{\delta_{n+1}^2}.$$

The following algorithm, based on Theorem 1, implements repeated doublings in terms of the affine coordinates of $P = (x, y)$.

Fig. 1. Algorithm 1: Repeated doubling points.

INPUT: $P = (x, y) \in E$ $i \geq 2$.
OUTPUT: $Q = 2^i P$.

Set $V \leftarrow x^2$, $D \leftarrow V$, $W \leftarrow y$, $T \leftarrow b$.
for $k = 1$ to $i - 1$ do
    Set $V \leftarrow V^2 + T$.
    Set $W \leftarrow D\cdot T + V\cdot(aD + W^2 + T)$.
    if $k \neq i - 1$ then
        $V \leftarrow V^2$, $D \leftarrow D^2$, $T \leftarrow bD^2$, $D \leftarrow D\cdot V$.
    fi
od
Set $D \leftarrow D\cdot V$.
Set $M \leftarrow D^{-1}\cdot(V^2 + W)$.
Set $x \leftarrow D^{-1}\cdot V^2$.
Set $x_i \leftarrow M^2 + M + a$, $y_i \leftarrow x^2 + M\cdot x_i + x_i$.
return$(Q = (x_i, y_i))$.

Note that the correctness of this algorithm follows directly from the proof of Theorem 1 and formula (2).

Corollary 1 *Assume that P is an elliptic point of order larger than 2^i. Then Algorithm 1 performs $3i - 1$ general field multiplications, $i - 1$ multiplications by the fixed constant b, and $5i - 4$ field squarings.*

4.1   Complexity Comparison

Since Algorithm 1 cuts down the number of field inversions at the expense of more field multiplications, the computational advantage of Algorithm 1 over repeated doubling (using the standard point doubling formula (2)) depends on r, the cost-ratio of inversion to multiplication. Assuming that adding and squaring is fast, we conclude, from Corollary 1, that Algorithm 1 outperforms the computation of five consecutive doublings when $r > 2$. Table 2 shows the number of field multiplications needed for computing $2^5 P + Q$ for several methods and for different values of r. Note that the standard algorithm and Guajardo and Paar's formulas do not use the elliptic coefficient b, whereas Algorithm 1 does.

Table 2. Comparison of Algorithm 1 with other algorithms.

Ratio	Algorithm 1		Schroeppel [7]		G.P. [2]	Standard (2)
r	b sparse	b random	b sparse	b random	b random	b random
2.5	21	25	18.5	22.5	27	27
3	22	26	21.5	25.5	28	30
3.5	23	27	24.5	28.5	29	33
4	24	28	27.5	31.5	30	36

Algorithm 1 obtains its best performance for field implementations when r is at least three. If the elliptic curve is selected at random, then we expect Algorithm 1 to be up to 22% faster than the standard algorithm. For field implementations where $r < 3$, (for example [6,9]), Schroeppel's method [7] outperforms Algorithm 1.

5 A New Kind of Projective Coordinates

When field inversion in $GF(2^n)$ is relatively expensive, then it may be of computational advantage to use fractional field arithmetic to perform elliptic curve additions, as well as, doublings. This is done with the use of projective coordinates.

5.1 Basic Facts

A *projective plane* P^2 is defined to be the set of equivalence classes of triples (X, Y, Z), not all zero, where (X_1, Y_1, Z_1) and (X_2, Y_2, Z_2) are said to be equivalent if there exists $\lambda \in GF(2^n)$, $\lambda \neq 0$ such that $X_1 = \lambda X_2, Y_1 = \lambda^2 Y_2$ and $Z_1 = \lambda Z_2$. Each equivalence class is called a *projective point*. Note that if a projective point $P = (X, Y, Z)$ has nonzero Z, then P can be represented by the projective point $(x, y, 1)$, where $x = X/Z$ and $y = Y/Z^2$. Therefore, the projective plane can be identified with all points (x, y) of the ordinary (affine) plane plus the points for which $Z = 0$.

Any equation $f(x, y) = 0$ of a curve in the affine plane corresponds to an equation $F(X, Y, Z) = 0$, where F is obtained by replacing $x = X/Z, y = Y/Z^2$, and multiplying by a power of Z to clear the denominators. In particular, the *projective equation* of the affine equation $y^2 + xy = x^3 + ax^2 + b$ is given by

$$Y^2 + XYZ = X^3 Z + aX^2 Z^2 + bZ^4 .$$

If $Z = 0$ in this equation, then $Y^2 = 0$, i.e., $Y = 0$. Therefore, $(1,0,0)$ is the only projective point that satisfies the equation for which $Z = 0$. This point is called *the point at infinity* (denoted $\mathcal{O}$).

The resulting projective elliptic equation is

$$E = \{(x, y, z) \in P^2, y^2 + xyz = x^3 z + a x^2 z^2 + b z^4\} \ .$$

To convert an affine point (x, y) to a projective point, one sets $X = x$, $Y = y$, $Z = 1$. Similarly, to convert a projective point (X, Y, Z) to an affine point, we compute $x = X/Z$, $y = Y/Z^2$. The projective coordinates of the point $-P(X, Y, Z)$ are given by $-P(X, Y, Z) = (X, XZ + Y, Z)$. The algorithms for adding two projective points are given below.

5.2   Projective Elliptic Arithmetic

In this section we present new formulas for adding elliptic curve points in projective coordinates.

Projective Elliptic Doubling

The projective form of the doubling formula is

$$2(X_1, Y_1, Z_1) = (X_2, Y_2, Z_2) \ ,$$

where

$$Z_2 = Z_1^2 \cdot X_1^2 \ ,$$
$$X_2 = X_1^4 + b \cdot Z_1^4 \ ,$$
$$Y_2 = b Z_1^4 \cdot Z_2 + X_2 \cdot (a Z_2 + Y_1^2 + b Z_1^4) \ .$$

Projective Elliptic Addition

The projective form of the adding formula is

$$(X_0, Y_0, Z_0) + (X_1, Y_1, Z_1) = (X_2, Y_2, Z_2) \ ,$$

where

$$
\begin{aligned}
&A_0 = Y_1 \cdot Z_0^2 \ , &&D = B_0 + B_1 \ , &&H = C \cdot F \ , \\
&A_1 = Y_0 \cdot Z_1^2 \ , &&E = Z_0 \cdot Z_1 \ , &&X_2 = C^2 + H + G \ , \\
&B_0 = X_1 \cdot Z_0 \ , &&F = D \cdot E \ , &&I = D^2 \cdot B_0 \cdot E + X_2 \ , \\
&B_1 = X_0 \cdot Z_1 \ , &&Z_2 = F^2 \ , &&J = D^2 \cdot A_0 + X_2 \ , \\
&C = A_0 + A_1 \ , &&G = D^2 \cdot (F + a E^2) \ , &&Y_2 = H \cdot I + Z_2 \cdot J \ .
\end{aligned}
$$

These formulas can be improved for the special case $Z_1 = 1$:

$$(X_0, Y_0, Z_0) + (X_1, Y_1, 1) = (X_2, Y_2, Z_2),$$

where

$$A = Y_1 \cdot Z_0^2 + Y_0 \ ,$$
$$B = X_1 \cdot Z_0 + X_0 \ ,$$
$$C = Z_0 \cdot B \ ,$$
$$D = B^2 \cdot (C + aZ_0^2) \ ,$$
$$Z_2 = C^2 \ ,$$

$$E = A \cdot C \ ,$$
$$X_2 = A^2 + D + E \ ,$$
$$F = X_2 + X_1 \cdot Z_2 \ ,$$
$$G = X_2 + Y_1 \cdot Z_2 \ ,$$
$$Y_2 = E \cdot F + Z_2 \cdot G \ .$$

5.3   Performance Analysis

The new projective doubling algorithm requires three general field multiplications, two multiplications by a fixed constant, and five squarings. Since doubling a point takes one general field multiplication less than the previous projective doubling algorithm given in [5], we obtain an improvement of about 20% for doubling a point, in general. For sparse coefficients b, we may obtain an improvement of up to a 25%.

The new projective adding algorithm requires 13 general multiplications, one multiplication by a fixed constant and six squarings. If $a = 0$ (or $a = 1$) and $Z_1 = 1$, then only nine general field multiplications and four squarings are required. Thus, we obtain one field multiplication less than the previous projective addition algorithm presented in [5]. The number of field operations required to perform an elliptic addition for various kinds of projective coordinates is listed in Table 3.

Now we can estimate the improvement of a scalar multiplication using the new projective coordinates. We will consider only the case $a = 0$ (or $a = 1$) and $Z_1 = 1$, since for this situation we obtain the best improvement. The number of field operations for computing $2^5 P + Q$ is given in Table 3. Using these values we can conclude that the computation of a scalar multiplication, based on the new projective coordinates, is on average 17% and 38% faster than the previous projective coordinates [4, 5].

Table 3. The number of field operations for $2^5 P + Q$ ($a = 0$ or 1, $Z_1 = 1$)

Projective	Doubling		Adding		Cost of $2^5 P + Q$	
coordinates	#Mult.	#Sqr.	#Mult.	#Sqr.	#Mult.	#Sqr.
$(x/z, y/z^2)$	4	5	9	4	29	29
$(x/z^2, x/z^3)$	5	5	10	4	35	29
$(x/z, y/z)$	7	5	12	1	47	26

6 Conclusions

We have presented improved methods for faster implementation of the arithmetic of an elliptic curve defined over $GF(2^n)$. Our methods are easy to implement and can be applied to all elliptic curves defined over fields of characteristic two, independently of the specific field representation. They favor sparse elliptic coefficients but also perform well for elliptic curves selected at random. In general, they should lead to an improvement of up to 20% in the computation of scalar multiplication.

7 Acknowledgments

We thank the referees for their helpful comments.

References

1. W. Diffie and M. Hellman, "New Directions in Cryptography," *IEEE Transactions in Informations Theory*, IT-22:644-654, November 1976.
2. J. Guajardo and C. Paar, "Efficient Algorithms for Elliptic Curve Cryptosystems", *Advances in Cryptology, Proc. Crypto'97, LNCS 1294*, B. Kaliski, Ed., Springer-Verlag,1997,pp. 342-356.
3. K Koyama and Y. Tsuruoka, "Speeding up elliptic cryptosystems by using a signed binary window method," *Advances in Cryptology, Proc. Crypto'92, LNCS 740*, E. Brickell, Ed., Springer-Verlag, 1993, pp. 345-357.
4. A. Menezes, *Elliptic curve public key cryptosystems,* Kluwer Academic Publishers, 1993.
5. IEEE P1363: Editorial Contribution to Standard for Public Key Cryptography, February 9, 1998.
6. R. Schroeppel, H. Orman, S. O'Malley and O. Spatscheck, "Fast key exchange with elliptic curve systems," *Advances in Cryptology, Proc. Crypto'95, LNCS 963*, D. Coppersmith, Ed., Springer-Verlag, 1995, pp. 43-56.
7. R. Schroeppel, "Faster Elliptic Calculations in $GF(2^n)$," preprint, March 6, 1998.
8. J. Solinas, "An improved algorithm for arithmetic on a family of elliptic curves," *Advances in Cryptology, Proc. Crypto'97, LNCS 1294*, B. Kaliski, Ed., Spring-Verlag, 1997, pp. 357-371.
9. E. De Win, A. Bosselaers, S. Vanderberghe, P. De Gersem and J. Vandewalle, "A fast software implementation for arithmetic operations in $GF(2^n)$," *Advances in Cryptology, Proc. Asiacrypt'96, LNCS 1163*, K. Kim and T. Matsumoto, Eds., Springer-Verlag, 1996, pp. 65-76.

8 Appendix

Algorithm 2: Projective Elliptic Doubling Algorithm

Input: the finite field $GF(2^m)$; the field elements a and $c = b^{2^{m-1}}(c^2 = b)$ defining a curve E over $GF(2^m)$; projective coordinates (X_1, Y_1, Z_1) for a point P_1 on E.
Output: projective coordinates (X_2, Y_2, Z_2) for the point $P_2 = 2P_1$.

1.	$T_1 \leftarrow X_1$	
2.	$T_2 \leftarrow Y_1$	
3.	$T_3 \leftarrow Z_1$	
4.	$T_4 \leftarrow c$	
5.	if $T_1 = 0$ or $T_3 = 0$ then	
	$\quad$ output $(1, 0, 0)$ and stop.	
6.	$T_3 \leftarrow T_3^2$	
7.	$T_4 \leftarrow T_3 \times T_4$	
8.	$T_4 \leftarrow T_4^2$	
9.	$T_1 \leftarrow T_1^2$	
10.	$T_3 \leftarrow T_1 \times T_3$	$= Z_2$
11.	$T_1 \leftarrow T_1^2$	
12.	$T_1 \leftarrow T_4 + T_1$	$= X_2$
13.	$T_2 \leftarrow T_2^2$	
14.	if $a \neq 0$ then	
	$\quad T_5 \leftarrow a$	
	$\quad T_5 \leftarrow T_3 \times T_5$	
	$\quad T_2 \leftarrow T_5 + T_2$	
15.	$T_2 \leftarrow T_4 + T_2$	
16.	$T_2 \leftarrow T_1 \times T_2$	
17.	$T_4 \leftarrow T_3 \times T_4$	
18.	$T_2 \leftarrow T_4 + T_2$	$= Y_2$
19.	$X_2 \leftarrow T_1$	
20.	$Y_2 \leftarrow T_2$	
21.	$Z_2 \leftarrow T_3$	

This algorithm requires 3 general field multiplications, 5 field squarings and 5 temporary variables. If also $a = 0$, then only 4 temporary variables are required.

Algorithm 3: Projective Elliptic Adding Algorithm

Input: the finite field $GF(2^m)$; the field elements a and b defining a curve E over $GF(2^m)$; projective coordinates (X_0, Y_0, Z_0) and $(X_1, Y_1, 1)$ for points P_0 and P_1 on E.

Output: projective coordinates (X_2, Y_2, Z_2) for the point $P_2 = P_0 + P_1$, unless $P_0 = P_1$. In this case, the triple $(0, 0, 0)$ is returned. (The triple $(0,0,0)$ is not a valid projective point on the curve, but rather a marker indicating that the Doubling Algorithm should be used, see [5].)

1.	$T_1 \leftarrow X_0$	
2.	$T_2 \leftarrow Y_0$	
3.	$T_3 \leftarrow Z_0$	
4.	$T_4 \leftarrow X_1$	
5.	$T_5 \leftarrow Y_1$	
6.	$T_6 \leftarrow T_4 \times T_3$	
7.	$T_1 \leftarrow T_6 + T_1$	$= B$
8.	$T_6 \leftarrow T_3^2$	
9.	if $a \neq 0$ the	
	$\quad T_7 \leftarrow a$	
	$\quad T_7 \leftarrow T_6 \times T_7$	
10.	$T_6 \leftarrow T_5 \times T_6$	
11.	$T_2 \leftarrow T_6 + T_2$	$= A$
12.	if $T_1 = 0$ then	
	$\quad$ if $T_2 = 0$ then output $(0,0,0)$ and stop.	
	$\quad$ else output $(1,0,0)$ and stop.	
13.	$T_6 \leftarrow T_1 \times T_3$	$= C$
14.	$T_1 \leftarrow T_1^2$	
15.	if $a \neq 0$ then	
	$\quad T_7 \leftarrow T_6 + T_7$	
	$\quad T_1 \leftarrow T_7 \times T_1$	$= D$
	else $T_1 \leftarrow T_6 \times T_1$	$= D$
16.	$T_3 \leftarrow T_6^2$	$= Z_2$
17.	$T_6 \leftarrow T_2 \times T_6$	$= E$
18.	$T_1 \leftarrow T_6 + T_1$	
19.	$T_2 \leftarrow T_2^2$	
20.	$T_1 \leftarrow T_2 + T_1$	$= X_2$
21.	$T_4 \leftarrow T_3 \times T_4$	
22.	$T_5 \leftarrow T_3 \times T_5$	
23.	$T_4 \leftarrow T_1 + T_4$	$= F$
24.	$T_5 \leftarrow T_1 + T_5$	$= G$
25.	$T_4 \leftarrow T_6 \times T_4$	
26.	$T_5 \leftarrow T_3 \times T_5$	
27.	$T_2 \leftarrow T_4 + T_5$	$= Y_2$
28.	$X_2 \leftarrow T_1$	
29.	$Y_2 \leftarrow T_2$	
30.	$Z_2 \leftarrow T_3$	

This algorithm requires 9 general field multiplications, 4 field squarings and 7 temporary variables. If also $a = 0$, then only 6 temporary variables are required.

Cryptanalysis of a Fast Public Key Cryptosystem Presented at SAC '97

Phong Nguyen and Jacques Stern

École Normale Supérieure, Laboratoire d'Informatique
45, rue d'Ulm, F – 75230 Paris Cedex 05
{Phong.Nguyen,Jacques.Stern}@ens.fr
http://www.dmi.ens.fr/{pnguyen,stern}/

Abstract. At SAC '97, Itoh, Okamoto and Mambo presented a fast public key cryptosystem. After analyzing several attacks including lattice-reduction attacks, they claimed that its security was high, although the cryptosystem had some resemblances with the former knapsack cryptosystems, since decryption could be viewed as a multiplicative knapsack problem. In this paper, we show how to recover the private key from a fraction of the public key in less than 10 minutes for the suggested choice of parameters. The attack is based on a systematic use of the notion of the orthogonal lattice which we introduced as a cryptographic tool at Crypto '97. This notion allows us to attack the linearity hidden in the scheme.

1 Introduction

Two decades after the discovery of public key cryptography, only a few asymmetric encryption schemes exist, and the most practical public key schemes are still very slow compared to conventional secret key schemes. Extensive research has been conducted on public-key cryptography based on the knapsack problem. Knapsack-like cryptosystems are quite interesting: they are easy to implement, can attain very high encrypt/decrypt rates, and do not require expensive operations. Unfortunately, all the cryptosystems based on the additive knapsack problem have been broken, mainly by means of lattice-reduction techniques. Linearity is probably the biggest weakness of these schemes.

To overcome this problem, multiplicative knapsacks have been proposed as an alternative. The idea of multiplicative knapsack is roughly 20 years old and was first proposed in the open literature by Merkle and Hellman [3] in their original paper. Merkle-Hellman's knapsack was (partially) cryptanalyzed by Odlyzko [8], partly because only decryption was actually multiplicative, while encryption was additive.

Recently, two public-key cryptosystems based on the multiplicative knapsack problem have been proposed: the Naccache-Stern cryptosystem [4] presented at Eurocrypt '97, and the Itoh-Okamoto-Mambo cryptosystem [1] presented at SAC '97. In the latter one, both encryption and decryption were relatively

S. Tavares and H. Meijer (Eds.): SAC'98, LNCS 1556, pp. 213–218, 1999.

fast. After analyzing several attacks including lattice-reduction attacks, Itoh, Okamoto and Mambo claimed that the security of their cryptosystem was high.

We present a very effective attack against this cryptosystem. In practice, one can recover the private key from the public key in less than 10 minutes for the suggested choice of parameters. The attack is based on a systematic use of the notion of the orthogonal lattice which we introduced as a cryptographic tool at Crypto '97 [5]. As in [5,7,6], this technique enables us to attack the linearity hidden in the keys generation process.

2   Description of the Cryptosystem

The message space is $\mathbb{Z}_M$, the ring of integers modulo an integer M. Let N be a product of two large primes P and Q. Let l and n be integers such that $l \leq n$. Select positive integers $q_1, \ldots, q_n$ less than $P^{1/l}$ and distinct primes $q_1', \ldots, q_n'$ such that:

- For all i, q_i' divides q_i.
- For all $i \neq j$, q_j' does not divide q_i/q_i'.

Choose an integer t in $\mathbb{Z}_N$ coprime with P, and integers $k_1, \ldots, k_n$ in $\mathbb{Z}_N$ satisfying the following congruence:

$$k_i \equiv tq_i \pmod{P}.$$

Finally, select random elements $e_1, \ldots, e_n$ in $\mathbb{Z}_M$.
    The public key consists of: the (e_i, k_i)'s, M, N, n and l.
    The secret key consists of: P, Q, t, the q_i's and the q_i''s.

2.1   Encryption

Let $s \in \mathbb{Z}_M$ be the plaintext. Alice chooses l integers $i_1, \ldots, i_l$ (not necessarily distinct) in $\{1, \ldots n\}$. The ciphertext is $(m, r) \in \mathbb{Z}_M \times \mathbb{Z}_N$ defined by:

$$m \equiv s + e_{i_1} + e_{i_2} + \cdots + e_{i_l} \pmod{M}$$
$$r \equiv k_{i_1} k_{i_2} \ldots k_{i_l} \pmod{N}$$

2.2   Decryption

Let (m, r) be the ciphertext. First, Bob computes $r' \equiv (t^l)^{-1} r \pmod{P}$, We have:

$$r' \equiv q_{i_1} q_{i_2} \ldots q_{i_l} \pmod{P}.$$

Since each q_i^l is strictly less than P, we actually have:

$$r' = q_{i_1} q_{i_2} \ldots q_{i_l}.$$

Eventually, Bob recovers s as follows:

1. Let $i = 1$.
2. If q_i' divides r', let $m := m - e_i \pmod{M}$ and $r' := r'/q_i$.
3. If $r' = 1$, Bob gets m as a plaintext. Otherwise, increment i and start again at Step 2.

2.3   Parameters

In their paper [1], Itoh, Okamoto and Mambo analyzed several possible attacks, including a lattice-reduction attack. They concluded that their cryptosystem was secure for the following choice of parameters:

- $N = 1024$ bits, $P = 768$ bits and $Q = 256$ bits.
- $n = 180$ and $l = 17$.
- $q_{max} = 2^{45}$ (6 bytes) and $q'_{max} = 2^{32}$ (4 bytes).

In this example, the public key takes 45 Kbytes and the private key takes 1.8 Kbytes. Compared to RSA-1024 with small exponent, encryption speed is similar, but decryption is about 50 times faster.

3   The Orthogonal Lattice

We recall a few useful facts about the notion of an orthogonal lattice, which was introduced in [5] as a cryptographic tool. Let L be a lattice in $\mathbb{Z}^n$ where n is any integer. The orthogonal lattice $L^\perp$ is defined as the set of elements in $\mathbb{Z}^n$ which are orthogonal to all the lattice points of L, with respect to the usual dot product. We define the lattice $\bar{L} = (L^\perp)^\perp$ which contains L and whose determinant divides the one of L. The results of [5] which are of interest to us are the following two theorems:

Theorem 1. *If L is a lattice in $\mathbb{Z}^n$, then* $\dim(L) + \dim(L^\perp) = n$ *and:*

$$\det(L^\perp) = \det(\bar{L}).$$

Thus, $\det(L^\perp)$ divides $\det(L)$. This implies that if L is a low-dimensional lattice in $\mathbb{Z}^n$, then a reduced basis of $L^\perp$ will consist of very short vectors compared to a reduced basis of L. In practice, most of the vectors of any reduced basis of $L^\perp$ are quite short, with norm around $\det(\bar{L})^{1/(n-\dim L)}$.

Theorem 2. *There exists an algorithm which, given as input a basis of a lattice L in $\mathbb{Z}^n$, outputs an LLL-reduced basis of the orthogonal lattice $L^\perp$, and whose running time is polynomial with respect to n, d and the size of the basis elements.*

In practice, one obtains a simple and very effective algorithm (which consists of a single lattice reduction, described in [5]) to compute a reduced basis of the orthogonal lattice, thanks to the celebrated LLL algorithm [2]. This means that, given a low-dimensional L in $\mathbb{Z}^n$, one can easily compute many short and linearly independent vectors in $L^\perp$.

4   Attacking the Scheme by Orthogonal Lattices

Let m be an integer less than n. Define the following vectors in $\mathbb{Z}^m$:

$$\mathbf{k} = (k_1, k_2, \ldots, k_m)$$
$$\mathbf{q} = (q_1, q_2, \ldots, q_m)$$

Note that an attacker knows $\mathbf{k}$, but not $\mathbf{q}$. By construction of the keys, we have the following congruence:

$$\mathbf{k} \equiv t\mathbf{q} \ (\text{mod} \ P).$$

This leads to a simple remark:

Lemma 3. *Let $\mathbf{u} \in \mathbb{Z}^m$. If $\mathbf{u} \perp \mathbf{k}$ then $\mathbf{u} \perp \mathbf{q}$ or $\|\mathbf{u}\| \geq P/\|\mathbf{q}\|$.*

Proof. We have: $t\mathbf{q}.\mathbf{u} \equiv 0 \ (\text{mod} \ P)$. Therefore $\mathbf{q}.\mathbf{u} \equiv 0 \ (\text{mod} \ P)$, and the result follows by Cauchy-Schwarz. □

This remark is interesting because $\|\mathbf{q}\|$ is much smaller than P. Indeed, since each $q_i < P^{1/l}$, we have:

$$\|\mathbf{q}\| < \sqrt{m}P^{1/l}.$$

Therefore, if $\mathbf{u} \in \mathbb{Z}^m$ is orthogonal to $\mathbf{k}$ then it is also orthogonal to $\mathbf{q}$ or satisfies

$$\|\mathbf{u}\| \geq \frac{P^{(l-1)/l}}{\sqrt{m}} \tag{1}$$

which implies that $\mathbf{u}$ is quite long.

Furthermore, from the previous section, one can expect to find many vectors orthogonal to $\mathbf{k}$, with norm around

$$\|\mathbf{k}\|^{1/(m-1)} \leq (P\sqrt{m})^{1/(m-1)}.$$

This quantity is much smaller than the right quantity of (1) when m is large enough, so that we make the following assumption:

Assumption 4. *Let $(\mathbf{b}_1, \mathbf{b}_2, \ldots, \mathbf{b}_{m-1})$ be a reduced basis of $\mathbf{k}^{\perp}$. Then the first $m - 2$ vectors $\mathbf{b}_1, \mathbf{b}_2, \ldots, \mathbf{b}_{m-2}$ are orthogonal to $\mathbf{q}$.*

Actually, one can prove that the first vector of an LLL-reduced basis satisfies the assumption, but this is not enough.

Now assume that the hypothesis holds. Then $\mathbf{q}$ belongs to the 2-dimensional lattice $L = (\mathbf{b}_1, \ldots, \mathbf{b}_{m-2})^{\perp}$. One expects the vectors $\mathbf{b}_1, \ldots, \mathbf{b}_{m-2}$ to have norm around $\|\mathbf{k}\|^{1/(m-1)}$. Therefore, the determinant of L should be around

$$\|\mathbf{k}\|^{(m-2)/(m-1)} \approx \|\mathbf{k}\|.$$

But $\mathbf{q}$ belongs to L and its norm is much smaller than $\|\mathbf{k}\|^{1/2}$. This leads to a more general assumption which is as follows:

Assumption 5. *Let $(\mathbf{b}_1, \mathbf{b}_2, \ldots, \mathbf{b}_{m-1})$ be a reduced basis of $\mathbf{k}^{\perp}$. Then $\mathbf{q}$ is a shortest vector of the 2-dimensional lattice $(\mathbf{b}_1, \mathbf{b}_2, \ldots, \mathbf{b}_{m-2})^{\perp}$.*

If this hypothesis holds, one can use the Gaussian algorithm for lattice reduction (which has worst-case polynomial time and average-case constant time) to recover $\pm\mathbf{q}$.

Next, we easily recover the secret factorization $P \times Q$ using the so-called differential attack described in [1]. More precisely, there exist integers $p_1, \ldots, p_n$ such that:

$$k_i \equiv p_i P + t q_i \pmod{N}.$$

Therefore, we have for all $i \neq j$:

$$q_j k_i - q_i k_j \equiv (p_i q_j - p_j q_i) P \pmod{N}.$$

It is likely that $gcd(q_j k_i - q_i k_j, N)$ is equal to P. And if it is not, we can try again with a different (i, j).

To sum up, the attack is the following:

1. Select an integer $m \leq n$.
2. Compute a reduced basis $(\mathbf{b}_1, \ldots, \mathbf{b}_{m-1})$ of the lattice $\mathbf{k}^\perp$.
3. Compute a reduced basis $(\mathbf{a}_1, \mathbf{a}_2)$ of the lattice $(\mathbf{b}_1, \ldots, \mathbf{b}_{m-2})^\perp$.
4. Compute a shortest vector $\mathbf{s}$ of the previous lattice.
5. Select integers $i \neq j$ in $\{1, \ldots, n\}$ and denote the coordinates of $\mathbf{s}$ by s_i.
6. If $gcd(s_j k_i - s_i k_j, N)$ is not a proper factor of N, restart at previous step.

In practice, we perform Steps 3 and 4 by a single LLL-reduction and take $\mathbf{a}_1$ as $\mathbf{s}$. Only Steps 2 and 3 take a little time. Note that we do not need to compute a complete reduced basis in Step 2 since the last vector is useless.

Once $\mathbf{q}$ and the secret factorization of N are found, it is not a problem to recover the rest of the secret key:

- t modulo P is given by $k_i \equiv t q_i \pmod{P}$.
- The q_i''s (or something equivalent) are revealed by the factors of the q_i's.

5 Experiments

We implemented the attack using the NTL package [9] which includes efficient lattice-reduction algorithms. We used the LLL floating point version with extended exponent to compute orthogonal lattices, since the entries of $\mathbf{k}$ were too large (about the size of N) for the usual floating point version.

In practice, the attack reveals the secret factorization as soon as $m \geq 4$ for the suggested choice of parameters. When $m \leq 20$, the total computation time is less than 10 minutes on a UltraSparc-I clocked at 167 MHz.

6 Conclusion

We showed that the cryptosystem presented by Itoh, Okamoto and Mambo at SAC '97 is not secure. The core of our attack is the notion of the orthogonal lattice which we introduced at Crypto '97, in order to cryptanalyze a knapsack-like cryptosystem proposed by Qu and Vanstone. The attack is very similar to the attack we devised against the so-called Merkle-Hellman transformations. This is because the congruence $\mathbf{k} \equiv t\mathbf{q} \pmod{P}$, which is used in the keys generation process, looks like a Merkle-Hellman transformation: in a Merkle-Hellman equation, we have an equality instead of a congruence.

We suggest that the design of multiplicative knapsack cryptosystems should avoid any kind of linearity. But this might be at the expense of efficiency.

References

1. K. Itoh, E. Okamoto, and M. Mambo. Proposal of a fast public key cryptosystem. In *Proc. of Selected Areas in Cryptography '97*, 1997. Available at http://adonis.ee.queensu.ca:8000/sac/sac97/papers/paper10.ps.
2. A. K. Lenstra, H. W. Lenstra, and L. Lovász. Factoring polynomials with rational coefficients. *Math. Ann.*, 261:515–534, 1982.
3. R. Merkle and M. Hellman. Hiding information and signatures in trapdoor knapsacks. *IEEE Trans. Inform. Theory*, IT-24:525–530, September 1978.
4. D. Naccache and J. Stern. A new public-key cryptosystem. In *Proc. of Eurocrypt '97*, volume 1233 of *LNCS*, pages 27–36. Springer-Verlag, 1997.
5. P. Nguyen and J. Stern. Merkle-Hellman revisited: a cryptanalysis of the Qu-Vanstone cryptosystem based on group factorizations. In *Proc. of Crypto '97*, volume 1294 of *LNCS*, pages 198–212. Springer-Verlag, 1997.
6. P. Nguyen and J. Stern. The Béguin-Quisquater server-aided RSA protocol from Crypto '95 is not secure. In *Proc. of Asiacrypt '98*, LNCS. Springer-Verlag, 1998.
7. P. Nguyen and J. Stern. Cryptanalysis of the Ajtai-Dwork cryptosystem. In *Proc. of Crypto '98*, volume 1462 of *LNCS*, pages 223–242. Springer-Verlag, 1998.
8. A. Odlyzko. Cryptanalytic attacks on the multiplicative knapsack cryptosystem and on Shamir's fast signature scheme. *IEEE Trans. Inform. Theory*, IT-30:594–601, 1984.
9. V. Shoup. Number Theory C++ Library (NTL) version 2.0. Can be obtained at http://www.cs.wisc.edu/~shoup/ntl/.

A Lattice-Based Public-Key Cryptosystem

Jin-Yi Cai* and Thomas W. Cusick

[1] Department of Computer Science
State University of New York at Buffalo, Buffalo, NY 14260
cai@cs.buffalo.edu
[2] Department of Mathematics
State University of New York at Buffalo, Buffalo, NY 14260
cusick@acsu.buffalo.edu

Abstract. Ajtai recently found a random class of lattices of integer points for which he could prove the following worst-case/average-case equivalence result: If there is a probabilistic polynomial time algorithm which finds a short vector in a random lattice from the class, then there is also a probabilistic polynomial time algorithm which solves several problems related to the shortest lattice vector problem (SVP) in any n-dimensional lattice. Ajtai and Dwork then designed a public-key cryptosystem which is provably secure unless the worst case of a version of the SVP can be solved in probabilistic polynomial time. However, their cryptosystem suffers from massive data expansion because it encrypts data bit-by-bit. Here we present a public-key cryptosystem based on similar ideas, but with much less data expansion.

Keywords: Public-key cryptosystem, lattice, cryptographic security.

1 Introduction

Since the origin of the idea of public-key cryptography, there have been many public-key techniques described in the literature. The security of essentially all of these depends on certain widely believed but unproven mathematical hypotheses. For example, the well-known RSA public-key cryptosystem relies on the hypothesis that it is difficult to factor a large integer n which is known to be a product of two large primes. This hypothesis has been extensively studied, but there is still no proof that for a typical such n, the prime factors cannot be found in less than k steps, where k is a very large number. From a computational complexity point of view, we generate a specific instance of a problem in NP (together with a solution, which is kept secret) and we rely on the belief that the problem is difficult to solve.

Apart from the lack of proof that any of these problems is really hard, i.e., there exists no efficient algorithm that will solve the problem in all cases, there is another serious issue. The mathematical hypothesis that these problems are difficult to solve really means difficult to solve in *the worst case*, but the security

---

* Research supported in part by NSF grant CCR-9634665 and an Alfred P. Sloan Fellowship.

S. Tavares and H. Meijer (Eds.): SAC'98, LNCS 1556, pp. 219–233, 1999.
© Springer-Verlag Berlin Heidelberg 1999

of the cryptographic algorithms depends more on the difficulty of *the average case*. For example, even if one day factoring is proved to be unsolvable in probabilistic polynomial time, to the users of the RSA system, there is no guarantee that the key they are actually using is hard to factor. To use these protocols, one must be able to generate specific instances of the problem which should be hard to solve. But typically there is no way to just generate known hard instances. One way to do this is to generate random instances of the problem, and hope that such instances are as hard on the average as in the worst case. However this property is known to be not true for a number of NP-hard problems.

Recently Ajtai [1] proved that certain lattice problems related to SVP have essentially the same average case and worst case complexity, and both are conjectured to be extremely hard. This development raises the possibility of public-key cryptosystems which will have a new level of security. Already Ajtai and Dwork [3] have proposed a public-key cryptosystem which has a provable worst-case/average-case equivalence. Specifically, the Ajtai-Dwork cryptosystem is secure unless the worst case of a certain lattice problem can be solved in probabilistic polynomial time.

Goldreich, Goldwasser and Halevi [11] have also given a public-key cryptosystem which depends on similar lattice problems related to SVP as in [1]. Unlike the work of [3], however, their method uses a trapdoor one-way function and also lacks a proof of worst-case/average-case equivalence.

The cryptosystems of [3] are unfortunately far from being practical. All of them encrypt messages bit-by-bit and involve massive data expansion: the encryption will be at least a hundred times as long as the message. (Note: In a private communication, Ajtai has informed us that this data expansion problem is being addressed by the authors of [3] as well.) In this paper we propose a public-key cryptosystem, based on the ideas of [1] and [3], which has much less data expansion. Messages are encrypted in blocks instead of bit-by-bit. We offer some statistical analysis of our cryptosystem. We also analyze several attacks on the system and show that the system is secure against these attacks. Whether there is a provable worst-case/average-case equivalence for this system is open.

2   Lattice problems with worst-case/average-case equivalence

Here we briefly define the terms for lattice problems, and describe the results of Ajtai [1] and some improvements.

Notation. $\mathbf{R}$ is the field of real numbers, $\mathbf{Z}$ is the ring of integers, $\mathbf{R}^n$ is the space of n-dimensional real vectors $a = \langle a_1, \ldots, a_n \rangle$ with the usual dot product $a \cdot b$ and Euclidean norm or length $\|a\| = (a \cdot a)^{1/2}$. $\mathbf{Z}^n$ is the set of vectors in $\mathbf{R}^n$ with integer coordinates, $\mathbf{Z}^+$ is the positive integers and $\mathbf{Z}_q$ is the ring of integers modulo q.

Definition. If $A = \{a_1, \ldots, a_n\}$ is a set of linearly independent vectors in $\mathbf{R}^n$, then we say that the set of vectors

$$\{\textstyle\sum_{i=1}^n k_i a_i : \ k_1, \ldots, k_n \in \mathbf{Z}\}$$

is a lattice in $\mathbf{R}^n$. We will denote the lattice by $L(A)$ or $L(a_1, \ldots, a_n)$. We call A a basis of the lattice. We say that a set in $\mathbf{R}^n$ is an n-dimensional lattice if there is a basis V of n linearly independent vectors such that $L = L(V)$. If $A = \{a_1, \ldots, a_n\}$ is a set of vectors in a lattice L, then we define the length of the set A by $\max_{i=1}^n \|a_i\|$. $\lambda_1(L) = \min_{0 \neq v \in L} \|v\|$.

A fundamental theorem of Minkowski is the following:

Theorem 1 (Minkowski). *There is a universal constant γ, such that for any lattice L of dimension n, $\exists v \in L$, $v \neq 0$, such that*

$$\|v\| \leq \gamma \sqrt{n} \det(L)^{1/n}.$$

The determinant $\det(L)$ of a lattice is the volume of the n-dimensional fundamental parallelepiped, and the absolute constant γ is known as Hermite's constant.

Minkowski's theorem is a pure existence type theorem; it offers no clue as to how to find a short or shortest non-zero vector in a high dimensional lattice. To find the shortest non-zero vector in an n-dimensional lattice, given in terms of a basis, is known as the Shortest Vector Problem (SVP). There are no known efficient algorithms for finding the shortest non-zero vector in the lattice. Nor are there efficient algorithms to find an approximate short non-zero vector, or just to approximate its length, within any fixed polynomial factor in its dimension n. This is still true even if the shortest non-zero vector v is unique in the sense that any other vector in the lattice whose length is at most $n^c \|v\|$ is parallel to v, where c is an absolute constant. In this case we say that v is unique up to a polynomial factor.

The best algorithm to date for finding a short vector in an arbitrary lattice in $\mathbf{R}^n$ is the L^3 algorithm of A.K. Lenstra, H.W. Lenstra and L. Lovász [14]. This algorithm finds in deterministic polynomial time a vector which differs from the shortest one by at most a factor $2^{(n-1)/2}$. C.P. Schnorr [16] proved that the factor can be replaced by $(1+\epsilon)^n$ for any fixed $\epsilon > 0$. However Schnorr's algorithm has a running time with $1/\epsilon$ in the exponent.

Regarding computational complexity, Ajtai [2] proved that it is NP-hard to find the shortest lattice vector in Euclidean norm, as well as approximating the shortest vector length up to a factor of $1 + \frac{1}{2^{n^k}}$. In a forthcoming paper [6], Cai and Nerurkar improve the NP-hardness result of Ajtai [2] to show that the problem of approximating the shortest vector length up to a factor of $1 + \frac{1}{n^\epsilon}$, for any $\epsilon > 0$, is also NP-hard. This improvement also works for all l_p-norms, for $1 \leq p < \infty$. Prior to that, it was known that the shortest lattice vector problem is NP-hard for the l_∞-norm, and the nearest lattice vector problem is NP-hard under all l_p-norms, $p \geq 1$ [12,17]. Even finding an approximate solution to within any constant factor for the nearest vector problem for any l_p-norm is NP-hard [4]. On the other hand, Lagarias, Lenstra and Schnorr [13] showed that the approximation problem (in l_2-norm) within a factor of $O(n)$ *cannot* be NP-hard, unless NP = coNP. Goldreich and Goldwasser showed that approximating the shortest lattice vector within a factor of $O(\sqrt{n/\log n})$ is not NP-hard assuming the polynomial time hierarchy does not collapse [9]. Cai showed that finding an

$n^{1/4}$-unique shortest lattice vector is not NP-hard unless the polynomial time hierarchy collapses [7].

What is most striking is a recent result of Ajtai [1] establishing the first explicit connection between the worst-case and the average-case complexity of the problem of finding the shortest lattice vector or approximating its length. The connection factor in the Ajtai connection has been improved in [5]. Ajtai defined a class of lattices in $\mathbf{Z}^m$ so that if there is a probabilistic polynomial time algorithm which finds a short vector in a random lattice from the class with a probability of at least $1/n^{O(1)}$, then there is also a probabilistic polynomial time algorithm which solves the following three lattice problems in *every* lattice in $\mathbf{Z}^n$ with a probability exponentially close to 1:

(P1) Find the length of a shortest non-zero vector in an n-dimensional lattice, up to a polynomial factor.

(P2) Find the shortest non-zero vector in an n-dimensional lattice where the shortest vector is unique up to a polynomial factor.

(P3) Find a basis in an n-dimensional lattice whose length is the smallest possible, up to a polynomial factor.

The lattices in the random class are defined modulo q (q is an integer depending only on n, as described below), that is, if two integer vectors are congruent modulo q then either both of them or neither of them belong to the lattice. More precisely, if $\nu = \{u_1, \ldots, u_m\}$ is a given set of vectors in $\mathbf{Z}_q^n$ then the lattice $\Lambda(\nu)$ is the set of all integer vectors $\langle h_1, \ldots, h_m \rangle$ so that

$$\sum_{i=1}^m h_i u_i \equiv 0 \pmod{q}.$$

For a fixed n, m and q, the probability distribution over the random class is defined by uniformly choosing a sequence of integer vectors $\langle u_1, \ldots, u_m \rangle$.

For a given n, the parameters m and q are defined by $m = \lceil c_1 n \rceil$ and $q = \lceil n^{c_2} \rceil$, where c_1 and c_2 are suitable constants.

The problem of finding a short vector in a lattice from the random class is a Diophantine problem. Questions of this type date back to Dirichlet's 1842 theorem on simultaneous Diophantine approximation. From this point of view the problem can be stated in the following way, which does not involve any explicit mention of lattices, as pointed out in [10].

(A1) Given $n, m = \lceil c_1 n \rceil, q = \lceil n^{c_2} \rceil$ and an n by m matrix M with entries in $\mathbf{Z}_q$, find a non-zero vector x so that $Mx \equiv 0 \pmod{q}$ and $\|x\| < n$.

Minkowski's theorem guarantees the existence of such short vectors x. Of course if the condition on $\|x\|$ is removed, then the linear system $Mx \equiv 0 \pmod{q}$ can be solved in polynomial time.

The theorem in [1] reduces the worst-case complexity of each of the problems **(P1)**, **(P2)**, **(P3)** to the average case complexity of **(A1)**. Currently the best bounds that can be achieved are stated below [5], [8]:

Theorem 2. *[5] For any constant $\epsilon > 0$, if there exists a probabilistic polynomial time algorithm $\mathcal{A}$ such that, for a given random lattice $\Lambda(\nu)$, where*

$\nu = (u_1, \ldots, u_m) \in \mathbf{Z}_q^{n \times m}$ is uniformly chosen, $q = \Theta(n^3)$ and $m = \Theta(n)$, $\mathcal{A}$ will find a vector of the lattice $\Lambda(\nu)$ of length $\leq n$ with probability $\frac{1}{n^{O(1)}}$, then, there also exists a probabilistic polynomial time algorithm $\mathcal{B}$ which for any given lattice $L = L(a_1, \ldots, a_n)$ by a basis $a_1, \ldots, a_n \in \mathbf{Z}^n$, outputs another basis for L, $b_1, \ldots, b_n$, so that,

$$\max_{i=1}^{n} \|b_i\| \leq \Theta(n^{3.5+\epsilon}) \min_{\text{all bases } b_1', \ldots, b_n' \text{ for } L} \max_{i=1}^{n} \|b_i'\|.$$

Theorem 3. *[8] Under the same hypothesis, there exists a probabilistic polynomial time algorithm $\mathcal{C}$ which for any given lattice $L = L(a_1, \ldots, a_n)$ by a basis will*

- *compute an estimate of $\lambda_1 = \lambda_1(L)$ up to a factor $n^{4+\epsilon}$, i.e., compute a numerical estimate $\tilde{\lambda}_1$, such that*

$$\frac{\lambda_1}{n^{4+\epsilon}} \leq \tilde{\lambda}_1 \leq \lambda_1;$$

- *find the unique shortest vector if it is an $n^{4+\epsilon}$-unique shortest vector.*

3   A new cryptosystem

Here we present the design of a new cryptosystem, which is based on the difficulty of finding or approximating SVP, even though no specific lattices are defined. The secret key in the new system is a vector u chosen with uniform distribution from the unit sphere $S^{n-1} = \{x \mid \|x\| = 1\}$, and a random permutation σ on $m+1$ letters. By allowing an exponentially small round-off error, we may assume that the coordinates of u are rational numbers whose denominators are bounded by some very large integer, exponential in n. Let $m = [cn]$ for a suitable absolute constant $c < 1$. For definiteness set $c = 1/2$. Let $H_i = \{v : v \cdot u = i\}$ denote the hyperplanes perpendicular to u. The public key in the new system is a parameter $b > 0$ and a set $\{v_{\sigma(0)}, \ldots, v_{\sigma(m)}\}$ of rational vectors, where each v_j is in one of the hyperplanes H_i for some $i \in \mathbf{Z}^+$, say $v_j \cdot u = N_j \in \mathbf{Z}^+$. We choose a sequence of numbers N_j so that it is superincreasing, that is

$$N_0 > b \text{ and } N_i > \sum_{j=0}^{i-1} N_i + b \text{ for each } i = 1, 2, \ldots, m.$$

Binary plaintext is encrypted in blocks of $m + 1$ bits. If $P = (\delta_0, \ldots, \delta_m)$ is a plaintext block ($\delta_i = 0$ or 1), then P is encrypted as a random perturbation of $\sum_{i=0}^{m} \delta_i v_{\sigma(i)}$. More precisely, the sender picks a uniformly chosen random vector r with $\|r\| \leq b/2$. Then the ciphertext is

$$\sum_{i=0}^{m} \delta_i v_{\sigma(i)} + r.$$

Decryption is accomplished by using the secret key u to compute the following inner product

$$
\begin{aligned}
S &= u \cdot \left(\sum_{i=0}^{m} \delta_i v_{\sigma(i)} + r \right) \\
&= \sum_{i=0}^{m} \delta_i (u \cdot v_{\sigma(i)}) + u \cdot r \\
&= \sum_{i=0}^{m} \delta_i N_{\sigma(i)} + u \cdot r \\
&= \sum_{i=0}^{m} \delta_{\sigma^{-1}(i)} N_i + u \cdot r.
\end{aligned}
$$

Since the N_i are superincreasing, we can use the greedy algorithm to efficiently recover the $\delta_{\sigma^{-1}(i)}$ from S, and then use the secret σ to recover δ_i. More precisely, if $\delta_{\sigma^{-1}(m)} = 1$, then $S \geq N_m - b/2$, and if $\delta_{\sigma^{-1}(m)} = 0$, then $S \leq N_0 + N_1 + \ldots + N_{m-1} + b/2$. Since $N_m > \sum_{i=0}^{m-1} N_i + b$, with the secret key one can discover whether $\delta_{\sigma^{-1}(m)} = 0$ or 1. Substituting S by $S' = S - \delta_{\sigma^{-1}(m)} N_m$, this process can be continued until all $\delta_{\sigma^{-1}(i)}$ are recovered. Then using the secret permutation σ, one recovers $\delta_0, \delta_1, \ldots, \delta_m$.

Thus decryption using u and σ involves an easy instance of a knapsack problem. As summarized in the article of Odlyzko [15], essentially all suggestions for cryptosystems based on knapsack problems have been broken. Here, however, the easy knapsack problem appears to have no bearing on the security of the system, since it appears that one must first search for the direction u.

The new cryptosystem has similarities with the third version of the Ajtai-Dwork cryptosystem (see [3]), but in the new system $m + 1 = O(n)$ bits of plaintext are encrypted to an n-dimensional ciphertext vector, instead of just one bit of plaintext.

We have not specified the distribution of the v_i, aside from its inner product with u being superincreasing. The following distribution has a strong statistical indistinguishability from $m + 1$ independent uniform samples of the sphere. Let M be a large integer, say, $M \gg 2^n$. Choose any $b' > b$. For analysis purposes we will normalize by denoting v_i/M as v_i. For each i, $0 \leq i \leq m$, let $v_i = \frac{2^i b'}{M} u + \sqrt{1 - \frac{2^{2i} b'^2}{M^2}} \rho_i$, where the ρ_i's are independently and uniformly distributed on the $(n - 2)$-dimensional unit sphere orthogonal to u. Note that each $\|v_i\| = 1$, after normalization. We denote this distribution by D. We note that

$$
u \cdot v_i - u \cdot \left(\sum_{j=0}^{i-1} v_j \right) = \frac{2^i b'}{M} > \sum_{j=0}^{i-1} \frac{2^j b'}{M} = \frac{b'}{M} > \frac{b}{M}.
$$

How secure is this new cryptosystem? We do not have a proof of worst-case/average-case equivalence. We can discuss several ideas for attacks that do not seem to work. The following discussion will also explain some of the choices made in the design of the cryptosystem.

We will first show that if we did not employ the random permutation σ, rather we publish as public key the unpermuted vectors $v_0, \ldots, v_m$, then there is an attack based on linear programming that will break the system in polynomial time.

The attack works as follows: From the given vectors $v_0, \ldots, v_m$ we are assured that the following set of inequalities defines a non-empty convex body containing the secret vector u.

$$v_0 \cdot x > b$$
$$v_1 \cdot x > v_0 \cdot x + b$$
$$v_2 \cdot x > (v_0 + v_1) \cdot x + b$$
$$\vdots \quad \vdots \quad \vdots$$
$$v_m \cdot x > (v_0 + v_1 + \cdots + v_{m-1}) \cdot x + b$$

Using linear programming to find a *feasible* solution to this convex body, we can compute in polynomial time a vector $\tilde{u}$ satisfying all the inequalities. Even though $\tilde{u}$ may not be equal to u, as along as $\tilde{u}$ satisfies the above set of inequalities, it is as good as u itself to decrypt the message $\sum_{i=0}^{m} \delta_i v_i + r$. Hence, the permutation σ is essential to the security of the protocol.

Next, let's consider the addition of the random perturbation r. This is to guard against an attack based on linear algebra, which works as follows.

Assume the message $w = \sum_{i=0}^{m} \delta_i v_{\sigma(i)}$ were sent without the perturbation vector r. Then this vector is in the linear span of $\{v_{\sigma(0)}, v_{\sigma(1)}, \ldots, v_{\sigma(m)}\}$, which is most likely to be linearly independent, by the way these v_i's are chosen. Then one can solve for the $m + 1 < n$ coefficients x_i in $w = \sum_{i=0}^{m} x_i v_{\sigma(i)}$. These coefficients are unique by linear independence, thus $x_i = \delta_i$, and we recover the plaintext.

The addition of the random perturbation r renders this attack ineffective, since with probability very near one, $w = \sum_{i=0}^{m} \delta_i v_{\sigma(i)} + r$ is not in the linear span of $\{v_{\sigma(0)}, v_{\sigma(1)}, \ldots, v_{\sigma(m)}\}$, which is of dimension at most $m + 1$. (If r were truly uniformly random from the ball $\|x\| \leq b/2$, then the probability that w belongs to the lower dimensional linear span is zero; if r is chosen with rational coordinates with exponentially large denominator, then this probability is exponentially small.) When the vector w is not in the linear span, to recover the coefficients δ_i appears to be no easier than the well known nearest lattice vector problem, which is believed to be intractable.

Finally if the lengths of the vectors v_i are not kept essentially the same, there can be statistical leakage of information (see Section 4). However, suppose the v_i are all roughly the same length, then the number of message bits $m = cn$ should be less than n. If m were equal to cn, for a constant $c > 1$, then there is the following cryptanalytic attack.

Suppose $\|v_i\| \approx V$ for each i and define numbers Q_i by

$$Q_i = \frac{|u \cdot v_i|}{\|u\|\|v_i\|} \approx \frac{N_i}{\|u\|V}.$$

Since the integers N_i are superincreasing, we can show that for all $i < m-3\log n$, $Q_i < Q_m/n^3$. In fact, let $m' = m - 3\log n$, then we can inductively prove that $Q_{m'+j} > 2^j Q_{m'-1} \geq 2^j Q_i$ for all $i < m'$. Thus for each $i < m'$ we have $Q_i < Q_m/n^3$. We will say that these Q_i are "unusually small" (compared to $\max Q_j$). Of course one cannot compute the Q_i's since one is given only the permuted ordering by σ and u is secret.

The attack begins with the selection of a random subset of $n - 1$ vectors v_i. If we get all $n - 1$ vectors having an unusually small dot product with the secret vector u, then the normal vector perpendicular to all these $n - 1$ vectors will be a good approximation to u. From this one can break the system. We show next that with non-trivial probability $1/n^{O(1)}$, all $n - 1$ vectors have an unusually small dot product with the secret vector u. This is at least

$$\frac{\binom{cn-3\log n}{n-1}}{\binom{cn}{n-1}} \geq \left(1 - \frac{3\log n}{cn - n}\right)^n \approx n^{-\frac{3}{c-1}}.$$

Thus one can try for a polynomial number of times, and with high probability one will find such a set of $n - 1$ vectors and break the system. This attack will not work if $m = n/2$.

4   Statistical analysis

It is clear from the discussion that the secret permutation σ as well as the random perturbation r are both necessary. With a secret permutation σ, however, an adversary may still attempt to find or approximate the secret vector u. In this section, the random perturbation r does not play an essential role in the analysis; it is easier to discard r in the following analysis, which is essentially the same with r, although a little less clean. Thus we will carry out the following analysis with $b = 0$ and $r = 0$.

A natural attack is to gather statistical information by computing some values associated with the vectors $v_{\sigma(0)}, \ldots, v_{\sigma(m)}$ which are invariant under the permutations. It is conceivable, for example, that $\sum_{i=0}^m v_i = \sum_{i=0}^m v_{\sigma(i)}$ might have a non-trivial correlation with the secret direction u since each v_i has a positive component in the u direction. We will show that if we did not choose our distribution for the v_i's carefully, then indeed this attack may succeed; but the distribution D appears to be secure against this attack.

Consider again the structural requirement that $v_0 \cdot u > 0$, and $v_i \cdot u > (v_0 + \cdots + v_{i-1}) \cdot u$. A natural distribution for the v_i's is to choose increment vectors w_i so that $v_i = (v_0 + \cdots + v_{i-1}) + w_i$, where w_i are independently and uniformly distributed on the $(n - 1)$-dimensional hemisphere $S_+^{n-1} = \{x \in$

$\mathbf{R}^n \mid ||x|| = 1, x \cdot u > 0\}$, which consists of all the unit vectors in $\mathbf{R}^n$ in the u direction. We will call this distribution F.

Let $s_i = v_0 + \cdots + v_i$, $0 \le i \le m$. Then $v_0 = w_0$, $v_i = s_{i-1} + w_i$, and $s_i = 2^i w_0 + \cdots 2^0 w_i$ by an easy induction. We need some preliminaries. Let β_n denote the n-dimensional volume of the unit n-ball, let δ_{n-1} denote the $(n-1)$-dimensional volume of the unit $(n-1)$-sphere, then

$$\beta_n = \int_0^1 \delta_{n-1} r^{n-1} dr = \frac{\delta_{n-1}}{n}, \quad n \ge 2.$$

And

$$\beta_n = \int_{-1}^1 \beta_{n-1}(\sqrt{1-h^2})^{n-1} dh = 2\beta_{n-1} I_n,$$

where the integral $I_n = \int_0^{\frac{\pi}{2}} \sin^n \theta d\theta = \frac{n-1}{n} I_{n-2} = \cdots = \frac{\sqrt{\pi}}{2} \frac{\Gamma\left(\frac{n+1}{2}\right)}{\Gamma\left(\frac{n+2}{2}\right)}$. $I_n \approx \sqrt{\frac{\pi}{2n}}$ asymptotically for large n. Also

$$\beta_n = \int_0^{2\pi} \int_0^1 \beta_{n-2}(\sqrt{1-h^2})^{n-2} r dr d\theta = \beta_{n-2} \frac{2\pi}{n} = \frac{\pi^{n/2}}{\Gamma\left(\frac{n+2}{2}\right)}.$$

We will use the uniform distribution U on sets such as the hemisphere S_+^{n-1}, namely the Lebesgue measure on S_+^{n-1}, and we will denote a random variable X uniformly distributed on such a set S by $X \in_U S$. The following analysis is carried out using the exact Lebesgue measure. In the actual cryptographic protocols, this must be replaced by an exponentially close approximation on the set of rational points with exponentially large (polynomially bounded in binary length) denominators. The errors are exponentially small and thus insignificant. For clarity of presentation, we will state all results in terms of the exact Lebesgue measure.

Lemma 1. *Let $w, w' \in_U S_+^{n-1}$ be independently and uniformly distributed on the unit (northern) hemisphere. Let u be the north pole. Then the expectation of the inner product*

$$\mathbf{E}[w \cdot u] = \frac{1}{(n-1)I_{n-2}} = \frac{2}{(n-1)\sqrt{\pi}} \frac{\Gamma\left(\frac{n}{2}\right)}{\Gamma\left(\frac{n-1}{2}\right)} \approx \sqrt{\frac{2}{\pi n}}.$$

Also,

$$\mathbf{E}[(w \cdot u)^2] = \frac{1}{n}.$$

$$\mathbf{E}[(w \cdot u)(w' \cdot u)] = (\mathbf{E}[w \cdot u])^2 \approx \frac{2}{\pi n}.$$

Proof For $w \in_U S_+^{n-1}$, the density function for the value of the inner product $h = w \cdot u$ is

$$p_{n-1}(h) = (\sqrt{1-h^2})^{n-3}/I_{n-2}.$$

Hence

$$\mathbf{E}[w \cdot u] = \int_0^1 h p_{n-1}(h)dh = \frac{1}{(n-1)I_{n-2}} = \frac{2}{(n-1)\sqrt{\pi}} \frac{\Gamma\left(\frac{n}{2}\right)}{\Gamma\left(\frac{n-1}{2}\right)} \approx \sqrt{\frac{2}{\pi n}}.$$

Similarly

$$\mathbf{E}[(w \cdot u)^2] = \int_0^1 h^2 p_{n-1}(h)dh = \frac{1}{n}.$$

We note in passing that $\mathbf{E}_{S^{n-1}}[(w \cdot u)^2]$ over the whole unit sphere S^{n-1} is $1/n$ as well, by symmetry $h \to -h$.

The last equality follows from independence of w and w',

$$\mathbf{E}[(w \cdot u)(w' \cdot u)] = \mathbf{E}[hh'] = \mathbf{E}[h]\mathbf{E}[h'] = (\mathbf{E}[h])^2 \approx \frac{2}{\pi n}.$$

$\square$

Lemma 2. *Let $w, w', w'' \in_U S_+^{n-1}$ be independently and uniformly distributed on the unit hemisphere. Then*

$$\mathbf{E}[w \cdot w'] = (\mathbf{E}[w \cdot u])^2 \approx \frac{2}{\pi n}.$$

$$\mathbf{E}[(w \cdot w')^2] = \frac{1}{n}.$$

$$\mathbf{E}[(w \cdot w')(w \cdot w'')] = (\mathbf{E}[w \cdot u])^2 \mathbf{E}[(w \cdot u)^2] \approx \frac{2}{\pi n^2}.$$

Proof Let $w, w', w'' \in_U S_+^{n-1}$. Choose a coordinate system so that u is the nth-coordinate. Then $w \cdot w' = \sum_{i=1}^n x_i(w)y_i(w')$. By linearity and independence $\mathbf{E}[w \cdot w'] = \sum_{i=1}^n \mathbf{E}[x_i]\mathbf{E}[y_i]$. For $i < n$, the symmetry $x_i \to -x_i$ implies that $\mathbf{E}[x_i] = 0$. For $i = n$, $x_n(w) = w \cdot u$, and similarly for $y_n(w')$. Then it follows that $\mathbf{E}[x_n] = \mathbf{E}[y_n] = \mathbf{E}[h]$, and

$$\mathbf{E}[w \cdot w'] = (\mathbf{E}[h])^2 \approx \frac{2}{\pi n}.$$

For $\mathbf{E}[(w \cdot w')^2]$, expand $(\sum_{i=1}^n x_i y_i)^2 = \sum_{i=1}^n x_i^2 y_i^2 + \sum_{1 \le i \ne j \le n} x_i y_i x_j y_j$. For $i \ne j$, at least one of i or j is not n, and by independence and the symmetry

$x_i \to -x_i$, we have $\mathbf{E}[x_i y_i x_j y_j] = \mathbf{E}[x_i x_j]\mathbf{E}[y_i y_j] = 0$. Thus the expectation of the second term is 0. By linearity and independence

$$\mathbf{E}[(w \cdot w')^2] = \sum_{i=1}^{n} \mathbf{E}[x_i^2]\mathbf{E}[y_i^2].$$

For $i = n$, it is $(\mathbf{E}[h^2])^2 = 1/n^2$. For $i < n$, by the symmetry $x_n \to -x_n$, it can be seen that $\mathbf{E}[x_i^2]$ is the same if we were to evaluate this expectation on the uniform distribution on the whole unit sphere. But on the whole sphere this is the same as $\mathbf{E}_{S^{n-1}}[x_n^2] = \mathbf{E}_{S^{n-1}}[h^2]$. This, however, by the symmetry $h \to -h$, is the same again if we were to evaluate it back on the hemisphere S_+^{n-1}. Hence ultimately $\mathbf{E}[x_i^2] = \mathbf{E}[h^2] = 1/n$, and $\mathbf{E}[x_i^2]\mathbf{E}[y_i^2] = 1/n^2$. It follows that

$$\mathbf{E}[(w \cdot w')^2] = 1/n.$$

Finally for $\mathbf{E}[(w \cdot w')(w \cdot w'')]$, we expand the product $(\sum_{i=1}^{n} x_i y_i)(\sum_{j=1}^{n} x_j z_j)$ $= \sum_{i=1}^{n} x_i^2 y_i z_i + \sum_{1 \leq i \neq j \leq n} x_i y_i x_j z_j$. For $i \neq j$, at least one of them is not n, so that either $\mathbf{E}[y_i] = 0$ or $\mathbf{E}[z_j] = 0$, thus $\sum_{1 \leq i \neq j \leq n} \mathbf{E}[x_i^2]\mathbf{E}[y_i]\mathbf{E}[z_j] = 0$. Then

$$\mathbf{E}[(w \cdot w')(w \cdot w'')] = \sum_{i=1}^{n} \mathbf{E}[x_i^2]\mathbf{E}[y_i]\mathbf{E}[z_i].$$

For $i < n$, $\mathbf{E}[y_i] = 0$ by symmetry as before. For $i = n$, it is $\mathbf{E}[h^2](\mathbf{E}[h])^2 \approx \frac{2}{\pi n^2}$.
□

For the distribution F, we will show that the secret information u is not safe. In fact we claim that s_m can be used to approximate the direction u. Consider $s_m \cdot u = 2^m(w_0 \cdot u) + \cdots + 2^0(w_m \cdot u)$.

$$\mathbf{E}_F[s_m \cdot u] = (2^m + \cdots + 2^0)\mathbf{E}[w \cdot u] \approx 2^{m+1}\sqrt{\frac{2}{\pi n}}.$$

Next we compute the variance $\mathbf{Var}_F[s_m \cdot u]$. First $(s_m \cdot u)^2 = 2^{2m}(w_0 \cdot u)^2 + \cdots + 2^0(w_m \cdot u)^2 + \sum_{0 \leq i \neq j \leq m} 2^{m-i} 2^{m-j}(w_i \cdot u)(w_j \cdot u)$. So

$$\mathbf{E}_F[(s_m \cdot u)^2] = (2^{2m} + \cdots + 2^0)\mathbf{E}[(w \cdot u)^2] + \sum_{0 \leq i \neq j \leq m} 2^{i+j}\mathbf{E}[(w \cdot u)(w' \cdot u)]$$

$$= \frac{4^{m+1} - 1}{3n} + \left[\sum_{0 \leq i,j \leq m} 2^{i+j} - \sum_{0 \leq i \leq m} 2^{2i}\right](\mathbf{E}[w \cdot u])^2$$

$$\approx \frac{4^{m+1}}{3n} + \frac{4^{m+2}}{3\pi n} = \frac{4^{m+1}}{3\pi n}(4 + \pi).$$

It follows that

$$\mathbf{Var}_F[s_m \cdot u] \approx \frac{4^{m+1}}{3\pi n}(\pi - 2).$$

We note that the normalized ratio

$$\frac{\mathbf{E}_F[s_m \cdot u]}{\sqrt{\mathbf{Var}_F[s_m \cdot u]}} \approx \sqrt{\frac{6}{\pi - 2}} \approx 2.2925564.$$

This indicates that s_m has a significant correlation with the hidden direction u, and hence u can not be considered secure under this distribution F.

More directly it can be shown that

$$\frac{\mathbf{E}_F[s_m \cdot (v_m - v_{m-1})]}{\mathbf{E}_F[||s_m||]\mathbf{E}_F[||v_m - v_{m-1}||]} \geq 1$$

asymptotically. Thus one can expect s_m to be used to distinguish v_m from the others.

We now return to our chosen distribution D, and show that in this distribution there is no easy statistical leakage. In this distribution, $v_i = \frac{2^i}{M}u + \sqrt{1 - \frac{2^{2i}}{M^2}}\rho_i$, and ρ_i are independently and uniformly distributed on the $(n-2)$-dimensional unit sphere orthogonal to u. Recall $||v_i|| = 1$. Let $s_i' = v_0 + \ldots + v_i$. We consider $||s_m'||^2$ and $s_m' \cdot u$. Clearly $||s_i'||^2 = (m+1) + \sum_{0 \leq i \neq j \leq m}(v_i \cdot v_j)$.

Lemma 3. *For $0 \leq i \neq j \leq m$,*

$$\mathbf{E}_D[v_i \cdot v_j] = \frac{2^{i+j}}{M^2}, \quad and \quad \mathbf{Var}_D[v_i \cdot v_j] = \frac{1}{n-1}\left(1 - \frac{2^{2i}}{M^2}\right)\left(1 - \frac{2^{2j}}{M^2}\right).$$

Proof We have $v_i \cdot v_j = \frac{2^{i+j}}{M^2} + \sqrt{1 - \frac{2^{2i}}{M^2}}\sqrt{1 - \frac{2^{2j}}{M^2}}(\rho_i \cdot \rho_j)$. By symmetry, $\mathbf{E}_{S^{n-2}}[\rho_i \cdot \rho_j] = 0$, so that $\mathbf{E}_D[v_i \cdot v_j] = \frac{2^{i+j}}{M^2}$. Thus,

$$\mathbf{Var}_D[v_i \cdot v_j] = \left(1 - \frac{2^{2i}}{M^2}\right)\left(1 - \frac{2^{2j}}{M^2}\right)\mathbf{Var}_D[\rho_i \cdot \rho_j].$$

We have $\mathbf{Var}_D[\rho_i \cdot \rho_j] = \mathbf{E}_D[(\rho_i \cdot \rho_j)^2] = \int_{-1}^1 h^2(p_{n-2}(h)/2)dh = \frac{1}{n-1}$. The lemma follows.   □

Now

$$\mathbf{E}_D[||s_m'||^2] = (m+1) + \sum_{0 \leq i \neq j \leq m} \frac{2^{i+j}}{M^2},$$

and $\sum_{0 \leq i \neq j \leq m} 2^{i+j} = \sum_{0 \leq i,j \leq m} 2^{i+j} - \sum_{i=0}^m 2^{2i} \approx \frac{8}{3}(2^m/M)^2$. Hence $\mathbf{E}_D[||s_m'||^2] \approx (m+1) + \frac{2^{2m+3}}{3}$. Ignoring the exponentially small term $2^{2m}/M^2$, $\mathbf{E}_D[||s_m'||^2] \approx m + 1$.

One should compare this with the uniform distribution U for which all v_i's are independently and uniformly distributed on S^{n-1}. In this case $\mathbf{E}_U[v_i \cdot v_j] = 0$, for $i \neq j$, and $\mathbf{E}_U[||s_m'||^2] = m + 1$.

We next evaluate the variance $\mathbf{Var}_D[||s'_m||^2]$.

$$\mathbf{Var}_D[||s'_m||^2] = \mathbf{E}_D\left[\left\{\sum_{0\le i\ne j\le m}(v_i\cdot v_j - E_D[v_i\cdot v_j])\right\}^2\right]$$

$$= 4\mathbf{E}_D\left[\left\{\sum_{0\le i<j\le m}\sqrt{1-\frac{2^{2i}}{M^2}}\sqrt{1-\frac{2^{2j}}{M^2}}(\rho_i\cdot\rho_j)\right\}^2\right]$$

$$= 4\mathbf{E}_D\left[\sum_{(i<j)}\left(1-\frac{2^{2i}}{M^2}\right)\left(1-\frac{2^{2j}}{M^2}\right)(\rho_i\cdot\rho_j)^2\right]$$

$$+ 4\mathbf{E}_D\left[\sum_{(i<j)\ne(i'<j')}c_{(ij)}c_{(i'j')}(\rho_i\cdot\rho_j)(\rho_{i'}\cdot\rho_{j'})\right].$$

For $(i < j) \ne (i' < j')$, there are two cases. If (i, j, i', j') are all distinct indices, then clearly $\rho_i\cdot\rho_j$ and $\rho_{i'}\cdot\rho_{j'}$ are independent. Thus $\mathbf{E}_D[(\rho_i\cdot\rho_j)(\rho_{i'}\cdot\rho_{j'})] = \mathbf{E}_D[\rho_i\cdot\rho_j]\mathbf{E}_D[\rho_{i'}\cdot\rho_{j'}] = 0$. If there are only 3 distinct indices among (i, j, i', j'), say $i = i'$, then by fixing ρ_i, the conditional distribution of $\rho_i\cdot\rho_j$ and $\rho_i\cdot\rho_{j'}$ over ρ_j and $\rho_{j'}$ are independent, and $\mathbf{E}_D[(\rho_i\cdot\rho_j)|\rho_i] = \mathbf{E}_D[(\rho_i\cdot\rho_{j'})|\rho_i] = 0$. Thus in any case $\mathbf{E}_D[(\rho_i\cdot\rho_j)(\rho_{i'}\cdot\rho_{j'})] = 0$, for $(i < j) \ne (i' < j')$, and

$$\mathbf{Var}_D[||s'_m||^2] = 4\sum_{0\le i<j\le m}\left(1-\frac{2^{2i}}{M^2}\right)\left(1-\frac{2^{2j}}{M^2}\right)\mathbf{E}_D\left[(\rho_i\cdot\rho_j)^2\right].$$

We have $\mathbf{E}_D\left[(\rho_i\cdot\rho_j)^2\right] = \mathbf{E}_{S^{n-2}}[h^2] = 1/(n-1)$. Ignoring exponentially small terms such as $2^m/M$, we get

$$\mathbf{Var}_D[||s'_m||^2] \approx \frac{4}{n-1}\binom{m+1}{2}.$$

This is to be compared to the uniform distribution U. Again $||s'_i||^2 = (m+1) + \sum_{0\le i\ne j\le m}(v_i\cdot v_j)$. But for the uniform distribution U, $\mathbf{E}_U[v_i\cdot v_j] = 0$, for $i \ne j$, and so

$$\mathbf{Var}_U[||s'_m||^2] = \mathbf{E}_U\left[\left\{\sum_{0\le i\ne j\le m}(v_i\cdot v_j)\right\}^2\right]$$

$$= 4\mathbf{E}_U\left[\sum_{0\le i<j\le m}(v_i\cdot v_j)^2 + \sum_{(i<j)\ne(i'<j')}(v_i\cdot v_j)(v_{i'}\cdot v_{j'})\right].$$

By the same argument, $\mathbf{E}_U[(v_i\cdot v_j)(v_{i'}\cdot v_{j'})] = 0$, for all $(i < j) \ne (i' < j')$. Hence $\mathbf{Var}_U[||s'_m||^2] = 4\sum_{0\le i<j\le m}\mathbf{E}_U[(v_i\cdot v_j)^2]$, where $\mathbf{E}_U[(v_i\cdot v_j)^2]$ is $\mathbf{E}[h^2]$

over $(n - 1)$-dimensional unit sphere, and thus equal to $1/n$ (see the proof of Lemma 1). It follows that

$$\mathbf{Var}_U[||s'_m||^2] = \frac{4}{n}\binom{m+1}{2}.$$

We conclude that at least in terms of the length of the sum $||s_m|| = ||v_0 + \ldots + v_m||$, our distribution D behaves very much like the uniform distribution U.

We return to the correlation between s'_m and u. It is easy to see that with distribution D

$$s'_m \cdot u = \sum_{i=0}^{m} \frac{2^i}{M} \approx \frac{2^{m+1}}{M},$$

which is exponentially small. Also since it is a constant $\mathbf{Var}_D[s'_m \cdot u] = 0$. For the uniform distribution U,

$$s'_m \cdot u = \sum_{i=0}^{m} v_i \cdot u,$$

and $\mathbf{E}_U[s'_m \cdot u] = 0$. For the variance

$$\mathbf{Var}_U[s'_m \cdot u] = \mathbf{E}_U[(s'_m \cdot u)^2]$$

$$= \mathbf{E}_U\left[\left(\sum_{i=0}^{m}(v_i \cdot u)\right)^2\right]$$

$$= \mathbf{E}_U\left[\sum_{i=0}^{m}(v_i \cdot u)^2 + \sum_{0 \leq i \neq j \leq m}(v_i \cdot u)(v_j \cdot u)\right].$$

By independence $\mathbf{E}_U[(v_i \cdot u)(v_j \cdot u)] = \mathbf{E}_U[v_i \cdot u]\mathbf{E}_U[v_j \cdot u] = 0$ for $i \neq j$. Also $\mathbf{E}_U[(v_i \cdot u)^2] = 1/n$. Hence $\mathbf{Var}_U[s'_m \cdot u] = m/n$. Therefore statistically one can not deduce much from $s'_m \cdot u$ in the distribution D, since it is exponentially small, and well within the range in which this value would have been under the uniform distribution, where $\mathbf{E}_U[s'_m \cdot u] = 0$ and $\mathbf{Var}_U[s'_m \cdot u] = \Omega(1)$.

In fact, suppose $u' \in S^{n-1}$ is any unit vector, $u' \neq \pm u$. The estimates of $\mathbf{E}_U[s'_m \cdot u'] = 0$ and $\mathbf{Var}_U[s'_m \cdot u'] = m/n = \Omega(1)$ are still valid. Let Π be the 2-dimensional plane spanned by u and u'. Let $u' = (\cos\theta)u + (\sin\theta)u^\perp$, where the unit vector $u^\perp \perp u$. Then we can choose a coordinate system such that $u^\perp$ is the $(n-1)$st coordinate for ρ_i, and $\mathbf{E}_D[\rho_i \cdot u^\perp] = \mathbf{E}_{S^{n-2}}[h] = 0$. Therefore $\mathbf{E}_D[s'_m \cdot u'] = (\cos\theta)\mathbf{E}_D[s'_m \cdot u]$. Thus $|\mathbf{E}_D[s'_m \cdot u']| \leq 2^{m+1}/M$, which is exponentially small. This implies that s'_m has correlation with no particular direction, the same as under the uniform distribution U.

References

1. M. Ajtai. Generating hard instances of lattice problems. In *Proc. 28th Annual ACM Symposium on the Theory of Computing*, 1996. Full version available from ECCC, *Electronic Colloquium on Computational Complexity* TR96-007, at http://www.eccc.uni-trier.de/eccc/.
2. M. Ajtai. The shortest vector problem in L_2 is NP-hard for randomized reductions. *Electronic Colloquium on Computational Complexity*, TR97-047 at http://www.eccc.uni-trier.de/eccc/.
3. M. Ajtai and C. Dwork. A public-key cryptosystem with worst-case/average-case equivalence. 1996. Available from ECCC, *Electronic Colloquium on Computational Complexity* TR96-065, at http://www.eccc.uni-trier.de/eccc/.
4. S. Arora, L. Babai, J. Stern, and Z. Sweedyk. The hardness of approximate optima in lattices, codes, and systems of linear equations. In *Proc. 34th IEEE Symposium on Foundations of Computer Science (FOCS)*, 1993, 724-733.
5. J-Y. Cai and A. Nerurkar. An Improved Worst-Case to Average-Case Connection for Lattice Problems. In *Proc. 38th IEEE Symposium on Foundations of Computer Science (FOCS)*, 1997, 468–477.
6. J-Y. Cai and A. Nerurkar. Approximating the SVP to within a factor $\left(1 + \frac{1}{\dim^\epsilon}\right)$ is NP-hard under randomized reductions. Available from ECCC, *Electronic Colloquium on Computational Complexity* TR97-059, at http://www.eccc.uni-trier.de/eccc/.
7. J-Y. Cai. A Primal-Dual Relation for Lattices and the Complexity of Shortest Lattice Vector Problem. To appear in *Theoretical Computer Science*.
8. J-Y. Cai. A new transference theorem and applications to Ajtai's connection factor. *Electronic Colloquium on Computational Complexity* TR98-005, at http://www.eccc.uni-trier.de/eccc/.
9. O. Goldreich and S. Goldwasser. On the Limits of Non-Approximability of Lattice Problems. *Electronic Colloquium on Computational Complexity* TR97-031, at http://www.eccc.uni-trier.de/eccc/.
10. O. Goldreich, S. Goldwasser, and S. Halevi. Collision-free hashing from lattice problems. 1996. Available from ECCC, *Electronic Colloquium on Computational Complexity* TR96-042, at http://www.eccc.uni-trier.de/eccc/.
11. O. Goldreich, S. Goldwasser, and S. Halevi. Public-key cryptosystems from lattice reduction problems. 1996. Available from ECCC, *Electronic Colloquium on Computational Complexity* TR96-056, at http://www.eccc.uni-trier.de/eccc/.
12. J. C. Lagarias. The computational complexity of simultaneous diophantine approximation problems. *SIAM Journal of Computing*, Volume 14, page 196–209, 1985.
13. J. C. Lagarias, H. W. Lenstra, and C. P. Schnorr. Korkin-Zolotarev Bases and Successive Minima of a Lattice and its Reciprocal Lattice. *Combinatorica*, 10:(4), 1990, 333-348.
14. A. K. Lenstra, H. W. Lenstra, and L. Lovász. Factoring polynomials with rational coefficients. *Mathematische Annalen*, 261:515–534, 1982.
15. A.M. Odlyzko. The rise and fall of knapsack cryptosystems. in *Cryptology and Computational Number Theory*, American Mathematical Society, pp. 75-88, 1990.
16. C. P. Schnorr. A hierarchy of polynomial time basis reduction algorithms. *Theory of Algorithms*, pages 375–386, 1985.
17. P. van Emde Boas. Another NP-complete partition problem and the complexity of computing short vectors in lattices. Technical Report 81-04, Mathematics Department, University of Amsterdam, 1981.

Fast DES Implementations for FPGAs and Its Application to a Universal Key-Search Machine *

Jens-Peter Kaps** and Christof Paar

Electrical and Computer Engineering Department
Worcester Polytechnic Institute
Worcester, MA 01605-2280
{kaps,christof}@ece.wpi.edu

Abstract. Most modern security protocols and security applications are defined to be *algorithm independent*, that is, they allow a choice from a set of cryptographic algorithms for the same function. Although an algorithm switch is rather difficult with traditional hardware, i.e., ASIC, implementations, Field Programmable Gate Arrays (FPGAs) offer a promising solution. Similarly, an ASIC-based key search machine is in general only applicable to one specific encryption algorithm. However, a key-search machine based on FPGAs can also be algorithm independent and thus be applicable to a wide variety of ciphers. We researched the feasibility of a universal key-search machine using the Data Encryption Standard (DES) as an example algorithm.
We designed, implemented and compared various architecture options of DES with strong emphasis on high-speed performance. Techniques like *pipelining* and *loop unrolling* were used and their effectiveness for DES on FPGAs investigated. The most interesting result is that we could achieve encryption rates beyond 400 Mbit/s using a standard Xilinx FPGA. This result is by a factor of about 30 faster than software implementations while we are still maintaining flexibility. A DES cracker chip based on this design could search 6.29 million keys per second.

1 Introduction

Most modern security protocols are defined algorithm independent and the protocol standards support a variety of algorithms. Although it is fairly easy to switch cryptographic algorithms in software, it is often painfully difficult in hardware. On the other hand, hardware solutions provide better performance and higher physical security. One answer to this problem is reconfigurable hardware, based on modern Field Programmable Gate Array, or FPGA, devices. Since FPGAs can switch algorithms they can be used to build *algorithm agile* applications. Although at a given time only one algorithm is configured, the FPGA can be reconfigured with a different algorithm. The following lists the main advantages of cryptographic algorithms on FPGAs

---

* This research was partially sponsored through NFS CAREER award CCR-9733246
** currently with GTE CyberTrust Solutions Incorporated

S. Tavares and H. Meijer (Eds.): SAC'98, LNCS 1556, pp. 234–247, 1999.
© Springer-Verlag Berlin Heidelberg 1999

- **Algorithm agility**, the same FPGA can be reprogrammed at run time to support different algorithms,
- **Scalable security**, through different versions of the same algorithm (e.g., Data Encryption Standard (DES) and triple-DES),
- **Alterable architecture parameters**, e.g., desirable features such as variable S-boxes, variable number of rounds, or different modes of operation can easily be realized,

Another interesting cryptographic application of FPGAs are key-search machines. Building a key-search machine in hardware is a major investment. Such a machine can be used to retrieve secrete keys for only one algorithm. Therefore it might be interesting to have a key-search machine which is also defined algorithm independent, supporting a variety of algorithms.

We designed a universal key-search machine and used a high speed DES implementation for FPGAs as an algorithm. DES is currently the most widely used private-key algorithm and it is also part of many other standards, e.g., IPSec protocols, ATM cell encryption, the Secure Socket Layer (SSL) protocol, and for various ANSI banking standards. Even though it is expected that DES will not be reapproved as a federal US standard this year, it is still important and will continue to play a major role for several years to come.

In Sect.2 we summarize previous relevant work. Section 3 explores different architecture options for DES like loop unrolling and pipelining and presents the architecture versions we decided to implement. In Sect. 4 we provide an overview of the projected universal key-search machine. Section 5 describes the design and implementation cycle and gives an overview of the hardware and software tools we used for our research. In Sect. 6 we present and compare the results of our implementations of the different architectures and extrapolate cost and speed of the proposed universal key-search machine running a DES Cracker chip.

2   Previous Work

Already one year after the Data Encryption Standard was released in 1976, Whitfield Diffie and Martin Hellman published a paper which describes in detail a key-search machine for DES [2]. They estimated that for $20 million a key-search machine could be built which recoveres a DES key within 12 hours. Two papers with quite different results appeared in 1993 [6] and [13]. Both papers describe a custom chip design and develop from there a key-search machine. Reference [13] calculates 3.5 hours time to break DES at $1 million. Their chip can test 50 million keys per second. The design shown in [6] uses 32 DES breaker (DESB) chips and breaks DES in 1.2 days, each DESB chip can test 5333 million keys per second. This number appears to be very high. The estimated cost including overhead is only $2500 which appears to be very low to us.

Modern custom hardware implementations of DES can achieve data rates of 1 Gbit/sec and beyond. References [4,3] were the first report of a custom chip employing modern GaAs technology to achieve 1 Gbit/sec. This design as well as many commercially available devices do not support a key change at full speed; i.e., a different key for every block of plain text.

The first paper to show an implementation of DES on FPGAs is [8]. Their approach generates key-specific circuitry for the Xilinx FPGAs. Therefore a binary image (bit-stream) for each key has to be precomputed before it can be used in the device. As for a key search machine the key has to be changed after every encryption, this chip is not suitable for this task. Their fastest implementation without decryption and adjusted to one key achieves a data rate of 26 Mbit/sec.

A group of cryptographers analyzed in 1996 the security of the key lengths of symetric ciphers based on current technology. In their report [1] employing FPGAs is highlited as an efficient approach for a brute-force attack against cryptographic systems. FPGAs are inexpensive, fast, and they need less inital investment than Application-Specific Integrated Circuits (ASIC). If we just take the plain FPGA cost for our design into account we would neet to invest three times as much money as [1] to recover keys at the same speed.

3   Architecture Options for the DES Algorithm

A high speed implementation of DES is crucial for an efficient DES cracker. But not only speed is an important factor, the design also must support a fast key change.

3.1   Basic DES

The DES algorithm possesses an iterative structure [10]. Data is passed through the Feistel network 16 times, each time with a different sub-key from the key transformation. This structure leads itself naturally to the block diagram shown in Fig. 1. The incoming data and key are passed through initial permutations. Then the data passes 16 times through the Feistel network and also 16 sub-keys are generated simultaneously. Both, the Feistel network operation and the sub-key generation is denoted in the block diagram as *Combinatorial Logic* (CLU, combinatorial logic unit). In order to be able to loop the output back to the input of the combinatorial logic unit we need *Registers* and *Multiplexers*. The multiplexer switches the input of the combinatorial logic unit between data from the previous round and new input data and key. The registers store the results of each loop and pass them on to the multiplexer. The output of the data register passes through the *Final Permutation*.

3.2   Loop Unrolling

Loop unrolling is the concatenation of combinatorial units in order to reduce the number of iterations if, for instance, two loops are unrolled then two rounds of DES will be calculated with one clock cycle. Figure 2 shows the block diagram. This block diagram differs from Fig. 1 only in the 2nd combinatorial logic unit.

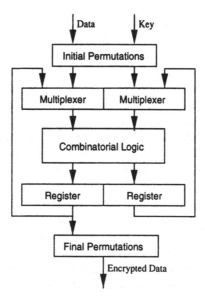

Fig. 1. DES block diagram

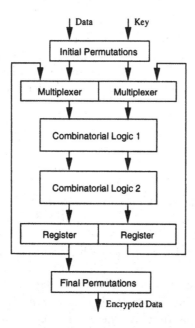

Fig. 2. Block diagram of DES with two unrolled loops

The initial and final permutations as well as the registers and multiplexers are the same.

Loop unrolling leads potentially to speed improvements for the following reason. In the not unrolled version, one iteration of DES has the following simple timing model: $T_{\mathrm{mux}} + T_{\mathrm{cl}} + T_{\mathrm{reg}}$, where T_{mux} denotes the time a signal needs to pass through a multiplexer, T_{cl} the delay introduced by the combinatorial logic, and T_{reg} the delay introduced by the register. Wiring delays are assumed to be included in the specified times for the elements. So for the whole 16 rounds this sums up to: $16 * T_{\mathrm{mux}} + 16 * T_{\mathrm{cl}} + 16 * T_{\mathrm{reg}}$.

The equation for one loop of the structure in Fig. 2 is thus: $T_{\mathrm{mux}} + 2*T_{\mathrm{cl}} + T_{\mathrm{reg}}$. This has to be executed 8 times, so that the over-all delay is now: $8 * T_{\mathrm{mux}} + 16 * T_{\mathrm{cl}} + 8 * T_{\mathrm{reg}}$. The same principle can be applied to four unrolled DES rounds resulting in: $4 * T_{\mathrm{mux}} + 16 * T_{\mathrm{cl}} + 4 * T_{\mathrm{reg}}$.

Obviously we can not reduce the delay introduced by the combinatorial logic units but we reduced the runs through the multiplexers and buffers. There is also another motivation for speed increase if modern design methods are applied. It is possible that the synthesis tools can optimize an unrolled design better, due to boundary optimization. Loop unrolling works thus well with a modern design process which can reduce the logic complexity and delay of the design.

3.3   Pipelining

Pipelining tries to achieve a speed improvement in a different way. Instead of processing one block of data at a time, a pipelined design can process two or more data blocks simultaneously. A design with two pipelines is shown in Fig. 3. The block diagram in Fig. 3 is very similar to the one with the two unrolled loops (Fig. 2). The only difference is the additional buffer between the combinatorial logic units.

The first block of data x_1 and the associated key k_1 are loaded and passed through the initial permutations and the multiplexer. The 1st combinatorial logic unit computes $x_{1,1}$ and $k_{1,1}$ which are stored into the 1st register block. On the next clock cycle $x_{1,1}$ and $k_{1,1}$ leave the 1st registers and the 2nd combinatorial logic unit computes $x_{1,2}$ and $k_{1,2}$ which are put into the 2nd register block. At the same time the second block of data x_2 and the key k_2 are loaded and passed through the initial permutations, and the multiplexer, and the 1st combinatorial unit computes $x_{2,1}$ and $k_{2,1}$ which are moved into the 1st register block.

Now the pipeline is filled and with each clock cycle another iteration for two pairs of data and key are computed. The data which has entered the pipeline first, will also exit it first. At that time the next data and key pair can be loaded.

The advantage of this design is that two or more data–key pairs can be worked upon at the same time. As there is still only one instance of the initial permutations, the multiplexer and the final permutation, the cost in terms of resources on the chip will not be twice as high as if we implemented two full non-pipelined DES designs. Also there has to be only one control logic which is just slightly more complicated than for a non-pipelined DES design. The maximum clock speed should be roughly the same since during one clock cycle the same amount of logic resources has to be traversed as in the non-pipelined design. It

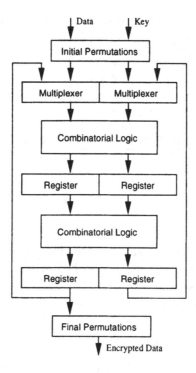

Fig. 3. Block diagram of DES with two pipeline stages

is also straightforward to design pipelines with more than two stages, e.g., with four.

3.4 Combination of Pipelining and Loop Unrolling

It is possible to combine both architecture acceleration techniques that we described above. For instance, each pipeline stage can contain two unrolled loops. The resulting block diagram looks similar to Fig. 3 except that each combinatorial logic unit is duplicated. During one clock cycle two iterations of two data–key pairs are computed: $x_{1,4}$ and $k_{1,4}$ are computed from $x_{1,2}$ and $k_{1,2}$, and $x_{2,2}$ and $k_{2,2}$ are computed from x_2 and k_2. An extension to four unrolled loops per pipeline stage is also possible.

3.5 Design Decisions

One major objective of this research was to obtain a realistic comparison of the different acceleration methods (loop unrolling, pipelining, combination of both) for FPGAs. Table 1 shows the architecture versions we decided to implement, results of which are described in Section 6.

Table 1. Implemented DES architectures

Name	Description
DES_ED16	standard DES (16 iterations)
DES_ED8	DES with 2 unrolled loops (8 iterations)
DES_ED4	DES with 4 unrolled loops (4 iterations)
DES_ED16x2	DES with 2 pipeline stages
DES_ED16x4	DES with 4 pipeline stages
DES_ED8x2	DES with 2 pipeline stages each containing 2 unrolled loops

4   Universal Key-Search Machine

The design of the Universal Key-Search Machine has to be algorithm independent. Therefore no assumption on the length of the plain text, cipher text, or key is made. The system architecture is described in the following subsection and the modifications necessary for our DES FPGA [7] to meet the I/O requirements for this architecture in the next subsection.

4.1   System Architecture

Our biggest concern is scalability. Therefore one design goal is to have a minimum amount of wiring and additional external logic (glue logic). A bus topology and an optimized addressing scheme was developed to meet this requirement. The addressing scheme requires no address decoding logic and there is no logical limit on the number of key-search chips connected to the bus.

Figure 4 gives an overview of the system architecture. When more key-search chips are attached additional buffers are needed to drive the bus. These bus drivers are left out of Fig. 4 for simplicity. The cracker-bus comprises an bidirectional 16 bit wide data bus, the control signals, **IE** (input enable) and **OE** (output enable), four chip select signals, **CS** (chip select), **CSI** (chip select in), **CSO** (chip select out), and **CSCLK** (chip select clock), and an indicator **done**. The function of the control chip is to interface the cracker-bus with the ISA bus of a PC. We use DESC, the DES Cracker chip, as an example.

The system supports sequential as well as parallel access. When CS is active all chips are selected and listening to the bus. The plain text and the cipher text can be send with IE to every chip at the same time. A different start key has to be sent to each chip. For this purpose we implemented an sequential addressing scheme. The signals CSI, CSO, and $CSCLK$ work with the chips as a long shift register. The control chip sends a logic '1' to the first chip. With each clock $CSCLK$ this '1' is passed from CSO to CSI of the next chip; the next chip is selected and the start key can be send with IE. The number of clock cycles $CSCLK$ it takes for the '1' to come back to the control chip is equal to the number of key-search chips put in. With OE the current key of the selected chip is clocked out. With this function the state of the key-search machine can be saved or the final key can be retrieved. The $done$ signal is a tristated signal

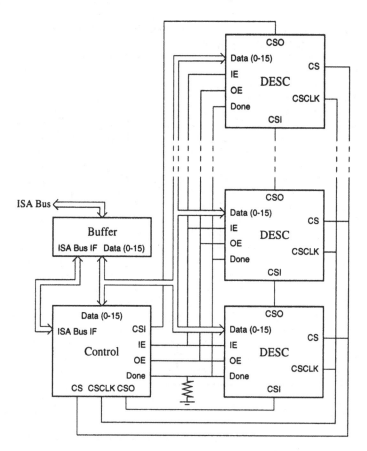

Fig. 4. System architecture

pulled to logic '0' with a resistor. When the key is found this signal goes to '1'.
Now the control chip addresses sequentially the chips. The chip which found the
key takes the '1' off as soon as it is selected. Through this scheme the controller
can identify the chip.

4.2 DES Cracker-Chip

The DES cracker chip (see Fig. 5) is based on our DES FPGA [7] denoted as
DES Core in the Figure. Each of the implemented DES architectures shown in
Table 1 can be used as a DES Core. Our DES FPGA is not optimized for key-
search but it supports a key change for every block of plain text at full speed.
The buffers for the cipher text and the plain text are 64 bits wide, the buffer
for the start key 56 bits. The bus can transfer 16 bits at a time, therefore the
buffers are split up in parts of maximum 16 bits. They are addressed through
a shift register (not shown in Fig. 5). When the chip is selected the signal *IE*
cycles through the input buffers. The 4×16 bits output register for the current

key is addressed via a 4 bits shift register, which in turn is operated through the *OE* signal when the chip is selected.

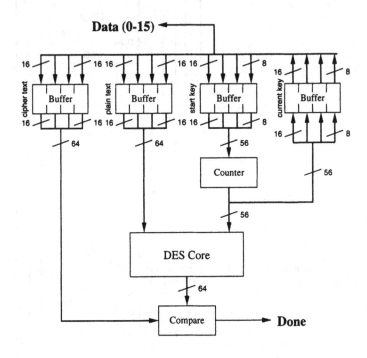

Fig. 5. DES Cracker-Chip

A 56 bit counter takes care of incrementing the key for each new encryption. A 56 bit compare unit compares the encrypted plain text with the cipher text. If both are the same the *done* signal goes high. The chip select circuitry and the control logic is not shown in the Figure.

5   Methodology

This section describes the design cycle and the tools used for our research of the DES core. The design procedure can roughly be broken down into the following stages:

1. Creating VHDL[1] descriptions of the DES design employing different architecture options and verifying each version.
2. Synthesis and logic optimization.

---

[1] VHDL is the VHSIC Hardware Description Language. VHSIC is an abbreviation for Very High Speed Integrated Circuit.

3. Place and Route for a specific device and back-annotated verification of the design.

Early in the design we decided upon Xilinx as FPGA vendor and a device family. That decision was based mainly on our previous work described in [5]. The entire design was implemented using VHDL and vendor specific macros. We used Synopsys version 1997.08 for the synthesis, logic optimization, design verification, and timing analysis.

Xilinx Alliance Series version M1.3.7 provided macro functions (LogiBLOX) and was used for the place and route. The Xilinx tools also perform a timing analysis after place and route which shows the **minimum clock period** for the given design. This clock period is guaranteed by Xilinx for the design and therefore is to be seen as rather pessimistic. We are using this timing result for our speed calculations.

6   Results

We implemented multiple versions of each architecture option listed in Table 1 in order to evaluate their effectiveness. In the following sections we compare the different designs. In most cases the designs are compared to the design *DES_ED16* which serves as our reference model.

The unit **CLB** stands for *combinatorial logic block* which is employed by Xilinx to measure the amount of logic resources on a device. We are using it here to compare the amount of logic resources used by a given design. The abbreviation **CLU** stands for *combinatorial logic unit* (see Chap. 3), i.e., one Feistel network round.

6.1   Loop Unrolling

We implemented two loop unrolled versions: *DES_ED8* and *DES_ED4*. The design *DES_ED8* contains two combinatorial logic units (*CLU*, see Sect. 3.2) and therefore encrypts or decrypts one data block in 8 clock cycles. The design *DES_ED4* contains four CLUs and provides the result after 4 clock cycles. Both designs are compared with the design *DES_ED16* in Table 2.

Table 2. Comparison of loop unrolled architectures

Design	Chip	CLBs used	CLBs per CLU	Min CLK in ns	Data Rate per CLU in Mbit/s	Data Rate in Mbit/s	Data Rate rel.
DES_ED16	XC4008E-3-PG223	262	262	40.4	99.1	99.1	1.00
DES_ED8	XC4013E-3-PG223	443	222	54.0	74.1	148.2	1.50
DES_ED4	XC4028EX-3-PG299	722	241	86.7	46.1	184.5	1.86

Our standard DES design *DES_ED16* allowed already for data rates of 99.1 Mbit/sec[2]. The design *DES_ED8* is with 148.2 Mbit/sec 50% faster than *DES_ED16* whereas the resource consumption (in CLBs) increases by 69%. The design *DES_ED4* is with 184.5 Mbit/sec only 25% faster than *DES_ED8*, the speed increase is only half as much as from the first unrolling. The resource consumption increases by 63%.

The number of CLBs divided by the number of CLUs indicates that the amount of logic resources consumed per unrolled CLU is almost constant. The speed divided by the number of CLUs shows that the speed for one CLU in the design *DES_ED4* is less then half the speed of *DES_ED16*. From this we can see that loop unrolling shows a non-linear speed improvement. It seems unlikely that a fourth loop unrolling would yield significantly higher performance.

6.2   Pipelining

We implemented two pipelined designs, *DES_ED16x2* and *DES_ED16x4*. The design *DES_ED16x2* contains two CLUs and therefore 2 pipeline stages and the design *DES_ED16x4* contains four CLUs and therefore 4 pipeline stages. The encryption or decryption of one block of data takes in both cases 16 clock cycles. Table 3 compares both designs with the design *DES_ED16*.

Table 3. Comparison of pipelined architectures

Design	Chip	CLBs used	CLBs per CLU	Min CLK in ns	Data Rate per CLU in Mbit/s	Data Rate in Mbit/s	Data Rate rel.
DES_ED16	XC4008E-3-PG223	262	262	40.4	99.1	**99.1**	1.00
DES_ED16x2	XC4013E-3-PG223	433	217	43.5	91.9	**183.8**	1.86
DES_ED16x4	XC4028EX-3-PG299	741	185	39.7	100.7	**402.7**	4.06

The design *DES_ED16x2* with two pipeline stages is with 183.8 Mbit/sec almost twice as fast as our reference design *DES_ED16*. And our design *DES_ED16x4* achieves 402.7 Mbit/sec which is four times faster. The speed divided by the number of CLUs shows that the speed per CLU stays almost constant for all designs. The lower speed for the design *DES_ED16x2* is caused by the lack of wiring resources on the device which results in a less efficient design.

The amount of logic resources consumed per implemented CLB is decreasing if we create more pipelines. This is due to the fact that the control unit does not become more complicated if we implement more pipelines. Also the multiplexers are implemented only once.

It is interesting to compare the pipelined designs with the loop unrolled designs. It can be seen that *DES_ED16x2* is both faster and smaller than the loop

---

[2] Mbit= 10^6 bit

unrolled *DES_ED8*. The difference is even more dramatically if the *DES_ED16x4* is compared with the *DES_ED4*. *DES_ED16* is more than twice as fast as *DES_ED4* and utilizes almost the same amount of CLBs.

6.3   Combination of Pipelining and Loop Unrolling

A design that contains loop unrolling as well as pipelining is in the simplest version already as large as the largest designs we have implemented so far which were *DES_ED16x4* and *DES_ED4*. Therefore we implemented only the design *DES_ED8x2* which contains 4 CLUs; 2 in each of the 2 pipeline stages. Table 4 compares this design with *DES_ED16x2* and *DES_ED8*.

Table 4. Comparison of a combined architecture with others

Design	Chip	CLBs	Min CLK in ns	Data Rate p. pipeline in Mbit/s	Data Rate in Mbit/s
DES_ED8x2	XC4028EX-3-PG299	733	48.0	166.5	**333.0**
DES_ED16x2	XC4013E-3-PG223	433	43.5	91.9	**183.8**
DES_ED8	XC4013E-3-PG223	443	54.0	148.2	**148.2**

It is not easy to compare this mixed design with the two other designs. The minimum clock period shows that the time it takes for two CLUs (loop unrolled) to execute in the design *DES_ED8x2* is faster than in the design *DES_ED8*. It is of course slower, but surprisingly not much, than one CLU in the design *DES_16x2*.

6.4   Extrapolation for Cracker

Our fastest design is DES_ED16x4 with a data rate of 402 Mbit/sec. This design could be the *DES Core* for our *DES Cracker-Chip* (DESC) as shown in Sect. 4.2. After the plain text, cipher text and start-key are loaded into the buffers the DES Core can start working at full speed. We expect that we can achieve the same data rate for the DESC chip as we achieve with DES_ED16x4. The data rate of 402 Mbit/sec corresponds to a key-search rate of 6.29 million keys per second. This means that a brute force attack on DES with 2^{56} keys would take 182 years on average. A fully unrolled design would theoretically search 25 million keys per second which is about half the speed of the key-search ASIC introduced in [13].

A cracker box comprising one control board, 16 boards each with four DESC chips (using DES_ED16x4 as the DES core) connected to a PC could search the entire key space of DES with a 40-bit key in one hour. The FPGA costs alone for such a box are about \$18.000[3]; when we include 50% overhead for the other

---

[3] Based on actual current (May 1998) pricing supplied by a vendor.

required circuits, boards, racks and an additional $1000 for the PC (a low end PC is fast enough) this cracker box would amount to cost $28.000.

Assume that eight such boxes could be attached to one PC and 160 PCs are connected via a network. This machine comprises 81,920 DESC chips and would cost about $35 million. It could find a DES key of 2^{56} bits length in one day on average. We want to notice that the prices for FPGAs are decreasing at a very high rate and that a key search machine at a considerable lower price might be feasible in the near future.

7 Comparison

We implemented all DES designs based on standard devices from Xilinx with a medium speed grade. With these devices we achieved speeds of up to 402.0 Mbit/sec using four pipelined stages and no loop unrolling. If we compare the reported DES speeds for high speed ASICs (1600 Mbit/sec) [11] and high speed software (13 Mbit/sec) [11], with our best result of 402.0 Mbit/sec we conclude that the speed-up factor from software to FPGAs is about 31, and from FPGAs to ASICs is about 4. It is difficult to provide a fair comparison to this respect but our results clearly show that FPGAs implementations of DES are very attractive for many applications. If we compare our result with the one from [8] we see that even though the same device (although at a slower speed grade) was used, the fastest implementation in the reference is by factor three slower and requires almost twice as much logic resources as our *DES_ED16*. It is to be noted that the fastest implementation in the reference is adjusted to one key.

Using our fastest design as a DES core for the DESC chip we can achieve a key-search rate of 6.29 million keys. For $38 million a machine can be built that breaks DES with 2^{56} keys in one day on average. Although it might very well be possible to build a DES-specific key search machine (much) cheaper, we would like to note that our design allows the application to a wide variety of block ciphers by simply downloading a different algorithm in to the FPGAs. Considering the significant cost of a key-search machine, this can be a major advantage if more than one algorithm is to be attacked.

References

1. M. Blaze, W. Diffie, R.L. Rivest, B. Schneier, T. Shimomura, E. Thompson, and M. Wiener. Minimal key lengths for symmetric ciphers to provide adequate commercial security. January 1996.
2. W. Diffie and M.E. Hellman. Exhaustive cryptanalysis of the NBS Data Encryption Standard. *Computer*, 10:74–84, June 1977.
3. H. Eberle. A high-speed DES implementation for network applications. In E.F. Brickell, editor, *Advances in Cryptology - CRYPTO '92. 12th Anual International Cryptology Conference Proceedings,* Lecture Notes in Computer Science, pages 521–539, Berlin, Germany, 1993. Springer-Verlag.

4. H. Eberle and C.P. Thacker. A 1 Gbit/second GaAs DES chip. In *Proceedings of the IEEE 1992 Custom Integrated Circuits Conference*, pages 19.7/1–4, New York, NY, USA, 1992. IEEE, IEEE.

5. G.M. Haskins. Securing Asynchronous Tranfer Mode networks. Masters thesis, Worcester Polytechnic Institute, Worcester, Massachusetts, USA, May 1997.

6. F. Hendessi and M. Aref. A successful attack against the DES. In T.A. Gulliver and N.P. Secord, editors, *Information Theory and Applications : Proceedings Third Canadian Workshop*, volume 793 of *Lecture Notes in Computer Science*, pages 78–90, Berlin, 1994. Springer Verlag.

7. Jens-Peter Kaps. High speed FPGA architectures for the Data Encryption Standard. Masters thesis, Worcester Polytechnic Institute, Worcester, Massachusetts, USA, May 1998.

8. J. Leonard and W.H. Magione-Smith. A case study of partially evaluated hardware circuits: Keyspecific DES. In W. Luk, P.Y.K. Cheung, and M. Glesner, editors, *Field-programmable Logic and Applications. 7th International Workshop, FPL '97*, Berlin, Germany, 1997. Springer-Verlag.

9. A.J. Menezes, S.A. Vanstone, and P.C. Van Oorschot. *Handbook of Applied Cryptography*. Discrete Mathematics and its Application. CRC Press, Florida, USA, 1997.

10. National Bureau of Standards FIPS Publication 46. *DES modes of operation*, 1977.

11. B. Schneier. *Applied Cryptography Second Edition: protocols, algorithms, and source code in C*. Wiley & Sons, New York, USA, 2nd edition, 1996.

12. D.R. Stinson. *Cryptography: Theory and Practice*. Discrete Mathematics and its Applications. CRC Press, Florida, USA, 1995.

13. M.J. Wiener. Efficient DES key search. Crypto '93 Rump Session Presentation, August 1993.

IDEA: A Cipher for Multimedia Architectures?

Helger Lipmaa

Küberneetika AS, Akadeemia 21, 12617 Tallinn, Estonia
helger@cyber.ee

Abstract. MMX is a new technology to accelerate multimedia applications on Pentium processors. We report an implementation of IDEA on a Pentium MMX that is 1.65 times faster than any previously known implementation on the Pentium. By parallelizing four IDEA's we reach an unprecedented 78 Mbits/s throughput per output block on a 166MHz MMX. In the light of rapidly increasing popularity of multimedia applications, causing more dedicated hardware to be built, and observing that most of the current block ciphers do not benefit from MMX, we raise the problem of designing block ciphers (and encryption modes) fully utilizing the basic operations of multimedia.

Keywords: block ciphers, fast implementations, IDEA, multimedia architectures, Pentium MMX.

1 Introduction

The second main objective besides security in designing cryptographic primitives is speed: even 10% difference in speed (by the same security level) may bias industry to prefer one cipher to another. Still, it is not an easy task to compare ciphers by virtue of speed. The reasons are manifold, depending on the human factor (the best known implementation may not be the best possible implementation) but also on the hardware available: ciphers optimized for 32-bit processors may not be optimal on 64-bit processors and vice versa. Application of new microprocessor techniques (DSP — Digital Signal Processing, VLIW – Very Long Instruction Word, SIMD — Single Instruction Multiple Data) in current general-purpose microprocessors will significantly sway our beliefs in the speed ratio of available ciphers [Cla97].

Because of the quickly increasing importance of multimedia, dedicated hardware will be commonplace tomorrow. Today's multimedia extensions (to name a few, Intel's MMX, Sun's VIS, HP's MAX-2, Cyrix's MMX, AMD's 3DNow!) are just the first flowers. New generations of multimedia enhanced processors will even more change our judgment of what it means to be "software" optimized.

MMX, incorporated in every new Intel processor (e.g., in the Pentium with MMX and the Pentium II), is a relatively new extension made to accelerate multimedia applications. Considering the worldwide spread of MMX capable computers, design and implementation of cryptographic primitives utilizing the basic operations of multimedia applications should be considered very seriously. Some work in this area has already been done by designing new hash functions

S. Tavares and H. Meijer (Eds.): SAC'98, LNCS 1556, pp. 248–263, 1999.
© Springer-Verlag Berlin Heidelberg 1999

and stream ciphers [HK97,Cla97,DC98]. Biham viewed a 64-bit processor as a SIMD parallel computer, which can compute 64 one-bit operations simultaneously, getting significant acceleration of DES [Bih97]. Using the same method ("*bit-slicing*"), several papers [SAM97,Kwa98] have later improved Biham's results.

There is a wide variety of block ciphers in more or less general use. The popularity of some of those ciphers is based on the trust in the design of the cipher, the popularity of some other ciphers is based on the high throughput in combination with reasonable security. In particular, the block cipher IDEA [LM90,LMM91] is believed to be very secure due to the proper interaction between three different group operations. Although, apart from DES, IDEA seems to be the most studied block cipher, no currently known attack (e.g., [BKR97], [DGV94] or [Haw98]) against the full IDEA performs better than exhaustive search. Interaction between three different group operations adds confidence in IDEA's security, but the frequent use of multiplication does not allow fast software implementations on common microprocessors (Table 1).

Block cipher	Block size	Cycles	Mbits/s
Square	128	244	87.1
Blowfish	64	158	67.2
RC5-32/16	64	199	53.4
CAST5	64	220	48.3
DES	64	340	31.2
SAFER (S)K-128	64	418	25.4
Shark	64	585	18.2
IDEA	64	590	18.0
3DES	64	928	11.4

Table 1. Performance in clock cycles per block of output and Mbits/s of several block ciphers on a 166MHz Pentium by Antoon Bosselaers [PRB98].

We describe an implementation of IDEA on MMX, that is significantly faster than the best *possible* implementation of IDEA on the standard Pentium. One attempt to optimize IDEA on MMX has already been taken: Masayasu Kumagai's implementation of non-standard IDEA [Kum97] encrypts three IDEA blocks in parallel, achieving 45.6 Mbits/s per individual encryption on a 200MHz Pentium MMX. Our implementation includes a fast version of standard IDEA and a parallel version that is about twice as fast as Kumagai's.

The MMX architecture was chosen for it being the *de facto* standard, IDEA was chosen because no other current "industry-standard" block cipher seems to benefit from the Pentium MMX and because of its practical importance. Moreover, in the following we demonstrate that IDEA utilizes only about one third of the Pentium MMX and is, additionally, easily parallelized without a significant parallelization overhead. The resulting parallel "4-way IDEA" is faster than any of the 64-bit block ciphers in Table 1; by doing this we transform a relative slow (and as generally believed, a *very* secure) cipher into a very fast (and still *very* secure) cipher. Observing that, we raise a question of designing new, multimedia optimized block ciphers.

Section 2 gives a background to MMX and multimedia extensions. Section 3 outlines the basics of the IDEA algorithm. Section 4 describes our implementation of IDEA on MMX. Section 5 describes shortly the fast parallel implementation of IDEA. Section 6 takes a more broad view of multimedia architectures and Sect. 7 gives a short description of "why can't most of the block ciphers be parallelized on the MMX" and raises the problem of designing new, multimedia-like constituted block ciphers. In Sect. 8 we outline the results and finally, Sect. 9 acknowledges the people who have to be acknowledged.

2   Introduction to MMX

At the time of writing this paper Intel's Pentium was the most widely used general purpose processor. We shall not present a detailed outline of Intel Pentium's architecture (an interested reader may turn to [Int97b] or [BGV96]).

MMX (MultiMedia eXtensions) is a relatively new technology to enhance performance of advanced media and communication applications. The MMX technology introduces new general-purpose instructions that operate in parallel on multiple data elements packed into 64-bit quantities (the 'SWAR' — SIMD Within A Register — architecture, [Die97]). These instructions accelerate the performance of multimedia applications such as motion video, combined graphics with video, image processing, audio synthesis, speech synthesis and compression, telephony, video conferencing, 2D graphics, and 3D graphics. These applications were broken down to identify the most compute-intensive routines, which were then analyzed in detail using advanced computer-aided engineering tools. The results of this extensive analysis showed many common, fundamental characteristics across these diverse software categories. The key attributes of these applications were:

- Small integer data types (for example: 8-bit graphics pixels, 16-bit audio samples).
- Small, highly repetitive loops.
- Frequent multiplies and accumulates.
- Compute-intensive algorithms.
- Highly parallel operations.

The new MMX instructions work on 8 new 64-bit registers called %mm0 ... %mm7. Some of the instructions have an 8-way parallel 8-bit, a 4-way parallel 16-bit, a 2-way parallel 32-bit and a 64-bit version but most of the operations (like multiplication and addition) have only versions corresponding to some subset of these possibilities. There are more operations for 8-bit and 16-bit data than for larger data types (the "small data types" paradigm).

All microprocessors in the Pentium family have another level of parallelism, called *super-scalar parallelism*. In particular, most of the MMX instructions can be executed in both U and V pipelines (in parallel with any other instruction), with the following exceptions.

- Multiplication requires three cycles (has *latency* 3) but can be pipelined, resulting in one multiplication operation every clock cycle (has *throughput* 1). Multiplication instructions cannot pair with other multiplication instructions.
- Shift, pack and unpack instructions cannot pair with each other.
- MMX instructions that access memory or integer registers can only execute in the U-pipe and cannot be paired with any instructions that are not MMX instructions.
- After updating an MMX register, one additional clock cycle must pass before that MMX register can be moved to either memory or to an integer register.

Throughput is 1 for every operation, latency is 1 for every operation but multiplication. It is important to understand the difference between the SIMD-parallelism provided by the MMX technology and the super-scalar parallelism. The first permits to execute *the same* operation on up to eight different data entities as one instruction, the second makes it possible to execute two *possibly different* instructions during the same machine cycle. Hence, the total level of parallelism inside a Pentium MMX can be up to 16.

Still, most of the applications do not benefit from MMX. Some of the limitations of MMX (and the Pentium family in general) are outlined below (cf [Int97b,Int97a] for more information):

Maximum two operands. Pentium/MMX instructions have the maximum of two operands, causing a high frequency of the move (**movq**) instructions in Pentium/MMX programs.

Lack of registers. There are only 8 MMX registers, which is rather insufficient for most of the compute-intensive applications.

Slow interaction with integer registers and memory. Data in memory has to be aligned to 64-bit boundaries (misalignment costs three cycles on the Pentium processor family) and arranged in a way that minimizes the number of cache misses. Correct data alignment may significantly expand the data structures (in the worst case, expanded data will not fit into the cache). The delay for a cache miss is at least eight internal clock cycles. Pairing limitations were already mentioned.

Limited number of instructions. MMX has only a limited set of specific operations. Because of the slow interaction between integer and MMX register sets, small programs using intensively both integer and MMX instructions will generally not benefit from MMX.

No flags register. The MMX command set does not change the flags register and therefore the wide variety of branch instructions available on the Pentium is not useable. The only two comparison operators on MMX (pcmpgt* and pcmpeq*; greater than, equal to) act on signed data and change the corresponding bits of the destination register to 1 (true) or 0 (false). Emulating different — especially unsigned — comparisons takes additional time.

No commands with immediate operands. Immediate operands have to be loaded from memory or generated by other means (e.g., by xoring or comparing a register to itself).

Only 16-bit signed multiplication. Applications intensively using the unsigned multiplication may become significantly slower. IDEA multiplication $\odot$ (Section 3), which is expensive to emulate using unsigned multiplication is even more expensive to emulate using only the signed multiplication (see Section 4). Emulation of $\odot$ using the available MMX instructions needs two multiplications: one to calculate the higher 16 bits of the result (pmulhw) and another to calculate the lower 16 bits (pmullw).

Standard reference for MMX optimization is [Int97a].

Definition 1. *Let the subscript s (resp. u) under a binary operator denote signedness (resp. unsignedness) of the corresponding operation. Let $*_s$ and $*_u$ be respectively the signed and unsigned multiplication operations from $\mathbb{Z}_{2^{16}}^2$ to $\mathbb{Z}_{2^{32}}$ ($*_u$ is the standard multiplication, expandable to $\mathbb{Z}_{2^{32}}^2$). Let $True(\phi)$ be $2^{16} - 1$ if ϕ is true and 0 otherwise. Next we define several basic operators corresponding one-to-one to the instructions of MMX. Actually the correspondence is 4-way, i.e., the MMX instructions execute four such operations in parallel. Let*

$$Cmpeq(a, b) := True(a = b)$$
$$Cmpgt(a, b) := True(a >_s b)$$
$$a \oplus b := a \text{ bitwise "xor" in } \mathbb{Z}_{2^{16}}$$
$$a \,\&\, b := a \text{ bitwise "and" in } \mathbb{Z}_{2^{16}}$$
$$a \boxminus b := a - b \mod 2^{16}$$
$$a \boxplus b := a + b \mod 2^{16}$$
$$Subus(a, b) := (a \boxminus b) \,\&\, True(a >_u b)$$
$$Mull(a, b) := (a *_s b) \,\&\, (2^{16} - 1)$$
$$Mulh(a, b) := \lfloor (a *_s b)/2^{16} \rfloor .$$

3   Introduction to IDEA

IDEA is — like most of the advanced block ciphers — an iterated cipher. IDEA consists of 8 identical rounds that map the 4-tuple of 16-bit round input, $(X_i^r)_{i=1}^4$, and the 6-tuple of 16-bit round subkeys $(Z_i^r)_{i=1}^6$ (expanded from the 128-bit key using the key expansion algorithm) into the $(X_i^{r+1})_{i=1}^4$. After eight rounds the output transformation will be executed. A round consists of several applications of three group operations, whole IDEA can be presented as a directed labeled graph with labels from the set $\{\oplus, \boxplus, \odot\}$ (Figure 1).

Technically, let $d : \mathbb{Z}_{2^{16}} \rightarrow \mathbb{Z}_{2^{16}+1}^*$, $d(x) = 2^{16}$ if $x = 0$ and $d(x) = x$ otherwise. The group operations used in IDEA are $a \oplus b$, $a \boxplus b$ and $a \odot b$, where

$$a \odot b := d^{-1}(d(a) \cdot d(b) \mod 2^{16} + 1).$$

In particular, these operations were chosen for no two of them to be distributive or associative to each other [LMM91]. This fact guarantees that all operations in the IDEA schematics must be executed in an order not contrary to the data dependencies. Operations not dependent on each other's output can be executed in parallel: M_1^r in parallel with M_2^r, A_1^r and A_2^r; E_1^r with E_2^r; E_3^r with E_4^r. On a SIMD architecture where only similar operations can be executed simultaneously, M_1^r and M_2^r cannot be performed in the same instruction as A_1^r and A_2^r.

IDEA satisfies most of the key attributes of multimedia applications used by designing MMX, therefore being an almost ideal candidate cipher to get benefit from MMX:

- IDEA has small integer data types (all the operations work on 16-bit data). Having only small data values enables to pack several of them into one register and thereafter process multiple plaintext blocks in parallel (one of the main factors in effective parallelization).
- IDEA processes the same data over and over without requiring random memory accesses, therefore needing less interaction with the slow memory. Additionally, IDEA lacks operations necessitating expensive, non-parallelizable, table lookups (another main factor in effective parallelization).
- IDEA is based on two 16-bit operations that are common in multimedia applications (16-bit multiplication and addition) and on exclusive or that is a primitive instruction in almost every microprocessor. Although IDEA's multiplication is not trivial to implement on MMX, MMX still provides *some* speedup (compared to the Pentium) per every multiplication (an important factor to get an overall speedup).

4   Fast Implementation

We have addressed all problems mentioned above and completed a fast implementation of IDEA on a Pentium MMX. Some of the tasks we had to solve are outlined below. We assume the plaintext to be in an MMX register and the pointer to the key schedule in an integer register. The ciphertext can be read afterwards from the same MMX register.

General optimization. Optimal use of registers, with minimized number of move instructions. Minimized use of memory: only constants and subkeys are read from memory. Subkeys and constants are correctly aligned to avoid time

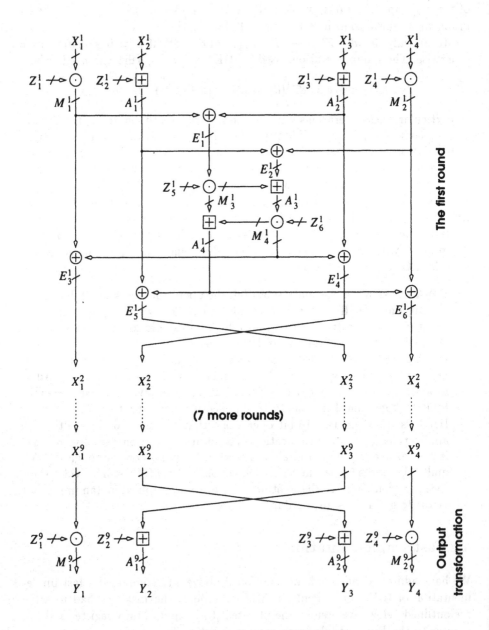

Fig. 1. The schematics of IDEA

penalties due to data misalignment (the key schedule has therefore expanded from 104 to 136 bytes). Data in memory is kept compactly and reused to reduce the number of cache misses. From the integer registers only one is used (as a pointer to the round subkeys). Nothing is written from MMX registers to memory or integer registers.

Effective use of super-scalar parallelism. In our implementation, 693 instructions are paired into 358 cycles. Excellent pairing (0.517 cycles per instruction) is a little miracle (cf to > 0.56 cycles per instruction got by Bosselaers when implementing hash functions for the Pentium, [Bos97]) and is definitely one of the sources of the effectiveness of our implementation.

Use of SIMD parallelism. M_1^r and M_2^r (resp, A_1^r and A_2^r, ...) are calculated in parallel by using the SWAR capability of MMX processors. This is another main factor in increasing the speed of IDEA.

Emulation of $\odot$, using the available MMX instructions, is done, as we believe, optimally. In the following we shortly explain how.

Lemma 1.

$$a *_u b = 2^{16} \cdot (Mulh(a, b) + (a \ \& \ Cmpgt(0, b)) + (b \ \& \ Cmpgt(0, a))) + Mull(a, b) \ .$$

Proof:

$a, b \geq_s 0$. In this case $a *_s b = a *_u b$.
$a \geq_s 0, 0 >_s b$. In this case, a is a positive and b is a negative number. Thus,
$\quad a *_s b = a *_u (b - 2^{16}) = a *_u b - 2^{16} *_u a$.
$b \geq_s 0, 0 >_s a$. Complementary to the previous case.
$0 >_s a, b$. In this case, $a *_s b = a *_u b - 2^{16} *_u a - 2^{16} *_u b - 2^{32} \equiv a *_u b - 2^{16} *_u a - 2^{16} *_u b$.

Results got by analyzing the four cases can be generalized by simple means to complete the proof. ∎

As already mentioned, MMX lacks unsigned comparison instructions. Our implementation needs one of them, which will be emulated using existing instructions. Let $Cmpleu(a, b) = True(a \leq_u b)$. It is easy to see that

$$Cmpleu(a, b) = Cmpeq(Subus(a, b), 0) \ .$$

Lemma 2. Let $a, b \in \mathbb{Z}_{2^{16}}$. Let $h := (a *_u b)/2^{16}$ and $l := (a *_u b) \ \& \ (2^{16} - 1)$ be calculated by the previous lemma. Then

$$a \odot b = ((1 \boxminus a \boxminus b) \ \& \ Cmpeq(h, l)) \boxplus$$
$$((1 \boxplus l \boxminus h \boxplus Cmpleu(h, l)) \ \& \ (Cmpeq(h, l) \oplus (2^{16} - 1))) \ .$$

Proof: The claim follows easily from Lemma 2 of [LM90] by noticing that $h = l$ iff $a *_u b = 0$. ∎

These formulas give a direct way to break down the $\odot$ operation into basic operations, corresponding one-to-one to MMX instructions. For example, $Cmpeq(h, l)$ corresponds to the instruction pcmpeqw, $Cmpgt(h, l)$ to pcmpgtw, $Subus(a, b)$ to psubusw, $Mull(a, b)$ to pmullw, $Mulh(a, b)$ to pmulhw. The given formula for emulation of 16-bit unsigned multiplication is, as far as we know, faster than any previously published algorithm for MMX and therefore interesting in itself.

Including also the necessary move instructions, the minimal number of MMX instructions needed to emulate $\odot$ by the procedure given above is 26. Additional highly processor (and algorithm) dependent mechanisms enable to get rid of three more instructions per IDEA multiplication, therefore resulting in 69 instructions per round (remember, we do M_1^r and M_2^r in parallel). Everything else (e.g., parallel adding, xoring) is accomplished in 13 instructions, hence a round takes 82 instructions. The output transform takes 29 instructions and the necessary endianness conversion takes 8 instructions, therefore the full IDEA has 693 instructions. After pairing, IDEA encryption takes 358 clock cycles or 29.7 Mbits/s on a 166MHz MMX. Note that our implementations are not subject to timing attacks [Koc96] due to the lack of jump instructions and any variable duration instructions.

Remark 1. Schneier and Whiting [SW97] have conjectured that there exists an IDEA implementation for the Pentium with ≈ 400 cycles, which is unrealistic for the next reason. Every round has four sequential emulations of $\odot$. The critical path of the $\odot$ operation contains integer multiplication (with latency 9) and at least 6 other instructions (moving one of the operands into the accumulator and afterwards converting the result of $*_u$ to the result of $\odot$) that cannot be paired with each other, therefore the multiplications take at least 60 cycles per round. The XOR and addition operators present in the IDEA schematics cannot be paired with emulations of $\odot$ and therefore take additional time. Adding the output transform and endianness conversion, there seems to be no obvious way to significantly better the implementation of Bosselaers.

5   Parallel Execution of Four IDEA-s

In MMX, four 16-bit multiplications are executed simultaneously. The same is true for every other instruction used to emulate $\odot$. During each round, three such 4-way multiplications are done, giving a 64-bit result, only a part of which is really used in the implementation described in Section 4 (Figure 2, left):

First multiplication (M_1^r, M_2^r). The first (bits $0 \ldots 15$) and the fourth (bits $48 \ldots 63$) word of the result are used. (This multiplication is also done during the output transformation.)

Second multiplication (M_3^r). The second word (bits $16 \ldots 31$) of the result is used.

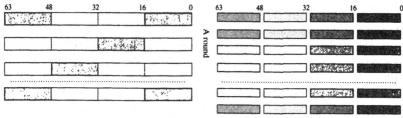

Fig. 2. Using of SWAR data during IDEA (unused fields are left blank). Left: one IDEA in parallel, right: four IDEA's in parallel.

Third multiplication (M_4^r). The third word (bits 32 ... 47) of the result is used.

We extended our implementation to encrypt four blocks in parallel, by fully using the results of all four multiplications at every step (Fig. 2, right. Note that such implementation will require unparallelizing the execution of M_1^r and M_2^r). Two 4×4 matrix transpositions (to (un)parallelize four 64-bit blocks), additional endianness conversions and extensive memory access (due to the lack of registers) will "slow" the implementation down to $\approx$ 135 cycles per IDEA encryption. A not-fully optimized implementation encrypts one IDEA block in 135.75 cycles (543 cycles/1056 instructions for 4-way IDEA), Table 2. This scales up to about 212 Mbits/s on a 450 MHz Pentium II, compared to the 300 Mbits/s of the fastest (known) hardware solution by Ascom.

Cipher	IDEA	4-way IDEA
166 MHz MMX, seconds	8.97 − 9.07	3.53 − 3.56
166 MHz MMX, Mbits/s	28.2 − 28.5	71.8 − 72.4
MMX, cycles (with overhead)	372 − 376	147 − 148
MMX, cycles (w/o overhead)	358	135.75
233 MHz Pentium II, seconds	7.78 − 7.96	2.38 − 2.43
233 MHz Pentium II, Mbits/s	32.2 − 32.9	105.1 − 107.2
Pentium II, cycles (with overhead)	453 − 464	139 − 142

Table 2. Test data. The "real life" throughput of IDEA-ECB on the Pentium MMX and on the Pentium II. Seconds - the time to encrypt four million 64-bit blocks.

6 Different Multimedia Extensions

If MMX had the unsigned multiplication instruction, the number of instructions per IDEA multiplication would decrease by 6. If MMX had the unsigned comparison instruction `pcmpgtuw`, the number of instructions per IDEA multiplication would decrease by 2. In the presence of both of these instructions, IDEA encryption on MMX machines could be done much faster than DES (we estimate $\approx 250 - 255$ cycles); 4-way IDEA would be faster than Square [DKR97] or any of the recently proposed AES candidate ciphers (we estimate $\approx 95 - 100$ cycles). Conditional move instructions, present in the Cyrix's — but not in the Intel's — version of MMX, would further speed up IDEA. If even such imperceptible changes fastened up a cipher significantly, what about the multimedia extensions that differ from MMX in major aspects?

Lately, in May 1998, Motorola unveiled their new multimedia architecture called AltiVec [Mot98], claimed to be much more powerful than any of the previously mentioned architectures. In particular, AltiVec has increased parallelism (128-bit vector registers) and a family of instructions to perform up to *eight* 16-bit (un)signed multiplications (with accumulate) in parallel. Additionally, AltiVec has a special inter-element byte permutation instruction and several vector rotation instructions and therefore allows to implement new fast ciphers using data-dependent rotations and byte permutations. One of the goals of AltiVec (unlike the MMX) was to accelerate data encryption algorithms [Mot98, page 1-4]. A short comparison between MMX and AltiVec is given in Table 3.

Architecture	MMX	AltiVec
Company	Intel	Motorola
Year	1997	1999
Endianness	little	both
Max no of operands	2	4
♯(vector registers)	8 (FP)	32 (separate)
Width of vector registers	64	128
8-bit parallelism	8	16
16-bit parallelism	4	8
32-bit parallelism	2	4
16×16-bit signed multiplication	Yes	Yes
16×16-bit unsigned multiplication	No	Yes
Signed comparison	Yes	Yes
Unsigned comparison	No	Yes
Data-dependent rotation	No	Yes

Table 3. Short comparison of MMX and AltiVec.

Vector processors provide even more parallelism. Krste Asanović has reported an implementation of 32-way IDEA on a 40 MHz T0 [AJ96] reaching 112 Mbits/s. No "industry-standard" block cipher has that level of inner parallelism.

7   Block Cipher Parallelization

Ciphers using S-boxes and/or lookup tables (e.g., DES, alleged RC4, SEAL, Blowfish, Khufu) do not take major advantage from the multimedia extensions of MMX (though they could benefit from the larger cache or word-size) as the MMX registers cannot be used as memory pointers. Parallelization of these ciphers would need accessing several "randomly" chosen memory cells simultaneously. RC5 [Riv95], which does not use S-boxes, does not benefit from MMX either because of the expensive non-parallelizable variable rotation involved.

It is interesting to note that some of the newest block ciphers, including the AES candidates MARS [BCD+98] and RC6 [RRSY98], rely on the 32-bit unsigned multiplication. The reasoning of the authors is that such multiplication is very cheap on nowadays common microprocessors. This claim is indeed true, but MMX technology cannot be used to accelerate these ciphers (and neither can AltiVec) because of the lack of a 32-bit parallel multiplication. There is a certain tradeoff (and even a contradiction) here. MARS and RC6 are optimized for the new 32-bit processors (mainly for the Pentium II), utilizing fully the 32-bit operations provided by such processors. At the same time, these ciphers ignore the multimedia extensions existing in the very same processors.

Further work can be done in trying to optimize different conventional ciphers for the Pentium MMX, but as it was pointed out, most of the commonly known block ciphers do not benefit from MMX. Still, in some cases interleaving Pentium integer and MMX instructions may result in some speedup. In particular, bit-slice MMX implementations of different block ciphers should be more than twice as fast because of the longer wordsize and additional logical operations.

One could think that MMX was designed "especially" to accelerate IDEA, but it would be more correct to say that IDEA is a cipher with key attributes very similar to those of multimedia applications (cf Sect. 3), by a loose definition of multimedia applications as applications benefiting from the Pentium MMX (different vendors have optimized their processors to be optimal for different subsets of multimedia applications).

A family of new block ciphers can be designed to take full advantage of MMX. A straightforward way would be to iteratively execute four copies of the IDEA round function in parallel and then mix their outputs in a suitable way. Would it be sufficient to apply a well chosen 8/16-bit word permutation to the 256-bit output of every round of this 4-way IDEA to get a secure cipher? A way providing more efficient diffusion would be to use Pseudo-Hadamard Transforms [Mas94,SKW+98]. Further research in this area is deferred to a future work. An interested reader may turn to [Cla97], where parallelized versions of the stream cipher Wake were proposed.

A more general task is to study design principles of secure ciphers based on the same basic operations (e.g., massively parallel 16-bit multiplication and addition of sequential data) as the existing multimedia applications. Such ciphers would perform well on nowadays microprocessors, therefore reducing the need for separate encryption and multimedia hardware (it can be compared to the approach of [BP97] that uses the same hardware for RSA and IDEA). Efficient confusion on such ciphers may be achieved by using 16-bit multiplication mixed with other 8-bit and 16-bit operations; diffusion may be achieved by additionally using 32-bit and 64-bit operations (e.g., shifts — but remember the "small data type" paradigm).

Yet another task is to study encryption modes allowing fast parallel encryption and decryption. The ECB mode can be used for both parallel encryption and decryption, but it has limited security in real life situations. The CBC mode can be used for parallel decryption but not for parallel encryption. The resulting throughput of IDEA encryption on a 233 MHz Pentium II would be 32 – 33 Mbits/s for encryption and 105 – 107 Mbits/s for decryption in standard CBC mode (Table 2). Encryption modes allowing both fast parallel encryption and decryption are needed. Note that such encryption modes are not only important for software but also for hardware architectures. The hardware solution mentioned before provides a throughput of 300 Mbits/sec in ECB mode, and a throughput of 100 Mbits/sec in the other modes. An example candidate is the counter mode [MOV96, Sect. 7.2.2] which allows parallel encryption/decryption while providing almost ideal security in the random oracle model [BDJR97] but which is not suited for use with differentially weak ciphers [BK98].

One could see the problem also from the viewpoint of a processor designer and ask what (minimal) extensions should be added to an existing general-purpose processor to achieve significant speedup of industry-standard cryptographic primitives. While the general answer seems to be out of our reach due to the diversity of cryptographic primitives, suggestions can be given to accelerate any fixed primitive (see discussion in the beginning of Sect. 6).

8   Conclusion

We have shown that it is possible to speed up the IDEA block cipher significantly by using the MMX extensions of Intel's Pentium processor. This is remarkable when taking into account the unfriendliness of the instruction set of MMX. Our fast implementation is

- 1.65 times faster than the best known assembler implementation on the Pentium by Antoon Bosselaers,
- ≈ 2.55 times faster than the C version on the Pentium in the popular library SSLeay v0.90b, when compiled with egcs 1.0.2 and full optimization.

By parallelizing four IDEA's, the encryption speed is increased by a factor of about 2.64 times, giving a total acceleration of 4.35 times compared to the implementation of Bosselaers. Implications (including the massive parallel key search) of using such parallel versions of conventional ciphers were already described in [Bih97] and were not repeated in this paper.

By noting that most of the nowadays "industry-standard" block ciphers do not benefit from MMX, we raise the problem of designing block ciphers (and encryption modes) fully utilizing the basic operations of multimedia.

9   Acknowledgements

Complete development was done on a **Linux** machine, using **g++** compiler and **gas** assembler. We are thankful to Antoon Bosselaers for proof-reading and for providing the speed numbers in Table 1, and to Mark Tehver and anonymous referees for several valuable remarks.

References

[AJ96]   Krste Asanović and David Johnson. Torrent Architecture Manual. Technical report, The International Computer Science Institution, Berkley, December 1996. Technical report TR-96-056.

[BCD$^+$98] Carolynn Burwick, Don Coppersmith, Edward D'Avignon, Rosario Gennaro, Shai Halevi, Charanjit Jutla, Stephen M. Matyas Jr., Luke O'Connor, Mohammad Peyravian, David Safford, and Nevenko Zunic. MARS — A Candidate Cipher for AES. Available at http://www.research.ibm.com/security/mars.html, June 1998.

[BDJR97]   Mihir Bellare, Anand Desai, E. Jokipii, and Phil Rogaway. A Concrete Security Treatment of Symmetric Encryption: Analysis of the DES Modes of Operation. In *Proceedings of 38th Annual Symposium on Foundations of Computer Science*, 1997.

[BGV96]   Antoon Bosselaers, René Govaerts, and Joos Vandewalle. Fast Hashing on the Pentium. In Neal Koblitz, editor, *Advances in Cryptology — CRYPTO '96*, volume 1109 of *Lecture Notes in Computer Science*, pages 298–312. Springer-Verlag, 1996.

[Bih97]   Eli Biham. A Fast New DES Implementation in Software. In Eli Biham, editor, *Fast Software Encryption '97*, volume 1267 of *Lecture Notes in Computer Science*, pages 261–272. Springer-Verlag, 1997.

[BK98]   Alex Biryukov and Eyal Kushilevitz. From Differential Cryptanalysis to Ciphertext-Only Attacks. In Hugo Krawczyk, editor, *Advances in Cryptology — CRYPTO '98*, volume 1462 of *Lecture Notes in Computer Science*, pages 72–88. Springer-Verlag, 1998.

[BKR97]   Johan Borst, Lars R. Knudsen, and Vincent Rijmen. Two Attacks on Reduced IDEA. In Walter Fumy, editor, *Advances in Cryptology — EUROCRYPT '97*, pages 1–13. Springer-Verlag, 1997.

[Bos97]   Antoon Bosselaers. Even faster hashing on the Pentium. Presented at the rump session of Eurocrypt'97, 1997.

[BP97]   Ahto Buldas and Jüri Poldre. A VLSI implementation of RSA and IDEA encryption engine. In *NORCHIP '97*, 1997.

[Cla97]   Craig S. K. Clapp. Optimizing a Fast Stream Cipher for VLIW, SIMD, and Superscalar Processors. In Eli Biham, editor, *Fast Software Encryption '97*, volume 1267 of *Lecture Notes in Computer Science*, pages 273–287. Springer-Verlag, 1997.

[DC98]   Joan Daemen and Craig S. K. Clapp. Fast Hashing and Stream Encryption with PANAMA. In Serge Vaudenay, editor, *Fast Software Encryption '98*, volume 1372 of *Lecture Notes in Computer Science*, pages 60–74. Springer-Verlag, 1998.

[DGV94]   Joan Daemen, René Govaerts, and Joos Vandewalle. Weak Keys for IDEA. In Douglas R. Stinson, editor, *Advances in Cryptology — CRYPTO '93*, volume 773 of *Lecture Notes in Computer Science*, pages 224–231. Springer-Verlag, 1994.

[Die97]   Hans Dietz. Technical Summary: SWAR Technology. Technical report, School of Electrical and Computer Engineering, Purdue University, February 1997. Available at http://dynamo.ecn.purdue.edu/~hankd/SWAR/over.html.

[DKR97]   Joan Daemen, Lars Knudsen, and Vincent Rijmen. The block cipher Square. In Eli Biham, editor, *Fast Software Encryption*, volume 1267 of *Lecture Notes in Computer Science*, pages 149–165. Springer-Verlag, 1997.

[Haw98]   Philip Hawkes. Differential-Linear Weak Key Classes of IDEA. In Kaisa Nyberg, editor, *Advances in Cryptology — EUROCRYPT '98*, volume 1403 of *Lecture Notes in Computer Science*, pages 112–126. Springer-Verlag, 1998.

[HK97]   Shai Halevi and Hugo Krawczyk. MMH: Software Message Authentication in the Gbit/Second Rates. In Eli Biham, editor, *Fast Software Encryption '97*, volume 1267 of *Lecture Notes in Computer Science*, pages 172–189. Springer-Verlag, 1997.

[Int97a]   Intel. *Intel Architecture Optimization Manual*, 1997. Order Number 242816-003.

[Int97b]   Intel. *Intel Architecture Software Developer's Manual. Volume 1: Basic architecture*, 1997. Order Number 243190.

[Koc96]   Paul Kocher. Timing Attacks on Implementations of Diffie-Hellman, RSA, DSS, and Other Systems. In Neal Koblitz, editor, *Advances in Cryptology — CRYPTO '88*, volume 1109 of *Lecture Notes in Computer Science*, pages 104–113, 1996.

[Kum97]   Masayasu Kumagai. Implementation of IDEA on MMX. Available at http://www2s.biglobe.ne.jp/~kumagai/idea_mmx.zip, April 1997.

[Kwa98]   Michael Kwan. Bitslice DES. Unpublished. Information available from http://www.cs.mu.oz.au/ mkwan/bitslice/Welcome.html, May 1998.

[LM90]   Xuejia Lai and James Massey. A proposal for a new block encryption standard. In Ivan Bjerre Damgård, editor, *Advances in Cryptology — EUROCRYPT '90*, volume 473 of *Lecture Notes in Computer Science*, pages 389–404. Springer-Verlag, 1990.

[LMM91]   Xuejia Lai, James L. Massey, and Sean Murphy. Markov Ciphers and Differential Cryptanalysis. In D. W. Davies, editor, *Advances in Cryptology — EUROCRYPT '91*, volume 547 of *Lecture Notes in Computer Science*, pages 17–38. Springer-Verlag, 1991.

[Mas94]   J. Massey. SAFER K-64: A Byte-Oriented Block-Ciphering Algorithm. In Ross Anderson, editor, *Fast Software Encryption*, volume 809 of *Lecture Notes in Computer Science*, pages 1–17. Springer Verlag, 1994.

[Mot98]   Motorola. *AltiVec Technology Programming Environments Manual*, May 1998. A preliminary revision 0.2.

[MOV96]   Alfred J. Menezes, Paul C. Van Oorschot, and Scott A. Vanstone. *Handbook of Applied Cryptography*. CRC Press, 1996.

[PRB98]   Bart Preneel, Vincent Rijmen, and Antoon Bosselaers. Recent Developments in the Design of Conventional Algorithms. In B. Preneel, R. Govaerts, and J. Vandewalle, editors, *Computer Security and Industrial Cryptography, State of the Art and Evolution*, volume 1528 of *Lecture Notes in Computer Science*, pages 90–115. Springer-Verlag, 1998.

[Riv95]   Ronald L. Rivest. The RC5 Encryption Algorithm. In Bart Preneel, edi-
            tor, *Fast Software Encryption*, volume 1008 of *Lecture Notes in Computer
            Science*, pages 86–96. Springer-Verlag, 1995.
[RRSY98] Ronald L. Rivest, Matt J. B. Robshaw, R. Sidney, and Y. L. Yin. The RC6
            Block Cipher. Available at `http://theory.lcs.mit.edu/~rivest/rc6.ps`,
            June 1998.
[SAM97]   Takeshi Shimoyama, Seiichi Amada, and Shiho Moriai. Improved Fast Soft-
            ware Implementation of Block Ciphers. In *International Conference on In-
            formation and Communications Security '97*, volume 1334 of *Lecture Notes
            in Computer Science*, pages 269–273, September 1997.
[SKW+98] Bruce Schneier, John Kelsey, Doug Whiting, David Wagner, Chris Hall,
            and Niels Ferguson. Twofish: A 128-Bit Block Cipher. In *Selected Areas
            in Cryptography '98*, June 1998. Lecture Notes in Computer Science (these
            proceedings).
[SW97]   Bruce Schneier and Doug Whiting. Fast Software Encryption: Designing En-
            cryption Algorithms for Optimal Software Speed on the Intel Pentium Pro-
            cessor. In Eli Biham, editor, *Fast Software Encryption '97*, volume 1267 of
            Lecture Notes in Computer Science, pages 242–259. Springer-Verlag, 1997.

A Strategy for Constructing Fast Round Functions with Practical Security Against Differential and Linear Cryptanalysis

Masayuki Kanda[1], Youichi Takashima[2], Tsutomu Matsumoto[3], Kazumaro Aoki[1], and Kazuo Ohta[1]

[1] NTT Information and Communication Laboratories
[2] NTT Human Interface Laboratories
1-1 Hikarino-oka Yokosuka-shi Kanagawa 239-0847 Japan
{kanda, ohta}@sucaba.isl.ntt.co.jp, maro@isl.ntt.co.jp
yoh@mistral.hil.ntt.co.jp
[3] Yokohama National University
79-5 Tokiwadai Hodogaya-ku Yokohama-shi Kanagawa 240-8501 Japan
tsutomu@mlab.dnj.ynu.ac.jp

Abstract. In this paper, we study a strategy for constructing fast and practically secure round functions that yield sufficiently small values of the maximum differential and linear probabilities p, q. We consider mn-bit round functions with 2-round SPN structure for Feistel ciphers.

In this strategy, we regard a linear transformation layer as an $n \times n$ matrix P over $\{0,1\}$. We describe the relationship between the matrix representation and the actual construction of the linear transformation layer. We propose a search algorithm for constructing the optimal linear transformation layer by using the matrix representation in order to minimize probabilities p, q as much possible. Furthermore, by this algorithm, we determine the optimal linear transformation layer that provides $p \leq p_s^5$, $q \leq q_s^5$ in the case of $n = 8$, where p_s, q_s denote the maximum differential and linear probabilities of s-box.

1 Introduction

1.1 Background

Differential cryptanalysis (DC) [6] proposed by Biham and Shamir and linear cryptanalysis (LC) [15] proposed by Matsui are the most powerful approaches to attacking most symmetric block ciphers. Accordingly, the designer should evaluate the security of any new proposed symmetric cipher against DC/LC and prove that it is sufficient invulnerable against them. It is known that there are four measures in order to evaluate the security of a cipher against DC/LC.

Precise measure The maximum average of differential and linear probabilities. They are also called differential probability [14] and approximate linear hull [18].

S. Tavares and H. Meijer (Eds.): SAC'98, LNCS 1556, pp. 264–279, 1999.
© Springer-Verlag Berlin Heidelberg 1999

Lai et al. [14] and Nyberg [18] stated that the precise evaluation of the security of cipher against DC/LC should be done using this measure. Generally speaking, however, since it is infeasible to evaluate these probabilities, this measure is not practical.

Theoretical measure The upper bounds of the maximum average of differential and linear probabilities. This measure was applied to evaluate the security of MISTY [17] and the cipher $\mathcal{KN}$ [20].

These probabilities are evaluated as $2p^2$, $2q^2$ for an R-round Feistel cipher ($R \geq 4$), respectively [20], where p, q is the maximum differential and linear probabilities of the round function. Furthermore, for an R-round Feistel cipher with bijective round function ($R \geq 3$), they are evaluated as p^2, q^2, respectively [3]. Nyberg and Knudsen [20] stated that Feistel ciphers evaluated with this measure are **provably secure** against DC/LC, which means that they are theoretically invulnerable to DC/LC, if these probabilities are sufficiently small. However, when designers intend to construct provably secure ciphers, they have to construct round functions yielding extremely small values of the probabilities p, q. This is a very strong constraint on the design of round functions.

Heuristic measure The maximum differential and linear characteristic probabilities. This measure was applied to estimate the security of DES [15], FEAL [21,2] and so on.

Biham and Shamir [6] claimed that the larger differential characteristic probability is, the higher is the success rate of DC, since they exploited a single path between plaintexts and ciphertexts which holds with significant differential characteristic probability. Matsui [15] also claimed the same for LC. However, these probabilities only give the lower bounds of the maximum average of differential and linear probabilities for some ciphers, since there are multiple paths between the same plaintexts and ciphertexts in practice [14, 18]. Furthermore, it takes much time to estimate them, since this involves the use of path searching algorithms, e.g., [16,21,2].

Practical measure The upper bounds of the maximum differential and linear characteristic probabilities.

As shown (3) and (4) in Sect. 2.2 below, given the maximum differential and linear probabilities of round function p, q, these probabilities are evaluated to be decreasing functions with the number of rounds of a Feistel cipher. Knudsen [11] noted that Feistel ciphers evaluated with this measure are **practically secure** against DC/LC, i.e., that they were believed to be invulnerable against DC/LC if these probabilities are less than a secure criterion, e.g., 2^{-64} for 64-bit ciphers and 2^{-128} for 128-bit ciphers. This implies that designers can construct practically secure Feistel ciphers that consist of good round functions while also considering their invulnerability to other known attacks and implementation efficiency with a moderate number of rounds.

Heys and Tavares studied the upper bounds of the maximum differential and linear characteristic probabilities of traditional SPN [9]. They focused on diffusion layers constructed by one bit operations. However, we consider that the SPN structures with diffusion layers constructed around word operations, e.g., 8-bit length, should be studied, since bit operations are hard to implement in software. Of interest, Rijmen et al. [22] introduced the branch number B for the SPN cipher, which is a lower bound of the number of active s-boxes in two consecutive rounds of a non-trivial linear trail or a non-trivial differential characteristic. The branch number B is a very similar in concept to these probabilities, since the security of an SPN cipher against DC/LC is evaluated by piling up the branch numbers every two rounds.

Of course, new ciphers have to be invulnerable against other known attacks; e.g., higher order differential attack [13, 12, 10] and interpolation attack [10]. Unfortunately, all the above measures evaluate the security against only DC/LC. That is, these evaluations do not prove that the cipher is also secure against other known attacks. For example, the cipher $\mathcal{KN}$ with 6-round, which was evaluated by the theoretical measure, was broken by higher order differential attack [10], even though the maximum average of differential and linear probabilities is less than 2^{-60}, which means that the cipher is provably secure against DC/LC [20].

Moreover, it is known empirically that, for good round functions, increasing the number of rounds will make the cipher more secure against most known attacks. Therefore, we consider that the number of rounds is more important in order to construct a secure cipher than the values of the maximum differential and linear probabilities of round function p, q. Thus, we believe that the practical measure is more useful that the theoretical measure evaluating the security of a cipher against DC/LC.

1.2   Design

Later we discuss the round functions of Feistel ciphers that are practically secure against DC/LC. Our design strategy is as follows:

(a) To enable the maximum differential and linear probabilities of round function to be evaluated, and to prove our cipher to be sufficiently invulnerable against DC/LC.
(b) To realize a fast round function.
(c) To enable it to be efficiently implemented on multiple platforms, e.g., 8-bit, 16-bit, 32-bit, and 64-bit processors.

As mentioned above, we believe that the number of rounds is more important in constructing secure ciphers than the values of the maximum differential and linear probabilities of round function p, q. Unfortunately, though increasing the number of rounds makes the cipher more secure, it also slows the encryption speed. This means that it is very difficult to construct a fast cipher that has a sufficient number of rounds, even if the round function is fast. Thus, we believe that our cipher should have a reasonable number of rounds and suitable differential and linear probabilities of round function p, q.

Furthermore, to satisfy strategy (c), we intend to use s-boxes as non-linear transformations, since the approach of realizing s-boxes by table look-up has no limitation with respect to the selection of function, software instructions of arithmetic operations depend on architectures, etc.

Thus, in this paper, we consider an mn-bit round function with m-bit s-boxes[1] to construct a fast and practically secure Feistel cipher. It is known that, for Feistel ciphers, major current round functions with s-boxes are based on (A) *1-round SPN structure*, or (B) *Recursive structure*. Examples of former are DES [8], CAST [1] and LOKI91 [5]; an example of the latter is MISTY [17].

Next, we focus on the total number of s-boxes in a cipher, which means how many s-boxes the cipher contains, not how many different s-boxes there are in the cipher. Roughly speaking, the encryption time is proportional to the total number of s-boxes under the assumptions that the run time except s-box process is negligible and that parallel processing is left out of consideration. That is, in order to construct a fast cipher, it is desirable that the total number of s-boxes in it be as small as possible. On the other hand, in order to construct a practically secure cipher, we must ensure that there are as many number of active s-boxes in it as the designer specified.

(A) 1-round SPN Structure This structure has one non-linear transformation layer with n parallel m-bit s-boxes and one linear transformation (permutation) layer.

Since the minimum number of active s-boxes in this structure is 1, the maximum differential and linear probabilities of round function p, q are equal to the maximum differential and linear probabilities of s-box p_s, q_s; i.e., $p = p_s$, $q = q_s$. It is known that p_s, q_s are equal to or larger than 2^{-m+1} when m is odd or 2^{-m+2} when m is even [7]. This means that too many rounds are required in order to construct a practically secure Feistel cipher; e.g., 33 rounds are required for a 128-bit Feistel cipher with a 64-bit bijective round function using 8-bit s-boxes whose the maximum differential and linear probabilities are $p_s = q_s = 2^{-6}$.

By the way, Nyberg [19] investigated the upper bounds of the maximum differential and linear probabilities of a generalized Feistel cipher with 1-round SPN structure, where "generalized" means "generalized Feistel structure," not "generalized linear layer." That is, unlike our research, the structure of the round function is not discussed. Furthermore, it only showed that the upper bounds of the maximum differential and linear probabilities of a cipher are of order p_s^{2n}, q_s^{2n} in the case of $n \leq 4$.

(B) Recursive Structure The recursive structure consists of nested functions with different input bit lengths.

For example, consider the 3-round recursive structure. In this case, since the number of s-boxes triples with each recursion, it is difficult to construct fast round functions because the number of s-boxes is too large. Furthermore, increasing the number of rounds rapidly increases the total number of s-boxes. Generally speaking, though the maximum differential and linear probabilities of round function p, q are extremely small, the number of rounds must be at least 8 in order to provide invulnerability against other known attacks.

---

[1] In this paper, we do not consider expand permutations in round functions.

We consider that too many s-boxes would be required in a practically secure cipher using 1-round SPN structure, since the maximum differential and linear probabilities of round function p, q are not sufficiently small. On the other hand, we also consider that it is difficult to increase the number of rounds of a cipher that use the recursive structure, since the number of s-boxes in the round function is too large. Accordingly, we need a new round function structure that makes the probabilities p, q much smaller than those achieved with the 1-round SPN structure and that requires a lot fewer s-boxes than the recursive structure.

1.3   2-round SPN Structure Approach

Our strategy is based on using mn-bit round functions consisting of three-layers (see Fig. 1): "1st non-linear transformation layer with n parallel m-bit s-boxes, linear transformation layer, and 2nd non-linear transformation layer with n parallel m-bit s-boxes" in this order. We call this the **2-round SPN structure** hereafter. Each layer has the following feature.

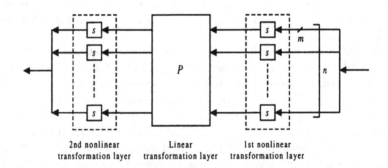

Fig. 1. 2-round SPN structural round function

1st non-linear transformation layer n parallel m-bit s-boxes are set in DES-like manner. In the following, we call this "1st non-linear layer."

2nd non-linear transformation layer n parallel m-bit s-boxes are set similar to the 1st non-linear layer. With the addition of this layer, the maximum differential and linear probabilities of round function become much smaller. In the following, we call this "2nd non-linear layer."

Linear transformation layer This makes the maximum differential and linear probabilities as small as possible for all non-zero input differences and all non-zero output mask values of round function[2]. Here, we consider that this is constructed only with bitwise XORs. That is, inputs are transformed

---

[2] Vaudenay proposed to use multipermutation as diffusion (permutation) [23]. Multipermutation is a good cryptographic primitive, but, it is very hard to satisfy the multipermutation requirements, and to make multipermutation with large bit size.

linearly to outputs per m-bits, especially per byte in the case of $m = 8$. In the following, we call this "linear layer."

By the way, all round subkeys are bitwise XORed with data before each s-box in order to form a Markov cipher [14]. Accordingly, we take no account of key addition layer here.

1.4   Result

For the 2-round SPN structure, an interesting issue is how to construct the optimal linear layer, since the two non-linear layers are already set. To construct the optimal linear layer, we represent a linear layer as an $n \times n$ matrix P over $\{0, 1\}$, and describe how to determine the matrix elements in order to minimize the differential and linear probabilities of round function p, q as much as possible.

In this paper, we propose a new search algorithm based on matrix representation for constructing the optimal linear layer. Furthermore, with this algorithm, we determine the optimal linear transformation layer that provides probabilities $p \leq p_s^5$, $q \leq q_s^5$ where $n = 8$, where p_s, q_s are the maximum differential and linear probabilities of the m-bit s-boxes in the non-linear layers.

As shown in (3) and (4) in Sect. 2.2 below, the number of rounds must be dependent on the maximum differential and linear probabilities of round function p, q to construct a practically secure cipher. We show that the round function with the 2-round SPN structure requires one-fourth as many rounds as the 1-round SPN structure to achieve the same differential and linear probabilities, while each round has double the number of s-boxes. Hence, the round function using the 2-round SPN structure is twice as efficient as that using the 1-round SPN structure.

1.5   Organization

This paper is organized as follows. In Sect. 2, we introduce some preliminary definitions. In Sect. 3, we describe the relationship between the matrix representation and the actual construction of the linear transformation layer. We propose a new search algorithm for constructing the optimal linear layer. In Sect. 4, we show the optimal linear layer for the case of $n = 8$. Finally, we conclude with a summary in Sect. 5.

2   Preliminaries

2.1   Notations

- $\#\{S\}$: number of elements in set S.
- $\Delta x, \Gamma x$: difference of x, mask value of x.
- $a \bullet b$: parity of bitwise product a and b.
- $P = [t_{ij}]$, $t_{ij} \in \{0, 1\}$, $0 \leq i, j < n$: matrix representation of a linear layer
- $z = {}^T[z_0, \ldots, z_{n-1}]$, $z_i \in GF(2)^m$, $0 \leq i < n$: an input to a linear layer
- $z' = {}^T[z'_0, \ldots, z'_{n-1}] = Pz$, $z'_i \in GF(2)^m$, $0 \leq i < n$: an output from a linear layer

2.2   Definitions

In this paper, we consider Feistel ciphers with mn-bit round function. We assume that a round key, which is used within one round, consists of independent and uniformly random bits and is bitwise XORed with data. Furthermore, we assume that an input also consists of independent and uniformly random bits and that all m-bit s-boxes work independently. Accordingly, we neglect the effect of the round key.

We use the following definitions in this paper.

Definition 1. *For given Δx, Γx and Δy, Γy, the differential and linear probabilities of an s-box are defined as:*

$$DP^s(\Delta x \rightarrow \Delta y) = \frac{\#\{x \in GF(2)^m | s(x) \oplus s(x \oplus \Delta x) = \Delta y\}}{2^m}$$

$$LP^s(\Gamma y \rightarrow \Gamma x) = \left(2 \times \frac{\#\{x \in GF(2)^m | x \bullet \Gamma x = s(x) \bullet \Gamma y\}}{2^m} - 1\right)^2$$

Definition 2. *The maximum differential and linear probabilities of an s-box, p_s, q_s, are defined as:*

$$p_s = \max_{\Delta x \neq 0, \Delta y} DP^s(\Delta x \rightarrow \Delta y)$$

$$q_s = \max_{\Gamma x, \Gamma y \neq 0} LP^s(\Gamma y \rightarrow \Gamma x)$$

Definition 3. *For given Δx, Γx and Δy, Γy, differential and linear probabilities of a round function with the 2-round SPN structure are defined as:*

$$p(\Delta x \rightarrow \Delta y) = \max_{\Delta z} \prod_{i=0}^{n-1} DP^s(\Delta x_i \rightarrow \Delta z_i) p(\Delta z \rightarrow \Delta z_i') DP^s(\Delta z_i' \rightarrow \Delta y_i)$$

$$q(\Gamma y \rightarrow \Gamma x) = \max_{\Gamma z'} \prod_{i=0}^{n-1} LP^s(\Gamma y_i \rightarrow \Gamma z_i') p(\Gamma z' \rightarrow \Gamma z_i) LP^s(\Gamma z_i \rightarrow \Gamma x_i)$$

where Δx denotes $(\Delta x_0, \ldots, \Delta x_{n-1})$. Also, Δy, Γx, and Γy are denoted in the same way as Δx. Δz, Γz denote the input difference and mask value of the linear layer, and yield $(\Delta z_0, \ldots, \Delta z_{n-1})$, $(\Gamma z_0, \ldots, \Gamma z_{n-1})$, respectively. Similarly, $\Delta z'$, $\Gamma z'$ denote the output difference and mask value of the linear layer, respectively. Here, Δz transforms for each i into $\Delta z_i'$ with the linear layer, $\{GF(2)^m\}^n \rightarrow GF(2)^m$; i.e., for each i, $\exists \Delta z_i'$ s.t. $p(\Delta z \rightarrow \Delta z_i') = 1$. Similarly, $\exists \Gamma z_i$ s.t. $p(\Gamma z' \rightarrow \Gamma z_i) = 1$.

Definition 4. *The maximum differential and linear probabilities of round function p, q are defined as:*

$$p = \max_{\Delta x \neq 0, \Delta y} p(\Delta x \rightarrow \Delta y)$$

$$q = \max_{\Gamma x, \Gamma y \neq 0} q(\Gamma y \rightarrow \Gamma x)$$

Definition 5. Active s-box *is defined as an s-box given a non-zero input difference or a non-zero output mask value.*
Note: When an s-box is bijective, the s-box given a non-zero output difference or a non-zero input mask value is also an active s-box.

Definition 6. *Assume that all s-boxes are bijective. The minimum number of active s-boxes n_d, n_l in a round function with the 2-round SPN structure for DC and LC are defined as:*

$$n_d = \min_{\Delta z \neq 0} [w_H(\Delta z) + w_H(\Delta z')]$$
$$n_l = \min_{\Gamma z' \neq 0} [w_H(\Gamma z) + w_H(\Gamma z')] \tag{1}$$

where $w_H(z)$ denotes the Hamming weight of z, which means the number of non-zero subblocks from $GF(2)^m$ of z; i.e. $w_H(z) = \#\{0 \leq i < n | z_i \neq 0\}$.

Theorem 1. *Let n_d, n_l denote the minimum number of active s-boxes in a round function for DC and LC, respectively. Then, the probabilities p, q hold for*

$$p \leq p_s^{n_d}, \quad q \leq q_s^{n_l} \tag{2}$$

Definition 7. *For an R-round Feistel cipher, assume that x_i $(0 \leq i \leq R+1)$, which is an input to the i-th round function, is an independent random variable. The maximum differential and linear characteristic probabilities $DCP_{max}^{(R)}$, $LCP_{max}^{(R)}$ are defined as:*

$$DCP_{max}^{(R)} = \max_{(\Delta x_0, \Delta x_1) \neq 0, (\Delta x_R, \Delta x_{R+1})} \prod_{i=1}^{R} p(\Delta x_i \to \Delta x_{i-1} \oplus \Delta x_{i+1})$$

$$LCP_{max}^{(R)} = \max_{(\Gamma x_1, \Gamma x_0), (\Gamma x_R, \Gamma x_{R+1}) \neq 0} \prod_{i=1}^{R} q(\Gamma x_i \to \Gamma x_{i-1} \oplus \Gamma x_{i+1})$$

Theorem 2. *For a Feistel cipher with $R = 2r$, $2r+1$ rounds, the upper bounds of the maximum differential and linear characteristic probabilities are estimated as follows [11]:*

$$DCP_{max}^{(R)} \leq p^r, \quad LCP_{max}^{(R)} \leq q^r \tag{3}$$

Theorem 3. *For a Feistel cipher with $R = 3r$, $3r+1$, $3r+2$ rounds and a bijective round function, the upper bounds of the maximum differential and linear characteristic probabilities are estimated as follows:*

In case of $R = 3r, 3r+1$: $\quad DCP_{max}^{(R)} \leq p^{2r}, \quad LCP_{max}^{(R)} \leq q^{2r}$
In case of $R = 3r+2$: $\quad DCP_{max}^{(R)} \leq p^{2r+1}, LCP_{max}^{(R)} \leq q^{2r+1}$ $\tag{4}$

Theorem 4. *Concatenation Rules [4, 16]*
When "$X \dashv (Y, Z)$" denotes that X branches into Y and Z, i.e., $X = Y = Z$, the following relations hold.

$X = Y \oplus Z \Rightarrow \Delta X = \Delta Y \oplus \Delta Z$, $\Gamma X = \Gamma Y = \Gamma Z$ (XOR operation)
$X \dashv (Y, Z) \Rightarrow \Delta X = \Delta Y = \Delta Z$, $\Gamma X = \Gamma Y \oplus \Gamma Z$ (BRANCH operation)

3   Relationship between Matrix Representation and Structure

We represent the linear layer as an $n \times n$ matrix P over $\{0,1\}$. This means that inputs are transformed linearly to outputs per m-bits, per byte in the case of $m = 8$, and the linear layer can be constructed with just bitwise XORs. For example,

$$z_i' = \bigoplus_{j=0}^{n-1} t_{ij} z_j = \bigoplus_{t_{ij}=1} \{z_j\}$$

where t_{ij} denotes the element of matrix P in row i and column j.

We assume that the maximum differential and linear probabilities of the s-box is p_s, q_s. From (2) in Theorem 1, it turns out that evaluating the upper bounds of the maximum differential and linear probabilities of round function p, q is equivalent to counting the minimum number of active s-boxes n_d, n_l. On the other hand, the optimal linear layer leads to the optimal round function; i.e., the upper bounds of the probabilities $p_s^{n_d}$, $q_s^{n_l}$ are minimum. Accordingly, constructing the optimal linear layer is equivalent to determining the matrix elements $P = [t_{ij}]$ yielding the maximum value of number n_d, n_l, because $p_s \leq 1$, $q_s \leq 1$. Here, we note that n_d, n_l is obviously equal to or less than $n + 1$, i.e., $n_d, n_l \leq n + 1$, from (1) in Definition 6, because $w_H(z) \leq n$.

For example, we consider the 4×4 matrix P_E such as

$$\begin{bmatrix} z_0' \\ z_1' \\ z_2' \\ z_3' \end{bmatrix} = \begin{bmatrix} 0\,1\,1\,1 \\ 1\,0\,1\,1 \\ 1\,1\,1\,0 \\ 1\,1\,1\,1 \end{bmatrix} \begin{bmatrix} z_0 \\ z_1 \\ z_2 \\ z_3 \end{bmatrix}$$

3.1   Relationship between Matrix Representation and Features of Round Function

The round function using the linear layer represented as matrix P_E has the following features. Here, we assume that all s-boxes are bijective.

- We obtain $z_0' = 0 \cdot z_0 \oplus 1 \cdot z_1 \oplus 1 \cdot z_2 \oplus 1 \cdot z_3 = z_1 \oplus z_2 \oplus z_3$. Similarly, z_1', z_2', z_3' can be represented by XORs among z_0, z_1, z_2, z_3. Furthermore, the differential characteristic can be represented in the same way.
- The matrix P_E represents which inputs of the linear layer are combined into each output of the linear layer. Each row corresponds to each output of the linear layer, and each column corresponds to each input of the linear layer, see Fig. 2.
- Invulnerability against DC and LC can be evaluated as the minimum number of active s-boxes n_d, n_l in the round function, respectively. n_d, n_l can be obtained with the search algorithm described in Sect. 3.2.
- The Hamming weight of the column vector denotes the avalanche effect. A large value of the Hamming weight means that the round function has a good avalanche effect.

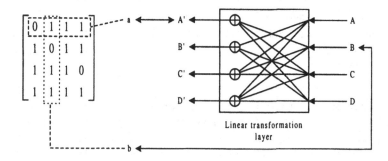

Fig. 2. Relationship between matrix representation and linear layer

As mentioned above, the preliminary features of the round function, i.e., invulnerability against DC/LC and the avalanche effect, are obtained by just matrix elements regardless of construction of linear layer.

3.2   Determination of Matrix Elements

Generally speaking, given matrix P, there are many constructions of the linear layer that correspond to matrix P. This is because matrix P denotes only the relationship between inputs and outputs of the linear layer, not the construction of the linear layer. That is, when several linear layers can be represented by the same matrix P, the round functions have same the features regardless of the constructions of the linear layers.

Accordingly, at first, we determine matrix elements in order to provide good invulnerability against DC/LC and good avalanche effect, and then realize the optimal linear layer.

Consideration of Differential Characteristic Based on the nature of differential characteristic, the matrix elements of $n \times n$ matrix P are determined by the following algorithm.

Search Algorithm

Step1 Define *security threshold* T $(2 \leq T \leq n)$.
Step2 Prepare a set of column vectors C whose Hamming weights are equal to or larger than $T - 1$.
Step3 Select a subset of n column vectors P_c from set C. Repeat the following steps3-1 and step3-2 until all subsets have been checked.
Step3-1 Compute the minimum number of active *s*-boxes n_d for subset P_c. This is represented as $n_d(P_c)$.
Step3-2 If $n_d(P_c) \geq T$, then accept the matrix consisting of subset P_c as a candidate matrix.
Step4 Output matrix P and $n_d(P)$ that yields the maximum value of n_d among the candidate matrices.

The security threshold T is used in order to restrict candidate subsets in Step 2 and Step 3-2. If candidate matrices are found by the above algorithm, then it can be proven that the minimum number of s-boxes is equal to or larger than T. That is, $p \le p_s^T$.

Consideration of Linear Expression Similarly, linear expression represents which output mask values of the linear layer are combined to yield each input mask value of the linear layer. The linear expression may be obtained with the concatenation rules in Theorem 4, when the construction of a linear layer is given. If each input mask value of the linear layer is represented by XORs among output mask values of the linear layer, the result is an other transformation matrix similar to that yielded by differential characteristic.

Here, based on our study of the relationship between the matrix for differential characteristic and the matrix for linear expression, we make the following two conjectures.

Conjecture 1. Given an $n \times n$ matrix P over $\{0, 1\}$ for a linear layer. The matrix for differential characteristic, i.e., relationship between input and output differences, is represented as the same matrix P, while the matrix for linear expression, i.e., relationship between output and input mask values, is represented as the transposed matrix $^T P$. That is, $\Delta z' = P \Delta z$, $\Gamma z = {}^T P \Gamma z'$.

Conjecture 2. The minimum number of active s-boxes n_d for differential characteristic represented as matrix P is equal to the minimum number of active s-boxes n_l for linear expression represented as the transposed matrix $^T P$.

Because of Conjecture 2, the minimum number of s-boxes n_l is also equal to or larger than T, when candidate matrices are found by the above algorithm. That is, $q \le q_s^T$. For example, matrix P_E holds the following relationship. It is proven that $n_d = 3$, $n_l = 3$ for matrices P_E and $^T P_E$, respectively.

$$\begin{bmatrix} 0 & 1 & 1 & 1 \\ 1 & 0 & 1 & 1 \\ 1 & 1 & 1 & 0 \\ 1 & 1 & 1 & 1 \end{bmatrix} \quad \Longleftrightarrow \quad \begin{bmatrix} 0 & 1 & 1 & 1 \\ 1 & 0 & 1 & 1 \\ 1 & 1 & 1 & 1 \\ 1 & 1 & 0 & 1 \end{bmatrix}$$

Matrix for differential characteristic P_E Matrix for linear expression $^T P_E$

3.3   Determining the Construction of Linear Layer

In this section, we describe how to determine the construction of a linear layer given matrix P.

Determination algorithm

Step 1 Choose two rows, and one row (row a) is XORed to the other row (row b) (called "primitive operation" hereafter).

Step 2 Transform the matrix P into the unit matrix I by the primitive operation, and count the number of the operations in order to find the transformation order yielding the minimum number of operations.

Step 3 Line A, which corresponds to row a, is XORed to line B, corresponds to row b, in reverse order of the transformation order found in Step 2.

For example, consider the matrix P_E. First, row 1 is XORed to row 2, and then row 4 is XORed to row 1. Repeat until matrix P_E is transformed into the unit matrix I. All states of transformation are described as follows.

$$\begin{bmatrix} 0\,1\,1\,1 \\ 1\,0\,1\,1 \\ 1\,1\,1\,0 \\ 1\,1\,1\,1 \end{bmatrix} \xrightarrow{(1\to 2)} \begin{bmatrix} 0\,1\,1\,1 \\ 1\,1\,0\,0 \\ 1\,1\,1\,0 \\ 1\,1\,1\,1 \end{bmatrix} \xrightarrow{(4\to 1)} \begin{bmatrix} 1\,0\,0\,0 \\ 1\,1\,0\,0 \\ 1\,1\,1\,0 \\ 1\,1\,1\,1 \end{bmatrix} \xrightarrow{(3\to 4)} \begin{bmatrix} 1\,0\,0\,0 \\ 1\,1\,0\,0 \\ 1\,1\,1\,0 \\ 0\,0\,0\,1 \end{bmatrix}$$

$$\xrightarrow{(2\to 3)} \begin{bmatrix} 1\,0\,0\,0 \\ 1\,1\,0\,0 \\ 0\,0\,1\,0 \\ 0\,0\,0\,1 \end{bmatrix} \xrightarrow{(1\to 2)} \begin{bmatrix} 1\,0\,0\,0 \\ 0\,1\,0\,0 \\ 0\,0\,1\,0 \\ 0\,0\,0\,1 \end{bmatrix}$$

For the above transformation, the number of operations is 5. That is, the linear layer can be constructed with 5 XORs.

In reverse order of the above transformation order, line 1 is XORed to line 2, and then line 2 is XORed to line 3, and so on. Finally, we construct the linear layer described in Fig. 3.

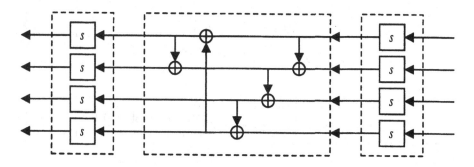

Fig. 3. Example of linear layer represented as the matrix P_E

4 Round Function with $n = 8$

In this section, we consider a round function with $n = 8$. Here, we assume that the linear layer and s-boxes are bijective, because this enables us to evaluate the invulnerability against DC/LC more precisely in the case of the bijective round function, described as (4) in Theorem 3 (Also, shown in [3]).

4.1   Determination for Matrix Elements

We determine an 8×8 matrix P $(n = 8)$ yielding the maximum value of n_d using the search algorithm of Sect. 3.2. At this time, we set the security threshold $T = 8, 7, \ldots$ in this order until we found candidate matrices. Furthermore, the following conditions were added within the algorithm.

Step3 Select subset of n column vectors P_c from the set C, and compute rank(P_c). If rank$(P_c) \neq 8$, then reject subset P_c.

Repeat the following steps until all subsets have been checked.

Step3-1 Compute the minimum number of active s-boxes n_d for P_c as follows.

- $n_d = \min\{n_{di} | 0 \leq i \leq 9\}$
- For any two columns (Columns a, b, $0 \leq i < 8$)
$$n_{d0} = 2 + \min_{(a,b)} \#\{(t_{ia}, t_{ib}) | t_{ia} \oplus t_{ib} \neq 0\}$$
- For any three columns (Columns a, b, c, $0 \leq i < 8$)
$$n_{d1} = 3 + \min_{(a,b,c)} \#\{(t_{ia}, t_{ib}, t_{ic}) | t_{ia} \oplus t_{ib} \oplus t_{ic} \neq 0\}$$
$$n_{d2} = 3 + \min_{(a,b,c)} \#\{(t_{ia}, t_{ib}, t_{ic}) | \text{Exception of } (0,0,0), (1,1,1)\}$$
- For any four columns (Columns a, b, c, d, $0 \leq i < 8$)
$$n_{d3} = 4 + \min_{(a,b,c,d)} \#\{(t_{ia}, t_{ib}, t_{ic}, t_{id}) | t_{ia} \oplus t_{ib} \oplus t_{ic} \oplus t_{id} \neq 0\}$$
$$n_{d4} = 4 + \min_{(a,b,c,d)} \# \left\{ (t_{ia}, t_{ib}, t_{ic}, t_{id}) \Big| \begin{array}{l} \text{Exception of } (0,0,0,0), \\ (1,1,0,0), (0,1,1,1), (1,0,1,1) \end{array} \right\}$$
$$n_{d5} = 4 + \min_{(a,b,c,d)} \# \left\{ (t_{ia}, t_{ib}, t_{ic}, t_{id}) \Big| \begin{array}{l} \text{Exception of } (0,0,0,0), \\ (1,0,1,0), (0,1,1,1), (1,1,0,1) \end{array} \right\}$$
$$n_{d6} = 4 + \min_{(a,b,c,d)} \# \left\{ (t_{ia}, t_{ib}, t_{ic}, t_{id}) \Big| \begin{array}{l} \text{Exception of } (0,0,0,0), \\ (1,0,0,1), (0,1,1,1), (1,1,1,0) \end{array} \right\}$$
$$n_{d7} = 4 + \min_{(a,b,c,d)} \# \left\{ (t_{ia}, t_{ib}, t_{ic}, t_{id}) \Big| \begin{array}{l} \text{Exception of } (0,0,0,0), \\ (0,1,1,0), (1,0,1,1), (1,1,0,1) \end{array} \right\}$$
$$n_{d8} = 4 + \min_{(a,b,c,d)} \# \left\{ (t_{ia}, t_{ib}, t_{ic}, t_{id}) \Big| \begin{array}{l} \text{Exception of } (0,0,0,0), \\ (0,1,0,1), (1,0,1,1), (1,1,1,0) \end{array} \right\}$$
$$n_{d9} = 4 + \min_{(a,b,c,d)} \# \left\{ (t_{ia}, t_{ib}, t_{ic}, t_{id}) \Big| \begin{array}{l} \text{Exception of } (0,0,0,0), \\ (0,0,1,1), (1,1,0,1), (1,1,1,0) \end{array} \right\}$$

Intuitively, the equations from n_{d0} to n_{d9} denote the minimum number of total active s-boxes when the number of active s-boxes in the 1st non-linear layer is given.

For example, consider equation n_{d0}.

When there are two active s-boxes in the 1st non-linear layer, two input differences of the linear layer can be represented as $\Delta z_a \neq 0$, $\Delta z_b \neq 0$. Each output difference of the linear layer shows $[\Delta z_i'] = [t_{ia} \cdot \Delta z_a \oplus t_{ib} \cdot \Delta z_b]$ $(0 \leq i < 8)$. Here, when we assume $\Delta z_a = \Delta z_b$ as the relationship among two input differences, it shows $[\Delta z_i'] = [(t_{ia} \oplus t_{ib}) \cdot \Delta z_a]$ $(0 \leq i < 8)$. By the way, an active s-box in the 2nd non-linear layer means an s-box in the 2nd non-linear layer whose input difference is non-zero, i.e., $\Delta z_i' \neq 0$. Accordingly, the minimum number of active s-boxes in the 2nd non-linear layer is $\min_{(a,b)} \#\{\Delta z_i' | \Delta z_i' \neq 0, \ 0 \leq i < 8\} = \min_{(a,b)} \#\{(t_{ia} \oplus t_{ib}) | t_{ia} \oplus t_{ib} \neq 0, \ 0 \leq i < 8\}$. This yields equation n_{d0}.

Similarly, equations n_{d1} and n_{d2} are obtained with the relationships among three input differences, and the equations from n_{d3} and n_{d9} are obtained using the relationships among four input differences of the linear layer.

Using the above search algorithm, we find that there is no matrix with $n_d \geq 6$, and that there are 10080 candidate matrices with $n_d = 5$. Accordingly, the invulnerability of the round function with one of the candidate matrices against DC is evaluated as $p \leq p_s^5$. Conversely, its invulnerability against LC is evaluated as $q \leq q_s^5$ because of Conjecture 2 in Sect. 3.2.

Furthermore, for all 10080 candidate matrices, the total Hamming weight is equal to 44; each matrix consists of 4 column (row) vectors with the Hamming weight of six and 4 other column (row) vectors with the Hamming weight of five.

4.2   Construction of Linear Layer

Next, from the above 10080 candidate matrices, we construct the linear layer by the determination algorithm in Sect. 3.3. Since, however, the computation complexity of determining the construction is very large, there are $(8 \times 7)^{16} \approx 2^{93}$ candidates when the linear layer consists of 16 XORs, it is impossible to determine the construction exhaustively. Accordingly, we consider a linear layer that consists of four blocks with 8 m-bit inputs and 4 m-bit outputs, see Fig. 4(A). Each block consists of 4 XORs, and all inputs pass only one XOR, described in Fig. 4(B). The linear layer consists of 16 XORs, and the computation complexity is much lower with about $(4 \times 3 \times 2 \times 1)^4 \approx 2^{18}$ candidates.

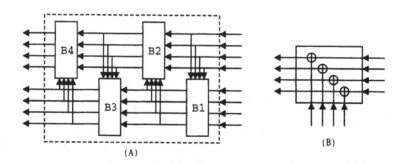

(A)   (B)

Fig. 4. Candidate construction of linear layer

This restricted search determines that 57 linear layers can be constructed from among the 10080 candidate matrices.

One of these matrices is shown below, and the linear layer with matrix P is shown in Fig. 5. Furthermore, it can be proven that, for this linear layer, the matrix for linear expression is represented as the matrix $^T P$, which is the matrix P transposed using the concatenation rules.

$$P = \begin{bmatrix} 0\,1\,1\,1\,1\,1\,1\,0 \\ 1\,0\,1\,1\,0\,1\,1\,1 \\ 1\,1\,0\,1\,1\,0\,1\,1 \\ 1\,1\,1\,0\,1\,1\,0\,1 \\ 1\,1\,0\,1\,1\,1\,0\,0 \\ 1\,1\,1\,0\,0\,1\,1\,0 \\ 0\,1\,1\,1\,0\,0\,1\,1 \\ 1\,0\,1\,1\,1\,0\,0\,1 \end{bmatrix} \iff {}^{T}P = \begin{bmatrix} 0\,1\,1\,1\,1\,1\,0\,1 \\ 1\,0\,1\,1\,1\,1\,1\,0 \\ 1\,1\,0\,1\,0\,1\,1\,1 \\ 1\,1\,1\,0\,1\,0\,1\,1 \\ 1\,0\,1\,1\,1\,0\,0\,1 \\ 1\,1\,0\,1\,1\,1\,0\,0 \\ 1\,1\,1\,0\,0\,1\,1\,0 \\ 0\,1\,1\,1\,0\,0\,1\,1 \end{bmatrix}$$

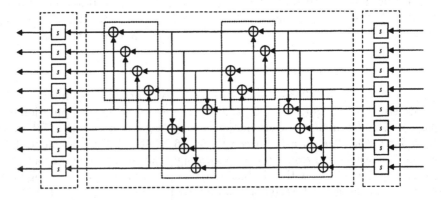

Fig. 5. Example of linear layer in the case of $n = 8$

5   Conclusion

This paper studied a strategy for constructing mn-bit round functions with the 2-round SPN structure that yield sufficiently small values of the maximum differential and linear probabilities p, q. This strategy regards a linear transformation layer as an $n \times n$ matrix P over $\{0,1\}$. We described the relationship between the matrix representation and the actual construction of the linear transformation layer. We proposed a search algorithm for constructing the optimal linear transformation layer by using the matrix representation.

Furthermore, with this algorithm, we determined the optimal linear transformation layer yielding probabilities $p \leq p_s^5$, $q \leq q_s^5$ in the case of $n = 8$.

References

1. C. M. Adams, "Simple and Effective Key Scheduling for Symmetric Ciphers," *Workshop on Selected Areas in Cryptology SAC'94*, 1994.
2. K.Aoki, K. Kobayashi, S. Moriai, "Best Differential Characteristic Search of FEAL," *Fourth International Workshop on Fast Software Encryption (FSE4)*, LNCS **1267**, 1997.

3. K. Aoki, K. Ohta, "Strict Evaluation of the Maximum Average of Differential Probability and the Maximum Average of Linear Probability," *IEICE Transactions Fundamentals of Electronics, Communications and Computer Sciences*, Vol. E80-A, No. 1, pp. 2-8, 1997.
4. E. Biham, "On Matsui's Linear Cryptanalysis," *Advances in Cryptology - EURO-CRYPT'94*, LNCS **950**, 1995.
5. L. Brown, M. Kwan, J. Pieprzyk, J. Seberry, "Improving Resistance to Differential Cryptanalysis and the Redesign of LOKI," *Advances in Cryptology - ASI-ACRYPT'91*, LNCS **739**, 1993.
6. E. Biham, A. Shamir, "Differential Cryptanalysis of DES-like Cryptosystems," *Journal of Cryptology*, Vol. 4 No. 1, pp. 3-72, 1991. (The extended abstract appeared at *CRYPTO'90*)
7. F. Chabaud, S. Vaudenay, "Links Between Differential and Linear Cryptanalysis," *Advances in Cryptology - EUROCRYPT'94*, LNCS **950**, 1995.
8. *Data Encryption Standard*, FIPS-PUB-46, 1977.
9. H. M. Heys, S. E. Tavares, "Substitution-Permutation Networks Resistant to Differential and Linear Cryptanalysis," *Journal of Cryptology*, Vol. 9 No. 1, pp. 1-19, 1996.
10. T. Jakobsen, L. R. Knudsen, "The Interpolation Attack on Block Ciphers," *Fourth International Workshop on Fast Software Encryption (FSE4)*, LNCS **1267**, 1997.
11. L. R. Knudsen, "Practically Secure Feistel Ciphers," *Cambridge Security Workshop on Fast Software Encryption (FSE1)*, LNCS **809**, 1994.
12. L. R. Knudsen, "Truncated and Higher Order Differentials," *Second International Workshop on Fast Software Encryption (FSE2)*, LNCS **1008**, 1995.
13. X. Lai, "Higher order derivatives and differential cryptanalysis," In *Proc. of Symposium on Communication, Coding, and Cryptography," in honor of James L. Massey on the occasion of his 60'th birthday*, Feb. 10-13, 1994, Monte-Verita, Ascona, Switzerland, 1994.
14. X. Lai, J. L. Massey, S. Murphy, "Markov Ciphers and Differential Cryptanalysis," *Advances in Cryptology - EUROCRYPT'91*, LNCS **547**, 1991.
15. M. Matsui, "Linear Cryptanalysis Method for DES Cipher," *Advances in Cryptology - EUROCRYPT'93*, LNCS **765**, 1994.
16. M. Matsui, "On Correlation Between the Order of S-boxes and the Strength of DES," *Advances in Cryptology - EUROCRYPT'94*, LNCS **950**, 1995.
17. M. Matsui, "New Block Encryption Algorithm MISTY," *Fourth International Workshop on Fast Software Encryption (FSE4)*, LNCS **1267**, 1997.
18. K. Nyberg, "Linear Approximation of Block Ciphers," *Advances in Cryptology - EUROCRYPT'94*, LNCS **950**, 1995.
19. K. Nyberg, "Generalized Feistel Networks," *Advances in Cryptology - ASI-ACRYPT'96*, LNCS **1163**, 1996.
20. K. Nyberg, L. R. Knudsen, "Provable Security Against a Differential Attack," *Journal of Cryptology*, Vol. 8 No. 1, pp. 27-37, 1995. (The extended abstract appeared at *CRYPTO'92*)
21. K. Ohta, S. Moriai, K. Aoki, "Improving the Search Algorithm for the Best Linear Expression.," *Advances in Cryptology - CRYPTO'95*, LNCS **963**, 1995.
22. V. Rijmen, J. Daemen, B. Preneel, A. Bosselaers, E. DcWin, "The Cipher SHARK," *Third International Workshop on Fast Software Encryption (FSE3)*, LNCS **1039**, 1996.
23. S. Vaudenay, "On the Need for Multipermutations: Cryptanalysis of MD4 and SAFER," *Second International Workshop on Fast Software Encryption (FSE2)*, LNCS **1008**, 1995.

The Nonhomomorphicity of Boolean Functions

Xian-Mo Zhang[1] and Yuliang Zheng[2]

[1] School of Info Tech & Comp Sci, the University of Wollongong, Wollongong NSW 2522, Australia. xianmo@cs.uow.edu.au
[2] School of Comp & Info Tech, Monash University, McMahons Road, Frankston, Melbourne, VIC 3199, Australia. yuliang@pscit.monash.edu.au
URL: http://www.pscit.monash.edu.au/links/

Abstract. We introduce the notion of *nonhomomorphicity* as an alternative criterion that forecasts nonlinear characteristics of a Boolean function. Although both *nonhomomorphicity* and *nonlinearity* reflect a "difference" between a Boolean function and all the affine functions, they are measured from different perspectives. We are interested in nonhomomorphicity due to several reasons that include (1) unlike other criteria, we have not only established tight lower and upper bounds on the nonhomomorphicity of a function, but also precisely identified the mean of nonhomomorphicity over all the Boolean functions on the same vector space, (2) the nonhomomorphicity of a function can be estimated efficiently, and in fact, we demonstrate a fast statistical method that works both on large and small dimensional vector spaces.
Key Words: Boolean Functions, Cryptography, Nonhomomorphicity, Nonlinear Characteristics.

1 Motivation of this Research

It is known that a function f on V_n is affine if and only if f satisfies such property that for any even k with $k \geq 4$,

$$f(u_1) \oplus \cdots \oplus f(u_k) = 0 \tag{1}$$

whenever $u_1 \oplus \cdots \oplus u_k = 0$.

In addition, it can be verified that f is affine if and only if there exists an even k with $k \geq 4$ such that (1) holds whenever $u_1 \oplus \cdots \oplus u_k = 0$. Therefore we regard (1) as a characteristic that is useful in telling a non-affine function from an affine one.

Now consider a non-affine function f on V_n. Let k be an even with $k \geq 4$ and $(u_1, \ldots, u_k)$ be a k-tuples with $u_1 \oplus \cdots \oplus u_k = 0$. If

$$f(u_1) \oplus \cdots \oplus f(u_k) = 0$$

then f satisfies the affine property at the particular vector $(u_1, \ldots, u_k)$. On the other hand, if

$$f(u_1) \oplus \cdots \oplus f(u_k) = 1$$

S. Tavares and H. Meijer (Eds.): SAC'98, LNCS 1556, pp. 280–295, 1999.
© Springer-Verlag Berlin Heidelberg 1999

then f behaves in a way that is against the affine property at $(u_1, \ldots, u_k)$.

The above observations motivate us to define the number of k-tuples of vectors in V_n, $(u_1, \ldots, u_k)$ with $u_1 \oplus \cdots \oplus u_k = 0$ such that the affine property (1) is satisfied, as the *homomorphicity* of f, and furthermore, the number of k-tuples of vectors in V_n, $(u_1, \ldots, u_k)$ with $u_1 \oplus \cdots \oplus u_k = 0$ such that the affine property (1) is not satisfied, as the *nonhomomorphicity* of f.

While nonhomomorphicity and nonlinearity are similar to each other in that they both reflect a "distance" between a Boolean function and all the affine functions, the former differentiates itself from the latter in the way the "distance" is measured. Nonhomomorphicity has several interesting properties suggesting that it can serve as a useful nonlinearity indicator: (1) unlike other criteria, we have not only established the tight lower and upper bounds on nonhomomorphicity, but also precisely identified the mean of nonhomomorphicity over all the Boolean functions with the same size, (2) the nonhomomorphicity of a function can be estimated efficiently. In fact, we show a fast statistical method for estimating the nonhomomorphicity of a function. The computing time of the statistical method is not relevant to the dimension (number of variables) of the function. This guarantees that we can use a computer program to analyze Boolean functions of higher dimensions efficiently.

2   Introduction to Boolean Functions

Denote by V_n the vector space of n tuples of elements from $GF(2)$. The *truth table* of a function f from V_n to $GF(2)$ (or simply functions on V_n) is a $(0, 1)$-sequence defined by $(f(\alpha_0), f(\alpha_1), \ldots, f(\alpha_{2^n-1}))$, and the *sequence* of f is a $(1, -1)$-sequence defined by $((-1)^{f(\alpha_0)}, (-1)^{f(\alpha_1)}, \ldots, (-1)^{f(\alpha_{2^n-1})})$, where $\alpha_0 = (0, \ldots, 0, 0)$, $\alpha_1 = (0, \ldots, 0, 1)$, ..., $\alpha_{2^{n-1}-1} = (1, \ldots, 1, 1)$. f is said to be *balanced* if its truth table contains an equal number of ones and zeros.

Given two sequences $\tilde{a} = (a_1, \cdots, a_m)$ and $\tilde{b} = (b_1, \cdots, b_m)$, their *component-wise product* is defined by $\tilde{a} * \tilde{b} = (a_1 b_1, \cdots, a_m b_m)$. In particular, if $m = 2^n$ and $\tilde{a}$, $\tilde{b}$ are the sequences of functions on V_n respectively, then $\tilde{a} * \tilde{b}$ is the sequence of $f \oplus g$.

Let $\tilde{a} = (a_1, \cdots, a_m)$ and $\tilde{b} = (b_1, \cdots, b_m)$ be two vectors (or sequences), the *scalar product* of $\tilde{a}$ and $\tilde{b}$, denoted by $\langle \tilde{a}, \tilde{b} \rangle$, is defined as the sum of the component-wise multiplications. In particular, when $\tilde{a}$ and $\tilde{b}$ are from V_m, $\langle \tilde{a}, \tilde{b} \rangle = a_1 b_1 \oplus \cdots \oplus a_m b_m$, where the addition and multiplication are over $GF(2)$, and when $\tilde{a}$ and $\tilde{b}$ are $(1, -1)$-sequences, $\langle \tilde{a}, \tilde{b} \rangle = \sum_{i=1}^{m} a_i b_i$, where the addition and multiplication are over the reals.

A $(1, -1)$-matrix H of order m is called a *Hadamard* matrix if $HH^t = mI_m$, where H^t is the transpose of H and I_m is the identity matrix of order m. A Sylvester-Hadamard matrix of order 2^n, denoted by H_n, is generated by the following recursive relation

$$H_0 = 1, \quad H_n = \begin{bmatrix} H_{n-1} & H_{n-1} \\ H_{n-1} & -H_{n-1} \end{bmatrix}, \quad n = 1, 2, \ldots. \tag{2}$$

Let ℓ_i, $0 \leq i \leq 2^n - 1$, be the i row of H_n. Then ℓ_i is the sequence of a linear function $\varphi_i(x)$ defined by the scalar product $\varphi_i(x) = \langle \alpha_i, x \rangle$, where α_i is the ith vector in V_n according to the ascending lexicographic order. (See for instance Lemma 2 of [7].)

Definition 1. *A function f on V_n is called an* affine *function if $f(x) = c \oplus a_1 x_1 \oplus \cdots \oplus a_n x_n$ where and each a_j and c are constant in $GF(2)$. In particular, f is called a* linear *function if $c = 0$.*

Definition 2. *The* Hamming weight *of a $(0, 1)$-sequence ξ is the number of ones in the sequence. Given two functions f and g on V_n, the* Hamming distance *$d(f, g)$ between them is defined as the Hamming weight of the truth table of $f(x) \oplus g(x)$, where $x = (x_1, \ldots, x_n)$. The* nonlinearity *of f, denoted by N_f, is the minimal Hamming distance between f and all the affine functions on V_n, i.e., $N_f = \min_{i=1,2,\ldots,2^{n+1}} d(f, \varphi_i)$ where $\varphi_1, \varphi_2, \ldots, \varphi_{2^{n+1}}$ are all the affine functions on V_n.*

It is known that the nonlinearity of a function f on V_n can be expressed as

$$N_f = 2^{n-1} - \frac{1}{2} \max\{ |\langle \xi, \ell_i \rangle|, 0 \leq i \leq 2^n - 1 \} \tag{3}$$

where ξ is the sequence of f and $\ell_0, \ldots, \ell_{2^n-1}$ are the rows of H_n, namely, the sequences of the linear functions on V_n. (For a proof of (3) see for instance Lemma 6 of [7].) In addition, the maximum nonlinearity of a function is $2^{n-1} - 2^{\frac{1}{2}n-1}$, namely, $N_f \leq 2^{n-1} - 2^{\frac{1}{2}n-1}$.

Given a function f on V_n, a $(1, -1)$ matrix defined by $M = ((-1)^{f(\alpha_i \oplus \alpha_j)})$, where $\alpha_i, \alpha_j \in V_n$ and $0 \leq i, j \leq 2^n - 1$, is called the $(1, -1)$ incidence matrix, or simply, the matrix of f. The following is attributed to R. L. McFarland [2]:

$$M = 2^{-n} H_n \operatorname{diag}(\langle \xi, \ell_0 \rangle, \langle \xi, \ell_1 \rangle, \ldots, \langle \xi, \ell_{2^n-1} \rangle) H_n \tag{4}$$

where ξ be the sequence of function f on V_n, ℓ_i be the ith row of H_n, and $\operatorname{diag}(a, b, \cdots, c)$ denotes the diagonal matrix whose entries on the diagonal are $a, b, \ldots, c$.

A function f on V_n is called a bent function [6] if $\langle \xi, \ell_i \rangle^2 = 2^n$ for every $i = 0, 1, \ldots, 2^n - 1$, where ξ is the sequence of f and ℓ_i is a row in H_n. A bent function on V_n exists only when n is a positive even number, and it achieves the highest possible nonlinearity $2^{n-1} - 2^{\frac{1}{2}n-1}$.

3   Homomorphicity and Nonhomomorphicity

The following lemma is important in this paper, as it explores a characteristic property of affine functions which will be useful in studying nonhomomorphicity.

Lemma 1. *Let f be a function on V_n. Then*

(i) f is an affine function if and only if f satisfies such property that for any even k with $k \geq 4$, $f(u_1) \oplus \cdots \oplus f(u_k) = 0$ whenever $u_1 \oplus \cdots \oplus u_k = 0$,

(ii) f is an affine function if and only if there exists an even k with $k \geq 4$ such that $f(u_1) \oplus \cdots \oplus f(u_k) = 0$ whenever $u_1 \oplus \cdots \oplus u_k = 0$.

Proof. Let f be a function on V_n. We first prove Part (ii) of the lemma.

Assume that f is affine. By using Definition 1, it is easy to verify that for any even k with $k \geq 4$, $f(u_1) \oplus \cdots \oplus f(u_k) = 0$ whenever $u_1 \oplus \cdots \oplus u_k = 0$. Conversely, assume that there exists an even k with $k \geq 4$ such that $f(u_1) \oplus \cdots \oplus f(u_k) = 0$ whenever $u_1 \oplus \cdots \oplus u_k = 0$. We now prove that f is affine.

Let u_1 and u_2 be any two vectors in V_n. Obviously, the k vectors u_1, u_2, $u_1 \oplus u_2, 0, \ldots, 0$ satisfy $u_1 \oplus u_2 \oplus (u_1 \oplus u_2) \oplus 0 \oplus \cdots \oplus 0 = 0$. From the assumption,

$$f(u_1) \oplus f(u_2) \oplus f(u_1 \oplus u_2) \oplus f(0) \oplus \cdots \oplus f(0) = 0 \qquad (5)$$

Consider two cases: $f(0) = 0$ and $f(0) = 1$.

Case 1: $f(0) = 0$. In this case $f(c\alpha) = cf(\alpha)$ holds for any vector $\alpha \in V_n$ and any value $c \in GF(2)$. Hence (5) can be rewritten as

$$f(u_1 \oplus u_2) = f(u_1) \oplus f(u_2) \qquad (6)$$

where u_1 and u_2 are arbitrary.

Let e_j denote the vector in V_n, whose the jth component is one and others are zero. For any fixed value x_j in $GF(2)$, $j = 1, \ldots, n$, from (6), $f(x_1 e_1 \oplus \cdots \oplus x_n e_n) = f(x_1 e_1) \oplus f(x_2 e_2 \oplus \cdots \oplus x_n e_n)$ Applying (6) repeatedly, we have $f(x_1 e_1 \oplus \cdots \oplus x_n e_n) = f(x_1 e_1) \oplus f(x_2 e_2) \oplus \cdots \oplus f(x_n e_n)$ Note that $f(0) = 0$ implies $f(c\alpha) = cf(\alpha)$ where c is any value in $GF(2)$ and α is any vector in V_n. Hence

$$f(x_1 e_1 \oplus \cdots \oplus x_n e_n) = x_1 f(e_1) \oplus \cdots \oplus x_n f(e_n) \qquad (7)$$

From the definition of e_j, $x_1 e_1 \oplus \cdots \oplus x_n e_n = (x_1, \ldots, x_n)$. On the other hand, if we write $f(e_j) = a_j$, $j = 1, \ldots, n$ then (7) can be rewritten as $f(x_1, \ldots, x_n) = a_1 x_1 \oplus \cdots \oplus a_n x_n$ This proves that f is linear.

Case 2: $f(0) = 1$. Set $g(x) = 1 \oplus f(x)$. Then g is linear. By using the result in Case 1, $g(x_1, \ldots, x_n) = b_1 x_1 \oplus \cdots \oplus b_n x_n$ where each $b_j \in GF(2)$. Hence $f(x_1, \ldots, x_n) = 1 \oplus b_1 x_1 \oplus \cdots \oplus b_n x_n$ This proves that f is affine.

We now prove Part (i) of the lemma. Assume that f is affine. From Definition 1, it is easy to check that for any even k with $k \geq 4$, $f(u_1) \oplus \cdots \oplus f(u_k) = 0$ whenever $u_1 \oplus \cdots \oplus u_k = 0$. Conversely, assume f satisfies such property that for any even k with $k \geq 4$, $f(u_1) \oplus \cdots \oplus f(u_k) = 0$ whenever $u_1 \oplus \cdots \oplus u_k = 0$. Then from Part (ii) of the lemma, f must be affine. □

From the characteristic property shown in Lemma 1, if a function f on V_n satisfies $f(u_1) \oplus \cdots \oplus f(u_k) = 0$ for a large number of k-tuples $(u_1, \ldots, u_k)$ of vectors in V_n with $u_1 \oplus \cdots \oplus u_k = 0$, then the function behaves more like an affine function. This leads us to introduce a new nonlinearity criterion.

Notation 1. *Let f be a function on V_n and k an even with $4 \leq k \leq 2^n$. For $c \in GF(2)$, denote by $\mathcal{H}_{f,c}^{(k)}$ the collection of ordered k-tuples $(u_1, \ldots, u_k)$ of* vectors in V_n with $u_1 \oplus \cdots \oplus u_k = 0$ satisfying $f(u_1) \oplus \cdots \oplus f(u_k) = c$ where $c \in GF(2)$ is constant.

Definition 3. Let f be a function on V_n and k an even with $4 \leq k \leq 2^n$. For $c \in GF(2)$, we call $\tilde{h}_{f,0}^{(k)} = \#\mathcal{H}_{f,0}^{(k)}$, the kth-order homomorphicity of f, and furthermore, $\tilde{h}_{f,1}^{(k)} = \#\mathcal{H}_{f,1}^{(k)}$, the kth-order nonhomomorphicity of f, where $\#S$ denotes the number of elements in a set S.

Note that there exist $2^{(k-1)n}$ k-tuples of vectors in V_n, $(u_1, \ldots, u_k)$, satisfying $\bigoplus_{j=1}^{k} u_j = 0$. Hence an interesting fact on $\tilde{h}_{f,c}^{(k)}$ follows:

Lemma 2. Let f be a function on V_n. Then $\tilde{h}_{f,0}^{(k)} + \tilde{h}_{f,1}^{(k)} = 2^{(k-1)n}$.

We note that Lemma 1 cannot be extended to the case of odd k. This explains why we have not defined homomorphicity or nonhomomorphicity for an odd order.

4 Calculations of Nonhomomorphicity

4.1 High Order Auto-Correlation

Recall that the auto-correlation of a function is defined as follows:

Definition 4. Let f be a function on V_n. For a vector $\alpha \in V_n$, denote by $\xi(\alpha)$ the sequence of $f(x \oplus \alpha)$. Thus $\xi(0)$ is the sequence of f itself and $\xi(0) * \xi(\alpha)$ is the sequence of $f(x) \oplus f(x \oplus \alpha)$. Let $\Delta(\alpha)$ be the scalar product of $\xi(0)$ and $\xi(\alpha)$. Namely

$$\Delta(\alpha) = \langle \xi(0), \xi(\alpha) \rangle$$

$\Delta(\alpha)$ is called the auto-correlation of f with a shift α.

Obviously, $\Delta(\alpha) = 0$ if and only if $f(x) \oplus f(x \oplus \alpha)$ is balanced, i.e., f satisfies the propagation criterion with respect to α. On the other hand, if $|\Delta(\alpha)| = 2^n$, then $f(x) \oplus f(x \oplus \alpha)$ is a constant and hence α is a linear structure of f.

Next we consider a generalization of the definition for auto-correlation. The generalization will turn out to be a useful tool in studying nonhomomorphic characteristics of functions.

Definition 5. Let f be a function on V_n and $\xi = (a_0, a_1, \ldots, a_{2^n-1})$ be the sequence of f. For a vector $\alpha \in V_n$ and an integer $k = 2, 3, \ldots$, the kth-order auto-correlation of f with a shift α, denoted by $\Delta^{(k)}(\alpha)$, is defined as

$$\Delta^{(2)}(\alpha) = \Delta(\alpha), \quad \Delta^{(k)}(\alpha) = \sum_{j=0}^{2^n-1} [a_j \Delta^{(k-1)}(\alpha_j \oplus \alpha)], \quad k = 3, 4, \ldots$$

where $\Delta(\alpha)$ is the auto-correlation of f as defined in Definition 4, and α_j is the vector corresponding to the integer j.

It is important to point out that nonhomomorphicity, high order auto-correlation and high order derivation introduced in [4] are three completely different concepts. Let f be a function on V_n. In [4], the *derivation* of f at vector β, denoted by $\Delta_\beta f(x)$, is defined as follows

$$\Delta_\beta f(x) = f(x) \oplus f(x \oplus \beta).$$

and the *kth-order derivation* of f at vectors $\beta_1, \ldots, \beta_k$, denoted by $\Delta^{(k)}_{\beta_1,\ldots,\beta_k} f(x)$, is defined recursively as

$$\Delta^{(k)}_{\beta_1,\ldots,\beta_k} f(x) = \Delta(\Delta^{(k-1)}_{\beta_1,\ldots,\beta_{k-1}} f(x)).$$

We can see the kth-order derivation of f at vectors $\beta_1, \ldots, \beta_k$, $\Delta^{(k)}_{\beta_1,\ldots,\beta_k} f(x)$, is itself a *function* on V_n. In contrast, both the kth-order nonhomomorphicity and the kth-order auto-correlation of f with a shift β are fixed integer values. To examine further how the three concepts differ, consider a bent function f of degree s. For k even with $k > s$, the kth-order derivation of f at vectors $\beta_1, \ldots, \beta_k$, $\Delta^{(k)}_{\beta_1,\ldots,\beta_k} f(x)$, is obviously the zero function. In contrast, for the kth-order auto-correlation of f, we have $\Delta^{(k)}(0) = 2^{-n} \sum_{i=0}^{2^n-1} \langle \xi, \ell_i \rangle^k = 2^{\frac{1}{2}nk}$ (which follows from Corollary 1 and Lemma 3 to be introduced later on), and for the kth-order nonhomomorphicity of f, we have $\tilde{h}^{(k)}_{f,1} = 2^{(k-1)n-1} - 2^{\frac{1}{2}nk-1}$, which follows from Theorem 3 in Section 5.

To examine the properties of the kth-order auto-correlation $\Delta^{(k)}(\alpha)$, we consider a matrix defined by $(\Delta^{(k)}(\alpha_i \oplus \alpha_j))$ where $i, j = 0, 1, \ldots, 2^n - 1$. Note that the diagonal of the matrix $(\Delta^{(k)}(\alpha_i \oplus \alpha_j))$ is composed of 2^n repetitions of $\Delta^{(k)}(0)$. By simple induction on k, we have the following result:

Theorem 1. *Let f be a function on V_n, M be the matrix of f and ξ be the sequence of f. Then*

$$(\Delta^{(k)}(\alpha_i \oplus \alpha_j)) = M^k = 2^{-n} H_n \, diag(\langle \xi, \ell_0 \rangle^k, \langle \xi, \ell_1 \rangle^k, \ldots, \langle \xi, \ell_{2^n-1} \rangle^k) H_n$$

where $\ell_0, \ell_1, \ldots, \ell_{2^n-1}$ are the rows of H_n.

This result shows that the two matrices, $(\Delta^{(k)}(\alpha_i \oplus \alpha_j))$ and

$$diag(\langle \xi, \ell_0 \rangle^k, \langle \xi, \ell_1 \rangle^k, \ldots, \langle \xi, \ell_{2^n-1} \rangle^k)$$

are similar in the sense that from the former one can easily find out the latter through the use of H_n, and vice versa. Furthermore, it is not hard to see that the sum of the entries on the diagonal of $(\Delta^{(k)}(\alpha_i \oplus \alpha_j))$ is identical to that of $diag(\langle \xi, \ell_0 \rangle^k, \langle \xi, \ell_1 \rangle^k, \ldots, \langle \xi, \ell_{2^n-1} \rangle^k)$. In other words,

$$\sum_{i=0}^{2^n-1} \Delta^{(k)}(\alpha_i \oplus \alpha_i) = 2^n \Delta^{(k)}(0) = \sum_{i=0}^{2^n-1} \langle \xi, \ell_i \rangle^k.$$

Hence we have proved

Corollary 1. *Let f be a function on V_n, M be the matrix of f and ξ be the sequence of f. Then $\Delta^{(k)}(0) = 2^{-n} \sum_{i=0}^{2^n-1} \langle \xi, \ell_i \rangle^k$.*

For $k = 2$, we have $\Delta^{(2)}(0) = 2^n$. This indicates that Corollary 1 embodies Parseval's equation (Page 416 of [5]) $\sum_{i=0}^{2^n-1} \langle \xi, \ell_i \rangle^2 = 2^{2n}$ as a special case in which $k = 2$.

4.2 Expression of Nonhomomorphicity by Other Indicators

Recall (3), the nonlinearity of a function f on V_n is related to the maximum $|\langle \xi, \ell_i \rangle|$, where ξ is the sequence of f and ℓ_i is the ith row of H_n. We give a precise expression of nonhomomorphicity by using the same indicator.

Theorem 2. *For a function f on V_n and k an even with $4 \leq k \leq 2^n$. $\tilde{h}_{f,0}^{(k)}$ and $\tilde{h}_{f,1}^{(k)}$ can be expressed as follows:*

(i) $\tilde{h}_{f,0}^{(k)} = 2^{(k-1)n-1} + \frac{1}{2}\Delta^{(k)}(0) = 2^{(k-1)n-1} + 2^{-n-1} \sum_{i=0}^{2^n-1} \langle \xi, \ell_i \rangle^k$
(ii) $\tilde{h}_{f,1}^{(k)} = 2^{(k-1)n-1} - \frac{1}{2}\Delta^{(k)}(0) = 2^{(k-1)n-1} - 2^{-n-1} \sum_{i=0}^{2^n-1} \langle \xi, \ell_i \rangle^k$

where ξ is the sequence of f and ℓ_i denotes the ith row of H_n.

Proof. We need only to prove that $\tilde{h}_{f,1}^{(k)} = 2^{(k-1)n-1} - \frac{1}{2}\Delta^{(k)}(0)$, as the rest part of the theorem follows from Corollary 1 and the fact that $\tilde{h}_{f,0}^{(k)} + \tilde{h}_{f,1}^{(k)} = 2^{(k-1)n}$.

Write $\xi = (a_0, a_1, \ldots, a_{2^n-1})$ where each $a_j = \pm 1$. Consider $u_j \in V_n$, $j = 1, \ldots, k$, and $\bigoplus_{j=1}^k u_j = 0$. Clearly, $\bigoplus_{j=1}^k f(u_j) = 1$ if and only if $\Pi_{j=1}^k a_{u_j} = -1$ where the subscript u_j in a_{u_j} is viewed as the integer representation of vector u_j. It is easy to verify

$$\frac{1}{2}(1 - \Pi_{j=1}^k a_{u_j}) = \begin{cases} 1 \text{ if } \bigoplus_{j=1}^k f(u_j) = 1 \\ \\ 0 \text{ if } \bigoplus_{j=1}^k f(u_j) = 0 \end{cases}$$

Hence

$$\tilde{h}_{f,1}^{(k)} = \frac{1}{2} \sum_{\bigoplus_{j=1}^k u_j = 0} (1 - a_{u_j} a_{u_2} \cdots a_{u_k})$$

$$= \frac{1}{2} \sum_{u_1, \ldots, u_{k-1} \in V_n} (1 - a_{u_1} a_{u_2} \cdots a_{u_{k-1}} a_{u_1 \oplus u_2 \oplus \cdots \oplus u_{k-1}})$$

$$= 2^{(k-1)n-1} - \frac{1}{2} \sum_{u_1, \ldots, u_{k-1} \in V_n} a_{u_1} a_{u_2} \cdots a_{u_{k-1}} a_{u_1 \oplus u_2 \oplus \cdots \oplus u_{k-1}}$$

$$= 2^{(k-1)n-1}$$

$$- \frac{1}{2} \sum_{u_1, \ldots, u_{k-2} \in V_n} a_{u_1} a_{u_2} \cdots a_{u_{k-2}} \sum_{u_{k-1} \in V_n} a_{u_{k-1}} a_{u_1 \oplus u_2 \oplus \cdots \oplus u_{k-2} \oplus u_{k-1}}$$

$$= 2^{(k-1)n-1} - \frac{1}{2} \sum_{u_1,\ldots,u_{k-2} \in V_n} a_{u_1} a_{u_2} \cdots a_{u_{k-2}} \Delta^{(2)}(u_1 \oplus u_2 \oplus \cdots \oplus u_{k-2})$$

$$= 2^{(k-1)n-1}$$

$$- \frac{1}{2} \sum_{u_1,\ldots,u_{k-3} \in V_n} a_{u_1} a_{u_2} \cdots a_{u_{k-3}} \sum_{u_{k-2} \in V_n} a_{u_{k-2}} \Delta^{(2)}(u_1 \oplus u_2 \oplus \cdots \oplus u_{k-2})$$

$$= 2^{(k-1)n-1} - \frac{1}{2} \sum_{u_1,\ldots,u_{k-3} \in V_n} a_{u_1} a_{u_2} \cdots a_{u_{k-3}} \Delta^{(3)}(u_1 \oplus u_2 \oplus \cdots \oplus u_{k-3})$$

$$\vdots$$

$$= 2^{(k-1)n-1} - \frac{1}{2} \sum_{u_1,u_2 \in V_n} a_{u_1} a_{u_2} \Delta^{(k-2)}(u_1 \oplus u_2)$$

$$= 2^{(k-1)n-1} - \frac{1}{2} \sum_{u_1 \in V_n} a_{u_1} \sum_{u_2 \in V_n} a_{u_2} \Delta^{(k-2)}(u_1 \oplus u_2)$$

$$= 2^{(k-1)n-1} - \frac{1}{2} \sum_{u_1 \in V_n} a_{u_1} \Delta^{(k-1)}(u_1) = 2^{(k-1)n-1} - \frac{1}{2} \Delta^{(k)}(0).$$

This completes the proof. □

5   Bounds on Nonhomomorphicity

First we introduce Hölder's Inequality [3] that will be used in our discussions on lower and upper bounds. It states that for real numbers $c_j \geq 0$, $d_j \geq 0$, $j = 1, \ldots, k$, p and q with $p > 1$ and $\frac{1}{p} + \frac{1}{q} = 1$, the following is true:

$$\left(\sum_{j=1}^{k} c_j^p\right)^{1/p} \left(\sum_{j=1}^{k} d_j^q\right)^{1/q} \geq \sum_{j=1}^{k} c_j d_j \tag{8}$$

where the quality holds if and only if there exists a constant $\nu \geq 0$ such that $c_j = \nu d_j$ for each $j = 1, \ldots, k$.

By using Hölder's Inequality, we can prove

Lemma 3. *Let f be a function on V_n and k an even integer with $k \geq 4$. Then*

$$\sum_{i=0}^{2^n-1} \langle \xi, \ell_i \rangle^k \geq 2^{n+\frac{1}{2}nk}$$

where the equality holds if and only if n is even and f is bent.

Armed with the above result, next we show a bound on nonhomomorphicity.

Theorem 3. *Let f be a function on V_n and k an even integer with $k \geq 4$. Then the following statements hold:*

(i) $\tilde{h}_{f,1}^{(k)}$ satisfies

$$2^{(k-1)n-1} - \frac{1}{2}(2^n - 2N_f)^k \leq \tilde{h}_{f,1}^{(k)} \leq 2^{(k-1)n-1} - 2^{\frac{1}{2}nk-1} \qquad (9)$$

where N_f denotes the nonlinearity of f,

(ii) An equality in (9) holds if and only if f is bent. In other words, f is bent if and only if

$$\tilde{h}_{f,1}^{(k)} = 2^{(k-1)n-1} - 2^{\frac{1}{2}nk-1}.$$

Recall that the nonlinearity of a function reaches the minimum nonlinearity if and only if the function is affine while the nonlinearity of a function reaches the maximum nonlinearity if and only if the function is bent. The above theorem shows there exists a consistent relationship between nonlinearity and nonhomomorphicity, especially when the order of nonhomomorphicity is large. Thus, if $\tilde{h}_{f,1}^{(k)}$ is large, we expect that f is closer to a bent function than to an affine one, and conversely if $\tilde{h}_{f,1}^{(k)}$ is small, then the function is closer to affine than to bent.

As $\tilde{h}_{f,0}^{(k)} + \tilde{h}_{f,1}^{(k)} = 2^{(k-1)n}$, we have the following complementary result:

Corollary 2. Let f be a function on V_n and k an even integer with $k \geq 4$. Then the following statements hold:

(i) $\tilde{h}_{f,0}^{(k)}$ satisfies

$$2^{(k-1)n-1} + 2^{\frac{1}{2}nk-1} \leq \tilde{h}_{f,0}^{(k)} \leq 2^{(k-1)n-1} + \frac{1}{2}(2^n - 2N_f)^k 2^{(k-1)n-1} \quad (10)$$

where N_f denotes the nonlinearity of f,

(ii) An equality in (10) holds if and only if f is bent. In other words, f is bent if and only if

$$\tilde{h}_{f,0}^{(k)} = 2^{(k-1)n-1} + 2^{\frac{1}{2}nk-1}.$$

A consequence of Theorem 3 and Corollary 2 is

Corollary 3. Let f be a function on V_n and k an even integer with $k \geq 4$. Then $\tilde{h}_{f,0}^{(k)} - \tilde{h}_{f,1}^{(k)} \geq 2^{\frac{1}{2}nk}$, and the equality holds if and only if f is bent.

An implication of the above corollary is that there exists no function on V_n such that $\tilde{h}_{f,0}^{(k)} = \tilde{h}_{f,1}^{(k)}$.

6   Comparing Nonhomomorphicity and Nonlinearity

A natural question on nonhomomorphicity is how it is related to other nonlinear characteristics, such as nonlinearity which indicates the minimum distance between a particular function and all the affine functions. It turns out that nonhomomorphicity and nonlinearity are two indicators that are not directly comparable. We demonstrate this by inspecting three specific functions f, g and h on V_{2s} with $s \geq 5$.

Recall that the rows in H_s, the Sylvester-Hadamard matrix of order 2^s, are denoted by ℓ_i, $i = 0, 1, \ldots, 2^s - 1$. The three functions are defined as follows:

1. f — the sequence of f is the concatenation of $\ell_1, \ell_2, \ldots, \ell_{2^s-1}$ with ℓ_1 being repeated twice, i.e., $\ell_1, \ell_1, \ell_2, \ldots, \ell_{2^s-1}$.
2. g — the sequence of g is composed of four repetitions of a bent sequence η of length 2^{2s-2}, i.e., η, η, η, η.
3. h — the sequence of f is the concatenation of $\ell_1, \ell_4, \ldots, \ell_{2^s-1}$ with ℓ_1 being repeated four times, i.e., $\ell_1, \ell_1, \ell_1, \ell_1, \ell_4, \ldots, \ell_{2^s-1}$.

By using (3), we know that the nonlinearities of the three functions are $N_f = N_g = 2^{2s-1} - 2^s$, and $N_h = 2^{2s-1} - 2^{s+1}$.

Consider k even with $k \geq 4$. By Theorem 2, we have the following nonhomomorphic characteristics for the three functions:

$$\tilde{h}_{f,1}^{(k)} = 2^{2(k-1)s-1} - 2^{-2s-1}(2^{sk+2s} - 2^{sk+s+1} + 2^{sk+k+s-1})$$

$$\tilde{h}_{g,1}^{(k)} = 2^{2(k-1)s-1} - 2^{-2s-1} \cdot 2^{sk+k+2s-2}$$

$$\tilde{h}_{h,1}^{(k)} = 2^{2(k-1)s-1} - 2^{-2s-1}(2^{sk+2s} - 2^{sk+s+2} + 2^{sk+2k+s-2})$$

Thus for these three functions f, g and h, their nonlinearities and nonhomomorphic characteristics are related as follows:

(i) f v.s. g: $N_f = N_g$, but $\tilde{h}_{f,1}^{(k)} > \tilde{h}_{g,1}^{(k)}$.
(ii) f v.s. h: $N_f > N_h$, and $\tilde{h}_{f,1}^{(k)} > \tilde{h}_{h,1}^{(k)}$.
(iii) g v.s. h: $N_g > N_h$, but $\tilde{h}_{g,1}^{(k)} < \tilde{h}_{h,1}^{(k)}$ if $k \leq s - 1$, and $\tilde{h}_{g,1}^{(k)} > \tilde{h}_{h,1}^{(k)}$ if $k \geq s$.

Properties of these three functions clearly show that nonlinearity and nonhomomorphicity are not comparable indicators. They, however, can be used to complement each other in studying cryptographic properties of functions.

The two functions g and h are of particular interest: while $\tilde{h}_{g,1}^{(k)} < \tilde{h}_{h,1}^{(k)}$ for $k \leq s - 1$, their positions are reversed for $k \geq s$. This motivates us to examine the behavior of nonhomomorphicity as k becomes large.

Theorem 4. *Let f and g be two functions on V_n. If $\tilde{h}_{f,1}^k \neq \tilde{h}_{g,1}^k$, then there is an even positive k_0, such that $\tilde{h}_{f,1}^k < \tilde{h}_{g,1}^k$ for every even k with $k \geq k_0$, or $\tilde{h}_{f,1}^k > \tilde{h}_{g,1}^k$ for every even k with $k \geq k_0$.*

Assume that $N_f > N_g$. Then from (3), we have

$$\max\{|\langle \xi, \ell_i \rangle|, 0 \leq i \leq 2^n - 1\} < \max\{|\langle \eta, \ell_i \rangle|, 0 \leq i \leq 2^n - 1\}.$$

Using a similar proof to that for the above theorem, we can show

Theorem 5. *Let f and g be two functions on V_n. If $N_f > N_g$, then there is an even positive k_0, such that $\hbar_{f,1}^k > \hbar_{g,1}^k$ for every even k with $k \geq k_0$.*

While Theorem 5 shows that nonhomomorphicity and nonlinearity are consistent when the dimension k is large, the three example functions f, g and h, together with Theorems 4 and 5, do indicate that nonhomomorphic characteristics of a function cannot be fully predicted by other cryptographic criteria, such as nonlinearity. Therefore, nonhomomorphicity can serve as another important indicator that forecasts certain cryptographically useful properties of the function.

Comparing (ii) of Theorem 2 and (3), we find that although both nonlinearity and nonhomomorphicity reflect non-affine characteristics, the former focuses on the maximum $|\langle \xi, \ell_i \rangle|$ while the latter is more concerned over all $|\langle \xi, \ell_i \rangle|$.

7   The Mean of Homomorphicity and Nonhomomorphicity

Let f be a function on V_n, χ denote an indicator (a criterion or a value), and χ_f denote the indicator of f. Note that there precisely 2^{2^n} functions on V_n. We are concerned with the mean of the indicator χ over all the functions on V_n, denoted by $\overline{\chi}$, i.e. $\overline{\chi} = 2^{-2^n} \sum_f \chi_f$.

The upper and lower bounds on χ_f cannot provide sufficient information on the distribution of χ of a majority of functions. For this reason, we argue that the mean of the indicator χ over all the functions on V_n, i.e. $\overline{\chi} = 2^{-2^n} \sum_f \chi_f$, should also be investigated. Note that there exist precisely 2^{2^n} functions with n variables.

Notation 2. *Let O_k (k is even) denote the collection of k-tuples $(u_1, \ldots, u_k)$ of vectors in V_n satisfying $u_{j_1} = u_{j_2}, \ldots, u_{j_{k-1}} = u_{j_k}$, where $\{j_1, j_2, \ldots, j_k\} = \{1, 2, \ldots, k\}$. Write $o_k = \#O_k$.*

It is easy to verify

Lemma 4. *Let n and k be positive integers and $u_1 \oplus \cdots \oplus u_k = 0$, where each u_j is a fixed vector in V_n. Then*

$$f(u_1) \oplus \cdots \oplus f(u_k) = 0$$

holds for every function f on V_n if and only if k is even and $(u_1, \ldots, u_k) \in O_k$.

Lemma 5. *In Notation 2, let k be an even with $2 \leq k \leq 2^n$. Then*

$$o_k = \sum_{t=1}^{k/2} \binom{2^n}{t} \sum_{p_1 + \cdots + p_t = k/2,\, p_j > 0} \frac{(k)!}{(2p_1)! \cdots (2p_t)!}$$

Proof. Let $(u_1, \ldots, u_k) \in O_k$. Then the multiple set $\{u_1, \ldots, u_k\}$ can be divided into t disjoint subsets $\Pi_1, \ldots, \Pi_t$ where (1) $1 \leq t \leq k$, (2) each Π_j is a $2p_j$ ($p_j > 0$) copy of a vector β_j i.e. $\Pi_j = \{\beta_j, \ldots, \beta_j\}$ and $|\Pi_j| = 2p_j$, (3) $\beta_j \neq \beta_i$, if $j \neq i$, (4) $\{u_1, \ldots, u_k\} = \Pi_1 \cup \cdots \cup \Pi_t$.

Note that there exist $\binom{2^n}{t}$ different choices of t distinguished vectors $\beta_1, \ldots,$ β_t from V_n. Arranging each multiple set $\{u_1, \ldots, u_k\}$, we obtain precisely $(k)!/$ $(2p_1)! \cdots (2p_t)!$ distinguished ordered sets. Note that $2p_1 + \cdots + 2p_t = k$ and $1 \le t \le k/2$. The proof is completed. $\square$

From Lemma 4, if $(u_1, \ldots, u_k) \in O_k$ then $f(u_1) \oplus \cdots \oplus f(u_k) = 0$ holds for every function f on V_n. Therefore, in this case $f(u_1) \oplus \cdots \oplus f(u_k) = 0$ with $u_1 \oplus \cdots \oplus u_k = 0$ does not really reflect an affine property. Hence we focus on $\mathcal{H}_{f,0}^{(k)} - O_k$ and $\mathcal{H}_{f,1}^{(k)}$.

Theorem 6. *Let k be an even with $2 \le k \le 2^n$. Then*

(i) the mean of $\tilde{h}_{f,0}^{(k)}$ over all the functions on V_n i.e. $2^{-2^n} \sum_f \tilde{h}_{f,0}^{(k)}$, satisfies

$$2^{-2^n} \sum_f \tilde{h}_{f,0}^{(k)} = \frac{1}{2} o_k + 2^{(k-1)n-1}$$

where o_k is given in Lemma 5.

(ii) the mean of $\tilde{h}_{f,1}^{(k)}$ over all the functions on V_n i.e. $2^{-2^n} \sum_f \tilde{h}_{f,1}^{(k)}$, satisfies

$$2^{-2^n} \sum_f \tilde{h}_{f,1}^{(k)} = -\frac{1}{2} o_k + 2^{(k-1)n-1}$$

Proof. To prove Part (i), we consider two cases for $(u_1, \ldots, u_k) \in \mathcal{H}_{f,0}^{(k)}$.

Case 1: $(u_1, \ldots, u_k) \in O_k$. From Lemma 4, $f(u_1) \oplus \cdots \oplus f(u_k) = 0$ holds for every function f on V_n.

Case 2: $(u_1, \ldots, u_k) \in \mathcal{H}_{f,0}^{(k)} - O_k$. Note that $f(u_1) \oplus \cdots \oplus f(u_k)$ takes the value zero and the value one with an equal probability of a half for a random function f on V_n. Therefore

$$2^{-2^n} \sum_f \tilde{h}_{f,0}^{(k)} = 2^{-2^n} \sum_f \#O_k + 2^{-2^n} \sum_f \#(\mathcal{H}_{f,0}^{(k)}(0) - O_k) = o_k + \frac{1}{2}[2^{(k-1)n} - o_k]$$

$$= \frac{1}{2} o_k + 2^{(k-1)n-1}$$

This proves (i) of the theorem.

Part (ii) can be proven in a similar way, once again by noting that $f(u_1) \oplus \cdots \oplus f(u_k)$ takes the value zero and the value one with an equal probability of a half, for a a random function f on V_n. $\square$

A function whose nonhomomorphicity is larger than the mean, namely $\tilde{h}_{f,1}^{(k)} >$ $2^{-2^n} \sum_f \tilde{h}_{f,1}^{(k)}$, indicates that the function is more nonlinear. The converse also holds.

8   Relative Nonhomomorphicity

The concept of relative nonhomomorphicity introduced in this section is useful for a statistical tool to be introduced later.

Notation 3. *Let k be an even with $k \geq 4$ and R_k denote the collection of ordered k-tuples $(u_1, \ldots, u_k)$ of vectors in V_n satisfying $u_1 \oplus \cdots \oplus u_k = 0$.*

We have noticed

$$\#R_k = 2^{(k-1)n} \quad \text{and} \quad \#(R_k - O_k) = 2^{(k-1)n} - o_k. \tag{11}$$

From the proof of Theorem 6, if $(u_1, \ldots, u_k) \in R_s - O_k$ then $f(u_1) \oplus \cdots \oplus f(u_k)$ takes the value zero and the value one with equal probability.

Definition 6. *Let f be a function on V_n and k be an even with $k \geq 4$. Define the kth-order relative nonhomomorphicity of f, denoted by $\rho_{f,1}^{(k)}$, as $\rho_{f,1}^{(k)} = \frac{\tilde{h}_{f,1}^{(k)}}{\#(R_k - O_k)}$, i.e. $\rho_{f,1}^{(k)} = \frac{\tilde{h}_{f,1}^{(k)}}{2^{(k-1)n} - o_k}$.*

From Theorem 6, we obtain

Corollary 4. *Let k be an even with $2 \leq k \leq 2^n$. Then the mean of $\rho_{f,1}^{(k)}$ over all the functions on V_n i.e. $2^{-2^n} \sum_f \rho_{f,1}^{(k)}$, satisfies $2^{-2^n} \sum_f \rho_{f,1}^{(k)} = \frac{1}{2}$.*

From Corollary 4,

$$\rho_{f,1}^{(k)} \begin{cases} \geq \frac{1}{2} & \text{then the nonhomomorphicity of } f \text{ is not smaller than the mean} \\ < \frac{1}{2} & \text{then the nonhomomorphicity of } f \text{ is smaller than the mean} \end{cases} \tag{12}$$

In practice, if $\rho_{f,1}^{(k)}$ is much smaller than $\frac{1}{2}$, then f should be considered cryptographically weak.

9   Estimating Nonhomomorphicity

As shown in Theorem 2, the nonhomomorphicity of a function can be determined precisely. In this section, however, we introduce a statistical method to estimate nonhomomorphicity. Such a method is useful in fast analysis of functions.

Denote a real-valued $(0,1)$ function on $R_k - O_k$, $t(u_1, \ldots, u_k)$, as follows

$$t(u_1, \ldots, u_k) = \begin{cases} 1, & \text{if } f(u_1) \oplus \cdots \oplus f(u_k) = 1 \\ 0, & \text{otherwise} \end{cases}$$

Hence from the definition of nonhomomorphicity we have

$$\tilde{h}_{f,1}^{(k)} = \sum_{(u_1, \ldots, u_k) \in R_k - O_k} t(u_1, \ldots, u_k)$$

Let Ω be a random subset of $R_k - O_k$. Write $\omega = \#\Omega$ and

$$\bar{t} = \frac{1}{\omega} \sum_{(u_1,\ldots,u_k)\in\Omega} t(u_1,\ldots,u_k) \tag{13}$$

Note that this is the "sample mean" [1]. In particular, $\Omega = R_n^{(k)} - O_k$, $\bar{t}$ is identified with the "true mean" or "population mean" [1], namely, $\rho_{f,1}^{(k)}$.

Now consider $\sum_{(u_1,\ldots,u_k)\in\Omega} (t(u_1,\ldots,u_k) - \bar{t})^2$. We have

$$\sum_{(u_1,\ldots,u_k)\in\Omega} (t(u_1,\ldots,u_k) - \bar{t})^2 = \sum_{(u_1,\ldots,u_k)\in\Omega} t^2(u_1,\ldots,u_k)$$

$$- 2\bar{t} \cdot \sum_{(u_1,\ldots,u_k)\in\Omega} t(u_1,\ldots,u_k) + \omega\bar{t}^2$$

Note that $t^2(u_1,\ldots,u_k) = t(u_1,\ldots,u_k)$. From (13),

$$\sum_{(u_1,\ldots,u_k)\in\Omega} (t(u_1,\ldots,u_k) - \bar{t})^2 = \omega\bar{t} - 2\omega\bar{t}^2 + \omega\bar{t}^2 = \omega\bar{t} - 2\omega\bar{t}^2 + \omega\bar{t}^2$$

$$= \omega\bar{t}(1 - \bar{t}) \tag{14}$$

Hence the quantity of $\sqrt{\frac{1}{\omega-1}\sum_{(u_1,\ldots,u_k)\in\Omega}(t(u_1,\ldots,u_k) - \bar{t})^2}$, which is called the "sample standard deviation" [1] and is usually denoted by μ, can be expressed as

$$\mu = \sqrt{\frac{1}{\omega - 1} \sum_{(u_1,\ldots,u_k)\in\Omega} (t(u_1,\ldots,u_k) - \bar{t})^2} = \sqrt{\frac{\omega\bar{t}(1 - \bar{t})}{\omega - 1}} \tag{15}$$

By using (4.4) in Section 4.B of [1], the "true mean" or "population mean", $\rho_{f,1}^{(k)}$, can be bounded by

$$\bar{t} - Z_{e/2}\frac{\mu}{\sqrt{\omega}} < \rho_{f,1}^{(k)} < \bar{t} + Z_{e/2}\frac{\mu}{\sqrt{\omega}} \tag{16}$$

where $Z_{e/2}$ denotes the value Z of a "standardized normal distribution" which to its right a fraction $e/2$ of the data, (16) holds with a probability of $(1-e)100\%$ [1]. For example,

when $e = 0.2$, $Z_{e/2} = 1.28$, and (16) holds with a probability of 80%,
when $e = 0.1$, $Z_{e/2} = 1.64$, and (16) holds with a probability of 90%,
when $e = 0.05$, $Z_{e/2} = 1.96$, and (16) holds with a probability of 95%,
when $e = 0.02$, $Z_{e/2} = 2.33$, and (16) holds with a probability of 98%,
when $e = 0.01$, $Z_{e/2} = 2.57$, and (16) holds with a probability of 99%,
when $e = 0.001$, $Z_{e/2} = 3.3$, and (16) holds with a probability of 99.9%.

From (13), $0 \leq \bar{t} < 1$ and it is easy to verify that μ in (15) satisfies $0 \leq \mu \leq \frac{1}{2}\sqrt{\frac{\omega}{\omega-1}}$, This implies that (16) can be simply replaced by

$$\bar{t} - \frac{Z_{e/2}}{2\sqrt{\omega - 1}} < \rho_{f,1}^{(k)} < \bar{t} + \frac{Z_{e/2}}{2\sqrt{\omega - 1}}, \tag{17}$$

where (17) holds with $(1 - e)100\%$ probability. Hence if ω i.e. $\#\Omega$ is large, then the lower bound and the upper bound on $\rho_{f,1}^{(k)}$ in (16) are closer to each other. On the other hand, if we choose $\omega = \#\Omega$ large enough then $Z_{e/2}\frac{\mu}{\sqrt{\omega}}$ is sufficiently small, and hence (16) and (17) will provide us with useful information. For instance, viewing Corollary 4 and (17), we can choose $\omega = \#\Omega$ such that $\frac{Z_{e/2}}{2\sqrt{\omega}-1} < 10^{-p}$. Hence $\omega \geq Z_{e/2} \cdot 10^{2p}$ is large enough. In this case (17) is specialized as

$$\bar{t} - 10^{-p} < \rho_{f,1}^{(k)} < \bar{t} + 10^{-p} \tag{18}$$

where (18) holds with $(1 - e)100\%$ probability.

In summary, we can analyze the nonhomomorphic characteristics of a function on V_n in the following steps:

1. we randomly fix even k with $k \geq 4$, for example, $k = 4, 6$ or 8, and randomly fix a large integer ω, for example, $\omega \geq Z_{e/2} \cdot 10^{2p}$, and randomly choose a subset of $R_k - O_k$, say Ω, with $\#\Omega = \omega$,
2. by using (13), we determine $\bar{t}$, i.e. "the sample mean",
3. by using (18), we determine the range of $\rho_{f,1}^{(k)}$ with a high reliability,
4. viewing $\rho_{f,1}^{(k)}$ in (18), from Corollary 4,

$$\rho_{f,1}^{(k)} \begin{cases} \geq \frac{1}{2} \text{ then } f \text{ is not less nonhomomorphic than the mean} \\ > \frac{1}{2} \text{ then } F \text{ is less nonhomomorphic than the mean} \end{cases} \tag{19}$$

where (19) holds with $(1 - e)\%$ probability,
5. if $\rho_{f,1}^{(k)}$ is much smaller than $\frac{1}{2}$ then f should be considered as cryptographically weak.

We have noticed that the statistical analysis has following advantages:

(1) the relative nonhomomorphicity, $\rho_{f,1}^{(k)}$ can be precisely identified by the use of "population mean" or "true mean",
(2) by using this method we do not need to search through the entire V_n,
(3) the method is highly reliable.

10   Extensions to S-boxes

Obviously, the concept of nonhomomorphicity of a Boolean function can be extended to that of an S-box in a straightforward way. Analysis of the general case of an S-box, however, has turned out to be far more complex. Nevertheless, we have obtained a number of interesting results on S-boxes, some of which encompass results presented in this paper. We will report the new results in a forthcoming paper. In the same paper we will also discuss how to utilize nonhomomorphic characteristics of an S-box employed by a block cipher in analyzing cryptographic weaknesses of the cipher.

11   Conclusions

Nonhomomorphicity is a new indicator for nonlinear characteristics of a function. It can complement the more widely used indicator of nonlinearity. Two useful properties of nonhomomorphicity are: (1) the mean of nonhomomorphicity over all the Boolean functions over the same vector space can be precisely identified, (2) the nonhomomorphicity of a function can be estimated efficiently, regardless of the dimension of the vector space.

12   Acknowledgment

The first author was supported by a Queen Elizabeth II Fellowship (227 23 1002).

References

1. Stephen A. Book. *Statistics*. McGraw-Hill Book Company, 1977.
2. J. F. Dillon. A survey of bent functions. *The NSA Technical Journal*, pages 191–215, 1972. (unclassified).
3. Friedhelm Erwe. *Differential And Integral Calculus*. Oliver And Boyd Ltd, Edinburgh And London, 1967.
4. X. Lai. Higher order derivatives and differential cryptanalysis. In *Proceedings of the Symposium on Communication, Coding and Cryptography, in the Honor of James L. Massey on the Occasion of his 60's Birthday*, pages 227–233. Kluwer Academic Publishers, 1994.
5. F. J. MacWilliams and N. J. A. Sloane. *The Theory of Error-Correcting Codes*. North-Holland, Amsterdam, New York, Oxford, 1978.
6. O. S. Rothaus. On "bent" functions. *Journal of Combinatorial Theory*, Ser. A, 20:300–305, 1976.
7. J. Seberry, X. M. Zhang, and Y. Zheng. Nonlinearity and propagation characteristics of balanced boolean functions. *Information and Computation*, 119(1):1–13, 1995.

Cryptanalysis of ORYX

D. Wagner[1], L. Simpson[2], E. Dawson[2], J. Kelsey[3],
W. Millan[2], and B. Schneier[3]

[1] University of California, Berkeley
daw@cs.berkeley.edu
[2] Information Security Research Centre,
Queensland University of Technology
GPO Box 2434, Brisbane Q 4001, Australia
{simpson,dawson,millan}@fit.qut.edu.au
[3] Counterpane Systems,
101 E Minnehaha Parkway, Minneapolis, MN 55419
{schneier,kelsey}@counterpane.com

Abstract. We present an attack on the ORYX stream cipher that requires only 25-27 bytes of known plaintext and has time complexity of 2^{16}. This attack directly recovers the full 96 bit internal state of ORYX, regardless of the key schedule. We also extend these techniques to show how to break ORYX even under a ciphertext-only model. As the ORYX cipher is used to encrypt the data transmissions in the North American Cellular system, these results are further evidence that many of the encryption algorithms used in second generation mobile communications offer a low level of security.

1 Introduction

The demand for mobile communications systems has increased dramatically in the last few years. Since cellular communications are sent over a radio link, it is easy to eavesdrop on such systems without detection. To protect privacy and prevent fraud, cryptographic algorithms have been employed to provide a more secure mobile communications environment. First generation mobile communications devices were analog. Analog cellphones rarely use encryption, and in any case analog encryption devices offered a very low level of security [2]. Over the last five years digital mobile communications systems have emerged, such as the Global Systems Mobile (GSM) standard developed in Europe and several Telecommunications Industry Association (TIA) standards developed in North America [6]. For these digital systems, a much higher level of security, using modern encryption algorithms, is possible. Unfortunately, algorithms which offer a high level of security have not been used in mobile telecommunications to date.

In the case of GSM telephony, it is shown in [4] that it may be possible to conduct a known plaintext attack against the voice privacy algorithm used in GSM telephones, the A5 cipher. More recently it was shown in [3] that it is

S. Tavares and H. Meijer (Eds.): SAC'98, LNCS 1556, pp. 296–305, 1999.

possible to clone GSM telephones by conducting a chosen-challenge attack on the COMP128 authentication algorithm.

The North American digital cellular standards designed by the TIA, including time division multiple access (TDMA) and code division multiple access (CDMA) both use roughly the same security architecture. The four cryptographic primitives used in these systems and described in the TIA standard [6] are:

- CAVE, for challenge-response authentication protocols and key generation.
- ORYX, a LFSR-based stream cipher for wireless data services.
- CMEA, a simple block cipher used to encrypt message data on the traffic channel.
- For voice privacy, TDMA systems use an XOR mask, or CDMA systems use keyed spread spectrum techniques combined with an LFSR mask.

The voice privacy algorithm in TDMA systems is especially weak since it is based on a repeated XOR mask. Such a system can be easily attacked using ciphertext alone [1]. The CMEA algorithm is susceptible to a known plaintext attack [7]. In this paper the security of the ORYX algorithm is examined.

ORYX is a simple stream cipher based on binary linear feedback shift registers (LFSRs) that has been proposed for use in North American digital cellular systems to protect cellular data transmissions [6]. The cipher ORYX is used as a keystream generator. The output of the generator is a random-looking sequence of bytes. Encryption is performed by XORing the keystream bytes with the data bytes to form ciphertext. Decryption is performed by XORing the keystream bytes with the ciphertext to recover the plaintext. Hence known plaintext-ciphertext pairs can be used to recover segments of the keystream. In this paper, the security of ORYX is examined with respect to a known plaintext attack conducted under the assumption that the cryptanalyst knows the complete structure of the cipher and the secret key is only the initial states of the component LFSRs.

For this attack, we assume that the complete structure of the cipher, including the LFSR feedback functions, is known to the cryptanalyst. The key is only the initial states of the three 32 bit LFSRs: a total keysize of 96 bits. There is a complicated key schedule which decreases the total keyspace to something easily searchable using brute-force techniques; this reduces the key size to 32 bits for export. However, ORYX is apparently intended to be a strong algorithm when used with a better key schedule that provides a full 96 bits of entropy. The attack proposed in this paper makes no use of the key schedule and is applicable to ORYX whichever key schedule is used.

2   The ORYX Cipher

The cipher ORYX has four components: three 32-bit LFSRs which we denote LFSR_A, LFSR_B and LFSR_K, and an S-box containing a known permutation L

of the integer values 0 to 255, inclusive. The feedback function for $LFSR_K$ is

$$x^{32} + x^{28} + x^{19} + x^{18} + x^{16} + x^{14} + x^{11} + x^{10} + x^9 + x^6 + x^5 + x + 1.$$

The feedback functions for $LFSR_A$ are

$$x^{32} + x^{26} + x^{23} + x^{22} + x^{16} + x^{12} + x^{11} + x^{10} + x^8 + x^7 + x^5 + x^4 + x^2 + x + 1$$

and

$$x^{32} + x^{27} + x^{26} + x^{25} + x^{24} + x^{23} + x^{22} + x^{17} + x^{13} + x^{11} + x^{10} + x^9 + x^8 + x^7 + x^2 + x + 1.$$

The feedback function for $LFSR_B$ is

$$x^{32} + x^{31} + x^{21} + x^{20} + x^{16} + x^{15} + x^6 + x^3 + x + 1.$$

The permutation L is fixed for the duration of a call, and is formed from a known algorithm, initialized with a value which is transmitted in the clear during call setup. Each keystream byte is generated as follows:

1. $LFSR_K$ is stepped once.
2. $LFSR_A$ is stepped once, with one of two different feedback polynomials depending on the content of a stage of $LFSR_K$.
3. $LFSR_B$ is stepped either once or twice, depending on the content of another stage in $LFSR_K$.
4. The high bytes of the current states of $LFSR_K$, $LFSR_A$, and $LFSR_B$ are combined to form a keystream byte using the combining function:

$$Keystream = \{High8_K + L[High8_A] + L[High8_B]\} \bmod 256$$

3   Attack Procedure

Since ORYX has a 96-bit keyspace, it is not feasible to simply guess the whole generator initial state and check if the guess is correct. However, if the generator initial state can be divided into smaller parts, and it is possible to guess one small part of the generator initial state, and incrementally check whether that guess is correct, the generator can be attacked. The attack presented in this paper uses this divide and conquer approach, and is a refinement of a method originally proposed in [8]. A feature of ORYX which contributes to the efficiency of the attack outlined in this paper is that the two stages of $LFSR_K$ whose contents control the selection of the feedback polynomial for $LFSR_A$ and the number of times $LFSR_B$ is stepped are both within the high eight stages of $LFSR_K$. Since the keystream bytes are formed from the contents of the high eight stages of each of the three LFSR states, we divide the keyspace and focus our attack on these 24 bits.

3.1   Attack Algorithm

Denote the high eight bits of the three LFSRs at the time the ith byte of keystream is produced by $High8_A(i)$, $High8_B(i)$ and $High8_K(i)$. The initial contents are $High8_A(0)$, $High8_B(0)$ and $High8_K(0)$, and all registers are stepped before the first byte of keystream, denoted $Z(1)$, is produced. To produce a keystream byte $Z(i+1)$ at time instant $i+1$, $LFSR_K$ is stepped once, then $LFSR_A$ is stepped once, then $LFSR_B$ is stepped either once or twice. The contents of $High8_A(i+1)$, $High8_B(i+1)$ and $High8_K(i+1)$ are then combined to form the keystream byte $Z(i+1)$. Therefore, there is no need to guess all 24 bits: if we guess the contents of $High8_A(1)$ and $High8_B(1)$ we can use the first byte of the known keystream $Z(1)$ and the combining function to calculate the corresponding contents of $High8_K(1)$. Thus the attack requires exhaustive search of only a 16 bit subkey: the contents of $High8_A(1)$ and $High8_B(1)$.

For a particular 16-bit guess of $High8_A(1)$ and $High8_B(1)$, we use $Z(1)$ and calculate the corresponding contents of $High8_K(1)$. After this calculation, the attack proceeds iteratively as we construct a path of guesses of $High8_A(i)$, $High8_B(i)$ and $High8_K(i)$ which are consistent with the known keystream. In each iteration a set of predictions for the next keystream byte is formed, and the guess evaluated by comparing the known keystream byte with the predicted values.

In the ith iteration, we exploit the fact that after stepping the three LFSRs to produce the next output byte, $High8_K(i+1)$ and $High8_A(i+1)$ effectively have one unknown bit shifting into them and, depending on $High8_K(i+1)$, $High8_B(i+1)$ has either one or two unknown bits, for each byte of output. We try all possible combinations of these new input bits, a total of 12 combinations, and compute the output byte for each case. At most, there will be 12 distinct output bytes which are consistent with the guess of $High8_A(i)$, $High8_B(i)$ and $High8_K(i)$. We compare the known keystream byte $Z(i+1)$ with the predicted output bytes.

If $Z(i+1)$ is the same as one of the predicted output bytes, for the case where there are 12 distinct outputs, then a single possible set of values exists for $High8_K(i+1)$, $High8_A(i+1)$ and $High8_B(i+1)$. We use these values in the next iteration of the attack.

Occasionally, where there are less than 12 distinct outputs, and the keystream byte is the same as the predicted output byte for more than one combination of new input bits, we must consider more than one possible set of values for $High8_K(i+1)$, $High8_A(i+1)$, and $High8_B(i+1)$. That is, the path of consistent guesses we are following may branch. In this situation we conduct a depth-first search.

If the keystream byte is not the same as any of the predicted output bytes, then the guessed contents of $High8_A(i)$ and $High8_B(i)$ were obviously incorrect. We go back along the path to the last branching point and start to trace out another path. If we search each possible path without finding a path of consistent guesses of length equal to the number of bytes of known keystream, then the guessed contents of $High8_A(1)$ and $High8_B(1)$ were obviously incorrect, and we make a new 16-bit guess and repeat the procedure.

When we find a sufficiently long path of consistent guesses, we assume that the values for $High8_A(1)$, $High8_B(1)$ and $High8_K(1)$ were correct. This provides knowledge of the contents of the high eight stages of each of the three LFSRs at the time that the first byte of keystream was produced. For the 24 consecutive guesses $High8_A(i)$, $High8_B(i)$ and $High8_K(i)$ for $2 \leq i \leq 25$, each set of values: $High8_K(i)$ and $High8_A(i)$ gives another bit in the state of LFSR$_K$ and LFSR$_A$, respectively, and $High8_B(i)$ gives either one or two bits in the state of LFSR$_B$. Once we reconstruct the 32-bit state of each LFSR, at the time the first keystream byte was produced the LFSR states can then be stepped back to recover the initial states of the three LFSRs: the secret key of the ORYX generator. Thus we recover the entire key using a minimum of 25 bytes of keystream, and at most 2^{16} guesses. We use the recovered initial states to produce a candidate keystream and compare to the known keystream. If the candidate keystream is the same as the known keystream, the attack ends, otherwise we make a new 16 bit guess and repeat the procedure.

In practice, we may occasionally need a few more than 25 keystream bytes to resolve ambiguities in the final few bits of the LFSR states. That is, we may need to chase down a few more false trails to convince ourselves we got the last few bits right.

3.2   Testing Procedure

The performance of the attack was experimentally analyzed to find the proportion of performed attacks for which the initial states can be successfully recovered, for various keystream lengths. The experiments use the following procedure: Nonzero initial states are generated for LFSR$_A$,LFSR$_B$ and LFSR$_K$. A keystream segment of length N, $\{Z(i)\}_{i=1}^{N}$ is produced using the ORYX cipher as outlined in Section 2. The attack, as described in Section 3.1, is launched on the produced segment of the keystream. An outline of the testing procedure follows:

- *Input:* The length of the observed keystream sequence, N.
- *Initialization:* $i = 1$, where i is the current attack index, Also define i_{max},the maximum number of attack trials to be conducted.
  LFSR$_A$ initial state seed index, j is the current LFSR$_B$ initial state seed index and k is the current LFSR$_K$ initial state seed index.
- *Stopping Criterion:* The testing procedure stops when the number of attacks conducted reaches i_{max}.
- *Step 1:* Generate pseudorandom initial state seeds ASEED$_i$, BSEED$_i$ and KSEED$_i$ for LFSR$_A$, LFSR$_B$ and LFSR$_K$, respectively. (pseudorandom number routine *drand48*, see [5] is used).
- *Step 2:* Generate pseudorandom LFSR initial states using ASEED$_i$, BSEED$_i$ and KSEED$_i$ for LFSR$_A$, LFSR$_B$ and LFSR$_K$, respectively.
- *Step 3:* Generate the keystream sequence of bytes $\{Z(i)\}_{i=1}^{N}$.

- *Step 4:* Apply the attack to $\{Z(i)\}_{i=1}^{N}$ to obtain the reconstructions of the initial states of the three LFSRs.
- *Step 5:* If $i \leq i_{max}$, increment i and go to Step 1.
- *Step 6:* Stop the procedure.
- *Output:* Reconstructed initial states of LFSR$_A$, LFSR$_B$ and LFSR$_K$.

4   Implementation Issues for the Attack

The attack procedure described in section 3.1 involves assuming that a particular guess of $High8_A(i)$, $High8_B(i)$ and $High8_K(i)$ is correct, using this guess to form predictions for the next keystream byte, and then comparing a known keystream byte with the predictions: if the keystream byte contradicts all predictions, we conclude that the guess was wrong. However, it is possible that the keystream byte $Z(i + 1)$ will be the same as one of the predicted output bytes, although the values for High8$_A(i)$ and High8$_B(i)$ are incorrect. We refer to such a situation as a false alarm.

4.1   The Probability of a False Alarm

For the attack to be effective, the probability of a false alarm occurring must be small. Therefore, we require a high probability that an incorrect guess will be detected (through comparison of the predicted outputs with the corresponding keystream byte). That is, we require the probability that no predicted output byte matches the actual keystream byte to be significantly greater than one half, given that the guessed contents of $High8_A(i)$, $High8_B(i)$ and $High8_K(i)$ are incorrect.

Consider the formation of the predicted output values. Each predicted output is formed from an 8-bit possible value for $High8_A(i + 1)$, an 8-bit possible value for $High8_B(i + 1)$ and an 8-bit possible value for $High8_K(i + 1)$. So there are a total of 2^{24} different input combinations. The predicted output has 8 bits. Therefore, given a particular output value, there exist multiple input combinations which result in this output. As the inputs are all non-negative integers less than 256 and the combining function is the modulo 256 sum of the three inputs, all output values are equally likely if the input combinations are equally likely. Thus each output value is produced by 2^{16} different input combinations. If one of these input combinations is the correct combination, then there are $2^{16} - 1$ other combinations which, although incorrect, produce the same value. The probability that a single incorrect input combination produces the same output as the correct combination is $\frac{2^{16}-1}{2^{24}-1} \approx 0.0039$. The probability that a single incorrect input combination produces a value different to the output of the correct combination is the complement of this; approximately 0.9961. We select a set of twelve input combinations. The probability that an incorrect guess will be detected (through comparison of the predicted outputs with the corresponding keystream byte) is the probability that none of the predicted outputs is the same as the known keystream value, given that all of the twelve possible input combinations are incorrect. The situation can be approximated by the binomial distribution. Therefore,

$$P(\text{incorrect guess detected}) \approx (.9961)^{12} = 0.9541$$

Since the probability that no predicted output byte matches the actual keystream byte, given that the guessed contents of $High8_A(i)$, $High8_B(i)$ and $High8_K(i)$ are incorrect, is 0.9541, the probability that at least one predicted output byte matches the actual keystream byte, given that the guessed contents of $High8_A(i)$, $High8_B(i)$ and $High8_K(i)$ are incorrect, is 0.0459. The probability of a false alarm is less than five percent, and the nice attribute of a false alarm is that once we are on the wrong track, we have a 0.9541 probability of detecting this at each step.

Using the binomial distribution to calculate approximate probabilities, given the guessed bits are incorrect,

$$P(\text{keystream byte matches 1 prediction}) \approx \binom{12}{1}(0.0039)^1(0.9961)^{11} = 0.0448$$

$$P(\text{keystream byte matches 2 predictions}) \approx \binom{12}{2}(0.0039)^2(0.9961)^{10} = 0.0010$$

$$P(\text{keystream byte matches} \geq 2 \text{ predictions}) \approx 0.0001$$

From this, we conclude that most of the time, we will generate very few false trails—typically just one or two. Thus, we perform a depth-first search of the possible states, but we seldom spend much time on a false trail.

Note that if we have the correct states for $High8_A(i)$, $High8_B(i)$ and $High8_K(i)$ we never mistakenly think we have the wrong state. Once we identify the correct $High8_A(1)$, $High8_B(1)$ and $High8_K(1)$, we can quickly find the correct states for $High8_A(i)$, $High8_B(i)$ and $High8_K(i)$, for $2 \leq i \leq n$ for some $n \geq 25$. From these we can reconstruct the initial states of the three LFSRs.

4.2   Effect of Length of Known Keystream

The minimum length of keystream required for this attack to be successful is 25 bytes; one byte to obtain the required eight bit value for $High8_K(1)$, giving a known eight bits in each of the three 32-bit LFSR initial states, and then one byte to recover each of the other 24 bits in the three LFSR initial states. The more keystream available, the more certain we are of successful reconstruction. However, if we have less than 25 bytes of known keystream, the attack can still be performed as outlined above to give a likely reconstruction of most of the LFSR states, and we use exhaustive search over the contents of the last few stages.

N	25	26	27
% Success	99.7	99.9	100.0

Table 1. Success rate (%) versus N.

5 Experimental Results

The performance of the attack was experimentally analyzed to find the the proportion of performed attacks for which the initial states can be successfully recovered, for the keystream lengths $N = 25$, 26, and 27. For each keystream length, the attack was performed one thousand times, using pseudorandomly generated LFSR initial states. The attack was considered successful if the reconstructed LFSR initial states were the same as the actual LFSR initial states. Table 1 shows as the success rate the proportion of attacks conducted which were successful, for each value of N.

From Table 1, we observe that even for the minimum keystream length, $N = 25$, the attack is usually successful. In a small number of cases, there exist multiple sets of LFSR initial states which produce the required keystream segment and the attack cannot identify the actual states used. However, as noted in Section 3.1 only a small increase in keystream length is required to eliminate these additional candidates.

6 Ciphertext-only Attacks

In many cases, the known-plaintext attack on ORYX can be extended to a ciphertext-only attack if some knowledge of the plaintext statistics is assumed. For example, when the plaintext is English text or other stereotyped data, ciphertext-only attacks are likely to be feasible with just hundreds or thousands of bytes of ciphertext.

To perform a ciphertext-only attack we start by identifying a probable string of at least seven characters; a word or phrase which is likely to appear in the plaintext. Examples of probable strings include "login:␣" and ".␣␣The␣". We then slide the probable string along each ciphertext position, hoping to find a "match" with the correct cleartext message.

If we align a probable plaintext string correctly, then we obtain a segment of the keystream with length equal to the length of the probable plaintext string. The known-plaintext attack described above can be performed on this keystream segment. If every path of guesses is ruled out by the end of the $N = 7$ bytes of known text, then we know the probable string does not match the cleartext at this position. Otherwise, we conclude that we have found a valid match; this may sound optimistic, but we show next that the probability of error is acceptably low.

With this procedure, false matches should be rare, because false paths of guesses are eliminated very quickly. After analyzing the first byte, 2^{16} possibilities for the high bytes of each register remain. From Section 4.1, only 0.0459 of the wrong possibilities remain undetected after the second byte; of those, the proportion which remain undetected after the third byte is 0.0459; and so on. This means that only $2^{16} \cdot (0.0459)^6 = 0.00061 \approx 2^{-10.7}$ wrong possibilities are expected to survive the tests after $N = 7$ bytes of known text are considered, on average. The probability of a false match being accepted is at most $0.00061 \approx 2^{-10.7}$. Therefore, with less than a thousand bytes of ciphertext, we expect to see no false matches, for probable strings of length $N = 7$. Using a slightly longer probable word will further reduce the probability of error.

The search for ciphertext locations which yield a valid match with the probable word can be performed quite efficiently. It should take about 2^{16} work, on average, to check each ciphertext position for a possible match. With less than a thousand bytes of ciphertext, the total computational effort to test a probable word is less than 2^{26}, and thus even a search with a dictionary of probable words is easily within reach.

Next, we describe how to use matches with the probable word to recover the ORYX key material. Each valid match provides $8 + (N - 1) = 14$ bits of information on the initial states of $LFSR_A$ and $LFSR_K$, and $8 + 1.5 \cdot (N - 1) = 17$ bits of information on the initial state of $LFSR_B$. Therefore, with three probable word matches, we will have accumulated about 42 bits of information on each of the 32 bit keys for $LFSR_A$ and $LFSR_K$, and 51 bits of information on the 32 bit key for $LFSR_B$. The key can be easily recovered by solving the respective linear equations over $GF(2)$. Alternatively, with two matches, we have 28 bits of information for $LFSR_A$ and $LFSR_K$, and 34 bits of information for $LFSR_B$. An exhaustive search over the remaining eight unknown bits should suffice to find the entire key with about 2^8 trials. For each key trial, we can decrypt the ciphertext, and check whether the result looks like plausible plaintext by using simple frequency statistics or more sophisticated techniques.

As long as we have enough ciphertext and can identify some set of probable words, it should be easy to find two or three matches and thus recover the entire ORYX key. In other words, it appears that even ciphertext-only attacks against ORYX have relatively low complexity, when some knowledge of the plaintext statistics is available. The computational workload and the amount of ciphertext required are modest, and these attacks are likely to be quite practical.

7   Summary and Conclusions

ORYX is a simple stream cipher proposed for use as a keystream generator to protect cellular data transmissions. The known plaintext attack on ORYX presented in this paper is conducted under the assumption that the cryptanalyst knows the complete structure of the cipher and the 96-bit secret key is only the initial states of the component LFSRs. The attack requires exhaustive search over 16 bits, and has over 99 percent probability of success if the cryptanalyst knows 25 bytes of the keystream. The probability of success is increased if the cryptanalyst has access to more than 25 bytes of the keystream. In our trials, a keystream length of 27 bytes was sufficient for the attack to correctly recover the key in every trial. Furthermore, we have shown how to extend this to a ciphertext-only attack which is likely to be successful with only hundreds or thousands of bytes of known ciphertext.

These results indicate that the ORYX algorithm offers a very low level of security. The results further illustrate the low level of security offered in most second generation mobile telephone devices. The authors are of the opinion that, in most cases, this is due to the lack of public scrutiny of the cryptographic algorithms prior to their adoption for widespread use. It is to be hoped that the past reliance on security through obscurity will not be repeated in the cryptographic algorithms to be used in the third generation of mobile communications systems, due for use early in the twenty-first century.

References

1. E. Dawson and L. Nielsen. Automated cryptanalysis of XOR plaintext strings. *Cryptologia*, volume XX Number 2, pages 165–181. April 1996.
2. B. Goldburg, E. Dawson and S. Sridharan. The automated cryptanalysis of analog speech scramblers. *Advances in Cryptology - EUROCRYPT'91*, volume 547 of *Lecture Notes in Computer Science*, pages 422–430. Springer-Verlag, 1991.
3. M. Briceno, I. Goldberg and D. Wagner. GSM cloning. 20 April, 1998. `http://www.isaac.cs.berkeley.edu/isaac/gsm.htm`
4. J. Dj. Golić. Cryptanalysis of alleged A5 stream cipher. *Advances in Cryptology - EUROCRYPT'97*, volume 1233 of *Lecture Notes in Computer Science*, pages 239–255. Springer-Verlag, 1997.
5. H. Schildt. *C the Complete Reference* Osborne McGraw-Hill, Berkeley, CA, 1990.
6. TIA TR45.0.A, *Common Cryptographic Algorithms* June 1995, Rev B.
7. D. Wagner, B. Schneier and J. Kelsey. Cryptanalysis of the cellular message encryption algorithm. *Advances in Cryptology - CRYPTO'97*, volume 1294 of *Lecture Notes in Computer Science*, pages 526–537. Springer-Verlag, 1997.
8. D. Wagner, B. Schneier and J.Kelsey. *Cryptanalysis of ORYX.* unpublished manuscript, 4 May 1997.

A Timing Attack on RC5

Helena Handschuh[1] and Howard M. Heys[2]

[1] ENST, Computer Science Department
46, rue Barrault, F-75634 Paris Cedex 13
handschu@enst.fr
GEMPLUS, Cryptography Department
34, rue Guynemer, F-92447 Issy-les-Moulineaux
handschuh@gemplus.com
[2] Memorial University of Newfoundland, Faculty of Engineering
St. John's, NF, Canada A1B 3X5
howard@engr.mun.ca

Abstract. This paper describes a timing attack on the RC5 block encryption algorithm. The analysis is motivated by the possibility that some implementations of RC5 could result in the data-dependent rotations taking a time that is a function of the data. Assuming that encryption timing measurements can be made which enable the cryptanalyst to deduce the total amount of rotations carried out during an encryption, it is shown that, for the nominal version of RC5, only a few thousand ciphertexts are required to determine 5 bits of the last half-round subkey with high probability. Further, it is shown that it is practical to determine the whole secret key with about 2^{20} encryption timings with a time complexity that can be as low as 2^{28}.

Keywords: Cryptanalysis, Timing Attacks, Block Cipher.

1 Introduction

RC5 is an iterative secret-key block cipher invented by R. Rivest [1]. It has variable parameters such as the key size, the block size, and the number of rounds. A particular (parameterized) RC5 algorithm is designated as RC5-$w/r/b$ where w is the word size (one block is made of two words), r is the number of rounds, and b is the number of bytes for the secret key. Our attack works for every choice of these parameters. However, we will focus on the "nominal" choice for the algorithm, RC5-32/12/16, which has a 64-bit block size, 12 rounds, and a 128-bit key.

The security of RC5 relies on the heavy use of data-dependent rotations. The application of the two powerful attacks of differential and linear cryptanalysis to RC5 is considered by Kaliski and Yin [2], who show that the 12-round nominal cipher appears to be secure against both attacks. In [3], Knudsen and Meier extend the analysis of the differential attacks of RC5 and show that, by searching for appropriate plaintexts to use, the complexity of the attack can be reduced by a factor of up to 512 for a typical key of the nominal RC5. As well, it is shown that keys exist which make RC5 even weaker against differential cryptanalysis. Recently, in [4], new differential cryptanalysis results imply that 16 rounds are required for the cipher with $w = 32$ to be secure. The results on linear cryptanalysis are refined by Selcuk in [5] and, in [6], it is shown that a small fraction of keys results in significant susceptibility to linear cryptanalysis. Despite these results, from a practical perspective RC5 seems to be secure against both differential and linear cryptanalysis.

In [7], Kocher introduces the general notion of a timing attack. The attack attempts to reveal the key by making use of information on the time it takes to encrypt. The applicability of the attack on asymmetric systems is demonstrated by examining timing variations for modular exponentiation operations. As noted by Kocher, implementations of RC5 on processors which do not execute the rotation in constant time are at risk from timing attacks. We will show that for implementations where the rotations take a variable amount of time, linear in the number of left shifts, RC5 is vulnerable to timing attacks which recover the extended secret key table with only 2^{20} ciphertexts from the sole knowledge of the total amount of rotations carried out during encryption.

2   Description of Cipher

RC5 works as follows: the secret key is first extended into a table of $2r + 2$ secret words S_i of w bits. We will assume that this key schedule algorithm is rather one-way and will therefore focus on recovering the extended secret key table and not the secret key itself. The description of the key schedule can be found in [1].

Let (L_0, R_0) denote the left and right halves of the plaintext. Then the encryption algorithm is given by:

$$
\begin{aligned}
&L_1 = L_0 + S_0 \\
&R_1 = R_0 + S_1 \\
&\text{for } i = 1 \text{ to } 2r \text{ do} \\
&\quad L_{i+1} = R_i \\
&\quad R_{i+1} = ((L_i \oplus R_i) \leftarrow R_i) + S_{i+1}
\end{aligned}
\tag{1}
$$

where "+" represents addition modulo-2^w, "$\oplus$" represents bit-wise exclusive-or, and "$X \leftarrow Y$" is the rotation of X to the left by the $\log_2 w$ least significant bits of Y. For example, if $w = 32$, X is rotated to the left by the number of positions indicated by the value of $Y \bmod 32$.

The ciphertext is (L_{2r+1}, R_{2r+1}). The transformation performed for a given i value is called a half-round: there are $2r$ half-rounds. Each half-round involves exactly one data-dependent rotation and one sub-key S_i.

To decrypt, the operations of the algorithm must be appropriately reversed to generate the data for each half-round by essentially going backwards through the encryption algorithm. For example, data is rotated right and the subkeys are applied by subtracting modulo-2^w from the data.

In section 3 hereafter we describe the foundations of the timing attack and give some preliminaries and in section 4 we describe our timing attack as it is used to obtain $\log_2 w$ bits of the last half-round subkey. In section 5, we discuss how to derive the remaining subkey bits and, in section 6, we present some experimental results on the likelihood of the success of the attack. Finally, in section 7, we discuss the complexity of the attack.

3   Preliminaries

In this section, we describe our assumptions for the timing attack and show how to correlate the total amount of rotations carried out during encryption with the value of the second last rotation.

3.1   Timing attacks

For a complete description of Kocher's timing attacks we refer to [7]. The main idea is to correlate the time measurements for processing different inputs with the secret exponent in RSA or Diffie-Hellman like systems. Since for every non-zero bit in the secret exponent the whole process performs an extra modular multiplication which depends on some computable intermediate value, correlations can be found between the variations of the time measurements on some sample space and the fact that an exponent bit is set or not. This way, as the distributions become more accurate, more and more bits can be derived and finally the whole secret key can be recovered.

In symmetric-key cryptosystems, things tend to get more complicated as usually only constant-time operations such as additions, exclusive-ors or table look-ups are performed. Nevertheless, under certain assumptions, given cryptosystems like RC5 can become vulnerable to timing attacks as well.

3.2   Hypothesis

Rivest notes that "on modern microprocessors, a variable rotation ... takes *constant time*" but there are certain types of processors which do not have this property. For instance for 8-bit microcontrollers on smart cards, or in other constrained environments, the rotation has to be performed step by step, one left or right shift at a time. It is not necessarily optimal to swap bytes or nibbles depending on the number of left shifts, as testing the rest modulo 16, 8 or 4 of this number may take more time than doing all the shifts no matter what. When machine cycles are measured during encryption, we can deduce from a certain amount of measurements how long the constant-time operations take, and how long the variable-time operations take: as these only concern rotations for RC5, we can deduce what the total amount of rotations is for every encryption. We believe a specific analysis can also endanger the algorithm if rotations are carried out in a non-linear but still variable amount of time. As well, it should also be noted, that a naive, straightforward hardware implementation could also result in rotation times that are a function of the cipher data and, hence, create susceptibility to the timing attack we describe in this paper.

In this paper, we focus on the case where we can assume that a left rotation by an amount y takes y times longer than a left rotation by an amount of 1. We show a ciphertext-only attack that recovers the extended secret key in a reasonable amount of time when the total number of rotation amounts is given. Kocher has already mentioned that "RC5 is at risk" and we show why. Note that the imperfectness and inherent randomness of timing measurements may cause the complexity to grow and the required number of ciphertexts to be much higher, but as the theoretical attack we present is of such low complexity, it should raise serious doubts about the strength of RC5, if implemented such that the rotation times are data dependent.

3.3   Foundation of the Attack

Let T_{2r} denote the total rotation amount for the encryption of a given plaintext. T_{2r} is given by:

$$T_{2r} = \sum_{i=1}^{2r} (R_i \bmod w) \tag{2}$$

We note that the amount of the last rotation is known because it is the value of the $\log_2 w$ least significant bits of L_{2r+1} which is the left half of the ciphertext. Therefore let us consider the total amount of rotations minus the last rotation. We denote this quantity by T_{2r-1} and

$$T_{2r-1} = \sum_{i=1}^{2r} (R_i \bmod w) - (L_{2r+1} \bmod w). \tag{3}$$

More generally, for a half-round k, we can define the intermediate amount of rotations so far and we have

$$T_k = \sum_{i=1}^{k} R_i \bmod w \tag{4}$$

and

$$T_{k-1} = \sum_{i=1}^{k} R_i \bmod w - (L_{k+1} \bmod w). \tag{5}$$

We also have the following:

$$T_{k-1} = T_{k-2} + R_{k-1} \bmod w \tag{6}$$

Now let us consider the way the total amounts of rotations are distributed over some large sample space. T_{k-2} can be represented by its mean value $(k-2) \times \frac{w-1}{2}$ added to some deviation. If this deviation (noise) is small, T_{k-1} and $R_{k-1} \bmod w$ are correlated at each half-round and the distribution of T_{k-1} gives us a good idea about what the distribution of $R_{k-1} \bmod w$ should look like.

This is the observation that leads to our timing attack. Knowledge of the second last rotation amount gives us some knowledge about the last subkey. We describe the attack in detail in section 4 hereafter.

4   Our Attack

In this section we describe the first part of our attack, which is how to derive the $\log_2 w$ least significant bits of the last subkey, S_{2r+1}. We shall describe two approaches to extracting the $\log_2 w$ least significant bits of the last subkey. We shall refer to these approaches as Method A and Method B. (Both methods were derived independently and are further elaborated in [8] for Method A and [9] for Method B.) The fundamental difference between the approaches is the indicator used as a metric to pick the most likely subkey.

4.1   Method A

In the last section, we showed how the total amount of rotations T_{k-1} and a given rotation $R_{k-1} \bmod w$ are correlated. Now we need an indicator to be able to qualify this correlation. A quite natural choice consists in using the following I correlation coefficient as an indicator:

$$I = E\{(T_{k-1} - \mu_{k-1})(R_{k-1} \bmod w - \frac{w-1}{2})\} \tag{7}$$

where μ_{k-1} is the mean of the distribution of the T_{k-1}s over some sample space.

In fact, it is even more convenient to use only the sign of $(R_{2k} \bmod w - \frac{w-1}{2})$ in order to partition the samples depending on the deviation from their mean value. The indicator we shall therefore be using is the following:

$$I = E\{(T_{k-1} - \mu_{k-1}) \times Sign(R_{k-1} \bmod w - \frac{w-1}{2})\} \tag{8}$$

In the case of two correlated distributions, this indicator is expected to have a higher absolute value than in the case of two independent distributions (when our guess about the second last rotation amount is wrong).

The first phase of the attack is a sample generation phase. We collect triplets corresponding to a plaintext encrypted under the unknown key and its ciphertext, and T_{2r-1}, the total amount of rotations minus the last one carried out during encryption. These samples are stored in a table and are ordered by the value of the $\log_2 w$ least significant bits of the left half of the ciphertext. Recall that this also is the value of the last rotation amount.

We denote by N the number of collected samples of total rotation amounts in each category. At each half-round we assume that the intermediate rotation amounts are uniformly distributed, independent random variables. For our analysis to work, we need the standard deviation of our sample space to be negligible over all half-rounds and all samples.

Let X_i be the rotation amount of the i-th sample at an intermediate half-round. Over all half-rounds and all samples, we have:

$$Var(2r \times X_i) = 2r \times Var(X_i) = 2r \times Var(X_0) \tag{9}$$

and

$$Var(2r \times \sum_{i=1}^{N} X_i) = 2r \times N \times Var(X_0) \approx 2r \times N \times \frac{w^2}{12} \tag{10}$$

The standard deviation σ is given as:

$$\sigma^2 = Var(\mu) = Var(\frac{2r \times \sum_{i=1}^{N} X_i}{N}) \approx \frac{2r}{N} \times \frac{w^2}{12} \tag{11}$$

An order of magnitude of the number N of samples needed is given by the condition:

$$\sigma \ll 1 \tag{12}$$

which gives us the following condition for N:

$$N \gg 2r \times \frac{w^2}{12} \tag{13}$$

Therefore in each category we need N to be much greater than 2^{11}. From a practical point of view, we implemented our attack with 2^{15} samples in each category ; there are w different categories, therefore the total number of samples required is $2^{15} \times w = 2^{20}$. We will keep this upper bound in our complexity evaluation in section 7 as it is convenient for practical implementations. (Actually, the practical attack requires quite more time and plaintexts than suggested by the theory, so this approximation fits much better to the experiments.)

Now we have:

$$R_{2r+1} = ((R_{2r-1} \oplus L_{2r+1}) \leftarrow L_{2r+1}) + S_{2r+1} \tag{14}$$

In this last equation, R_{2r+1} is the right half of the ciphertext and L_{2r+1} is the left half of the ciphertext. Therefore the right value of R_{2r-1} gives us the right value of S_{2r+1}.

We concentrate on the category of samples such that the $\log_2 w$ least significant bits of the left half of the ciphertext are equal to zero. In particular, this means that the last rotation amount is also equal to zero. Therefore we have the following equation:

$$(R_{2r+1} - S_{2r+1}) \bmod w = R_{2r-1} \bmod w \tag{15}$$

For each possible value of the $\log_2 w$ least significant bits of S_{2r+1}, we compute the supposed second last rotation amount $R_{2r-1} \bmod w$ for each sample in the category we concentrate on. This gives us a trial distribution over the 2^{15} samples.

Now divide the samples into two parts depending on the sign of the guessed rotations. Compute the correlation coefficients I^+ and I^- on each of the two parts. On the negative part, the correlation coefficient is supposed to be negative, and on the positive part, it is supposed to be positive. Finally compute the quantity $I^+ - I^-$. This indicator should have a higher value when the two distributions are correlated then when they are independent. The right value of the $\log_2 w$ least significant bits of the last subkey is suggested by the highest indicator.

4.2   Method B

In this section, we describe another approach to extracting the $\log_2 w$ least significant bits of the last subkey and, using a probabilistic model we get an estimate of the number of ciphertexts required to determine the bits with high probability. For convenience, we shall strictly focus our attention on the cipher with $w = 32$ and $r = 12$. Similar to the previous section, the model assumes that the rotations in each half-round are independent random variables that are uniformly distributed over $\{0, 1, 2, \ldots, 31\}$ with a mean of 15.5 and a variance of 85.25. Under these assumptions, the sum of the number of rotations for the first 22 half-rounds of the cipher, T_{22}, is a random variable with a mean of $\mu_{22} = 22 \cdot 15.5 = 341$ and variance of $\sigma_{22}^2 = 22 \cdot 85.25^2 = 1875.5$. As well, based on the central limit theorem, T_{22} is approximately Gaussian distributed.

To determine the correct value of the partial subkey $S_{25} \bmod 32$, a number of ciphertexts is used to test each possible value for the 5 bits and to determine which value is most consistent with the expected statistical distribution of the timing information. In particular, ciphertexts for which $L_{25} \bmod 32 = 0$ are used to compute an estimate of the variance of T_{22} based on the value of each candidate partial subkey: it is expected that the variance estimate when the correct value for the partial subkey is selected will be smaller than the estimate when an incorrect partial subkey is assumed.

Let K represent the actual key bits $S_{25} \bmod 32$ and let $\tilde{K}$ represent the guess of the partial subkey K. The candidate key $\tilde{K}$ can be represented by

$$\tilde{K} = (K + \tau) \bmod 32 \tag{16}$$

where $-15 \leq \tau \leq 16$. The estimate of the variance of T_{22} for a particular candidate key $\tilde{K}$ is given by

$$\phi_\tau = E\left\{e_{\tau,x}^2\right\} \tag{17}$$

where $e_{\tau,x}$ represents the difference between the measured number of rotations for the entire 24 half-rounds and the expected number of rotations given the assumed candidate key for a ciphertext with $R_{25} \bmod 32 = x$. The difference $e_{\tau,x}$ is composed as follows:

$$e_{\tau,x} = (T_{22} + R) - (\mu_{22} + \tilde{R}_{\tau,x}) \tag{18}$$

where R represents the actual value of the rotation in the 23rd half-round (i.e., $R = R_{23} \bmod 32$) and $\tilde{R}_{\tau,x}$ represents the guess of the rotation in the 23rd half-round (corresponding to the candidate key $\tilde{K}$) given $R_{25} \bmod 32 = x$. Using ciphertexts for which $L_{25} \bmod 32 = 0$, the size of the rotation in the 24th half-round is 0. Therefore, $T_{22} + R$ equals the total number of rotations, which, of course, can be derived from the timing information. The value of $\tilde{R}_{\tau,x}$ is determined from $\tilde{R}_{\tau,x} = (x - \tilde{K}) \bmod 32$. Hence, for a given ciphertext and candidate key, the value of $e_{\tau,x}$ can be calculated. We can also view equation (18) by letting $\Delta T = T_{22} - \mu_{22}$ and $\Delta R_{\tau,x} = R - \tilde{R}_{\tau,x}$ and we get

$$e_{\tau,x} = \Delta T + \Delta R_{\tau,x}. \tag{19}$$

Now assume that the cryptanalyst has available N ciphertexts for which $L_{25} \bmod 32 = 0$ and, hence, for large N, the number of ciphertexts corresponding to a particular value x for $R_{25} \bmod 32$ is given by $N_x \approx N/32$. For each ciphertext and candidate key $\tilde{K}$, $e_{\tau,x}$ is computed and the mean of the square of $e_{\tau,x}$ is calculated. The result is equivalent to

$$\phi_\tau = \frac{1}{N} \sum_{x=0}^{31} \sum_{i=1}^{N_x} [\Delta T_{x,i} + \Delta R_{\tau,x}]^2 \tag{20}$$

where $\Delta T_{x,i}$ represents the i-th value of ΔT for $R_{25} \bmod 32 = x$. For the correct guess of the key (i.e., $\tau = 0$), $\Delta R_{\tau,x} = 0$ since $\tilde{R}_{\tau,x} = R$ and, hence, $\phi_0 = (1/N) \sum_{x=0}^{31} \sum_{i=1}^{N_x} (\Delta T_{x,i})^2$. For incorrect candidate keys, for which $|\tau| \geq 1$, $\Delta R_{\tau,x} \neq 0$, and it can be shown that $E\{\phi_\tau\} > E\{\phi_0\}$. The cryptanalyst can therefore collect ciphertexts and timing information (implying rotation information) and determine the key K by picking the candidate key $\tilde{K}$ which minimizes ϕ_τ.

We now determine the probability that an incorrect key is selected over the correct key. For this to happen, we must have $\phi_\tau < \phi_0$ or, alternatively, $\phi_\tau - \phi_0 < 0$. Hence, the cryptanalyst must acquire enough ciphertext timings to ensure that $\phi_\tau - \phi_0 > 0$ for all $\tau \neq 0$. From (20), it can be seen that

$$\phi_\tau - \phi_0 = \frac{1}{N} \sum_{x=0}^{31} \sum_{i=1}^{N_x} [2\Delta R_{\tau,x} \Delta T_{x,i} + (\Delta R_{\tau,x})^2]. \tag{21}$$

We can consider $\phi_\tau - \phi_0$ to be a Gaussian distributed random variable with an expected value given by

$$E\{\phi_\tau - \phi_0\} \approx \frac{1}{32} \sum_{x=0}^{31} (\Delta R_{\tau,x})^2 \tag{22}$$

where we have used $N_x \approx N/32$ and $E\{\Delta T\} = 0$. As well, $\phi_\tau - \phi_0$ has a variance given by

$$Var\{\phi_\tau - \phi_0\} = \frac{1}{N^2} \sum_{x=0}^{31} \sum_{i=1}^{N_x} [(2\Delta R_{\tau,x})^2 \sigma_{22}^2] \approx \frac{\sigma_{22}^2}{8N} \sum_{x=0}^{31} (\Delta R_{\tau,x})^2. \tag{23}$$

It can be shown that

$$\sum_{x=0}^{31}(\Delta R_{\tau,x})^2 = |\tau|(32 - |\tau|)^2 + (32 - |\tau|)|\tau|^2 = 32|\tau|(32 - |\tau|) \qquad (24)$$

and, consequently, it can be easily verified that

$$\max_\tau P(\phi_\tau - \phi_0 < 0) = P(\phi_\omega - \phi_0 < 0) \qquad (25)$$

where $\omega = -1$ or $+1$. For $N = 2000$ ciphertexts for which $L_{25} \bmod 32 = 0$, based on the Gaussian distribution, we get $P(\phi_1 - \phi_0 < 0) = 0.0021$ and the probability of being able to determine the correct 5 bits of subkey, $S_{25} \bmod 32$, is given by

$$P(S_{25}[4\ldots 0] \text{ correct}) = 1 - P(\exists \tilde{K}|\tau \neq 0, \phi_\tau < \phi_0) \qquad (26)$$

where

$$P(\exists \tilde{K}|\tau \neq 0, \phi_\tau < \phi_0) < 31 \cdot P(\phi_1 - \phi_0 < 0) = 0.0651. \qquad (27)$$

Therefore, the probability of picking the correct 5 bits of subkey is greater than 93.5% with 2000 ciphertexts under the assumption that the rotations in all half-rounds are independent. Note that the ciphertexts must be chosen such that $L_{25} \bmod 32 = 0$, which is true on average for 1 in 32 random ciphertexts. Hence, the correct 5 bits of subkey can be derived with high probability using about 64000 random ciphertexts and their timings.

As we shall see in section 6, in fact, there are some dependencies in the rotations of different half-rounds which result in the probabilities of successfully deriving the key in practice being somewhat lower than expected from the model. Nevertheless, experimental evidence confirms that the approach works well when applied to the actual cipher.

5   Deriving the Remaining Subkey Bits

In the previous section, we illustrated how it is possible to determine $\log_2 w$ bits of the last half-round subkey S_{2r+1} with high probability using a set of random ciphertexts and their timing information. Fortunately, it is straightforward to apply the techniques on the same ciphertexts to determine the remaining bits of subkey S_{2r+1} and, with enough ciphertexts, it is possible to derive all the bits of the subkeys $S_i, 3 \leq i \leq 2r$, as well.

First, now that we have found the $\log_2 w$ least significant bits of S_{2r+1}, we have to derive the $w - \log_2 w$ other bits of the last subkey. We proceed the following way:

Based on the categories described in section 4.1, concentrate on one category of samples at a time, in increasing order. Depending on the value of the $\log_2 w$ least significant bits of the left ciphertext, the right half was rotated by some amount to the left. Therefore the right value of $R_{2r-1} \bmod w$ gives us the right value of the $\log_2 w$ bits of the last subkey which correspond to the $\log_2 w$ bit positions $i = [L_{2r+1} \bmod w]$ through $i = [L_{2r+1} \bmod w + \log_2 w - 1]$.

However, we proceed bit by bit in order to take advantage of the knowledge we already have on the least significant bits of the subkey. For each new rotation amount from 1 to $w - \log_2 w$, try each of the two possible values of the $\log_2 w$ concerned bits of the subkey: take the $\log_2 w - 1$ bits you already know and guess 0 or 1 for the next bit. This gives you only two possible distributions for $R_{2r-1} \bmod w$ in each category of samples. Using the indicator of either Method A or Method B determine the targeted key bit. The right value of the $\log_2 w$ concerned bits of the last subkey is suggested by the corresponding best indicator.

Once S_{2r+1} is derived, the remaining subkeys associated with each half-round $i, 2 \leq i \leq 2r - 1$, may be determined using the same set of ciphertexts. Once the subkey for a half-round i is determined, the ciphertext may be partially decrypted for one half-round so that the output of half-round $i - 1$ is known. Correspondingly, the timing of the partial encryption of the first $i-1$ half-rounds may be determined by subtracting the time to execute the i-th half-round from the time to encrypt the first i half-rounds of the cipher. The new ciphertext and timing information may then be used to extract the subkey for half-round $i - 1$ in exactly the same manner as for the subkey of half-round i.

The remaining subkeys, S_0, S_1, and S_2, are applied to the cipher by addition to the plaintext left half, plaintext right half, and the output of the rotation operation in the first half-round. All three of these subkeys cannot be determined using timing information but are trivially determined using only a modest number of known plaintexts and ciphertexts: S_1 is simply determined using one known plaintext and using the relationship $S_1 = L_2 - R_0$, S_2 can be determined with a modest number of known plaintexts using, for example, linear cryptanalysis [2], and S_0 can be easily derived once S_1 and S_2 are determined using $S_0 = [((R_2 - S_2) \rightarrow L_2) \oplus L_2] - L_0$.

6   Experimental Results

In this section we present the experimental results which validate the effectiveness of the attack. Both Methods A and B assume that the values of the rotations in different half-rounds are uniformly distributed, independent random variables. This assumption, however, is not strictly correct. Consider, for example, the following scenario for $w = 32$ and $r = 12$: $S_{24} = 0$ and $S_{25} \bmod 32 = 0$. Suppose the cryptanalyst is attempting to determine $S_{25} \bmod 32$ and is therefore considering ciphertexts for which $L_{25} \bmod 32 = 0$. If $R_{25} \bmod 32 = x = 16$, then $R_{22} \bmod 32 = (L_{25} \rightarrow 16) \oplus 16$ and, since $L_{25} \rightarrow 16$ is a uniformly distributed random variable, $R_{22} \bmod 32$ behaves as anticipated by the model. However, if $R_{25} \bmod 32 = x = 0$, then $R_{22} \bmod 32 = 0$ always and $R_{22} \bmod 32$ is not a uniformly distributed random variable as suggested by the model.

These discrepancies from the model add inaccuracies to the process of statistically deriving the subkeys for both Method A and Method B. However, experimental results for Method B demonstrate that the cryptanalytic technique

Number of	Probability of Success		
Random Ciphertexts	S_{25} mod 32	S_{25}	$S_3 \ldots S_{25}$
10^4	0.611	0.083	0.000
10^5	0.893	0.794	0.009
10^6	0.901	0.827	0.024

Table 1. Experimental Results for 1000 Keys on RC5-32/12/16 (Method B)

is still very effective and the statistical model of section 4.2 provides a rough approximation of the effectiveness of the attack to determine 5 key bits of the last half-round subkey. Using Method B, for 2000 ciphertexts with L_{25} mod 32 = 0 (equivalent to about 64000 random ciphertexts), experiments on the nominal RC5 for 1000 random keys resulted in an 86.2% chance of the partial subkey S_{25} mod 32 being correctly determined. Using 64000 random ciphertexts, the complete subkey S_{25} was correctly determined for 69.7% of the keys.

The effectiveness of the timing attack as determined in experiments using Method B is further illustrated in Table 1. It is clear from the table that few random ciphertexts are required to determine the bits of the last half-round subkey with a high probability. The correct derivation of all subkeys $S_3 \ldots S_{25}$ does not occur with nearly as high a probability: even modest deviations in probability from 1 when determining subkeys significantly reduces the probability that all 23 subkeys will be successfully determined. However, it is apparent that the attack can be very effective in determining subkeys for a large fraction of keys and should be seriously considered to ensure that an implementation of the cipher is not vulnerable. Note that the probability of the complete success for the attack can be improved upon using a key path search approach described in the following section.

7   Complexity of the attack

In this section, we discuss the complexity of the attack in the context of Method A (although much of the discussion applies equally well to Method B). The first part of the attack is the sample generation phase. We need around 2^{15} ciphertexts in each category. There are w categories. Therefore the complexity of this phase is $2^{15} \times w$ encryptions. The ordering has no extra cost.

The second part of the attack is divided into four steps as mentioned in section 4. From a complexity point of view, there are $2r - 2$ half-rounds to be considered. For each half-round:

- start by computing the mean of the total rotation amounts in each category and by subtracting this mean to each total timing. This step takes $2^{15} \times w$ operations.
- then, computing the $\log_2 w$ least significant bits of a subkey takes $2^{15} \times w$ complexity.

- each further bit takes $2^{15} \times 2$ operations, and there are $w - \log_2 w$ such bits.
- finally, before entering the next half-round, decrypt one half of the ciphertext with the subkey just found, and reorder all the categories by the $\log_2 w$ least significant bits of this new half of the ciphertext. At the same time, subtract the value of the current last rotation from the total rotation amount for each ciphertext. This whole step takes $4 \times 2^{15} \times w$ operations.

Thus the total complexity for the last $2r - 2$ half-rounds is:

$$C = (2r - 2)[8 \times w \times 2^{15}] \tag{28}$$

As an example, for RC5-32/12/16 this complexity equals $C = 22 \times 2^{23}$. As for the last four subkeys, the cost is almost negligible.

In conclusion, the attack can be carried out with the encryption of only $2^{15} \times w$ plaintexts and the time complexity of the analysis phase is roughly equivalent to searching the last $2r - 2$ subkeys. The total complexity is:

$$C = (2r - 2)[8 \times w \times 2^{15}] = (2r - 2) \times 2^{18 + \log_2 w} \tag{29}$$

These results were all checked by computer experiments based on Method A. In fact, the complexity is slightly higher because the right value of the $\log_2 w$ least significant subkey bits is not always suggested by the best indicators. Sometimes the second or third-best indicator corresponds to the right subkey. Thus, when implementing the attack, we had to make some complexity tradeoffs. However, when the top indicators are quite close, trying the 8 most likely paths is still possible. This only applies while searching the $\log_2 w$ least significant subkey bits; all other bits are always correctly guessed.

Experiments show that, using Method A, when the guess of the $\log_2 w$ least significant bits is wrong, along the next 4 half-rounds the indicators tend towards a characteristical pattern. Therefore the right key path can still be made out with little extra effort.

Considering RC5-32/12/16, for example, if there is no "wrong" indicator, the best complexity is around 2^{28} for one key search. On average, no more than 8 subkeys are expected to lead to a partial exhaustive path search. For every such key, searching through the 8 keys associated to the 8 best indicators over two half-rounds leads to the right result. Nevertheless, for the subkeys S_7 through S_4, this path search cannot apply anymore. Therefore we exhaustively search through the 8 best subkeys for these four half-rounds. The overall extra work factor should not exceed $8 \times 8^2 + 8^4 = 2^9 + 2^{12}$. However, this leads to an upper bound on the complexity which is far from being optimal.

In summary, the overall complexity of our attack does not exceed 2^{40} operations for RC5-32/12/16. On average, the complexity is quite lower though. It is more realistic to consider the best case where the analysis takes only 2^{28} operations.

In the general case, the complexity of the attack does not exceed:

$$C_{max} = (2r - 2) \times 2^{30 + \log_2 w} \tag{30}$$

On average it is closer to the theoretic complexity

$$C = (2r - 2) \times 2^{18 + \log_2 w} \qquad (31)$$

and does not require more than $N_{max} = 2^{15 + \log_2 w}$ ciphertexts and their associated timings, as well as a few known plaintexts to derive the last subkeys.

8   Conclusion

We have shown in some detail how to derive the extended secret key table of RC5-32/12/16 by a timing attack using only about 2^{20} ciphertext timings and in time complexity 2^{28} in the best case, and 2^{40} in the worst case. This confirms Kocher's statement that RC5 is at some risk on platforms where rotations take a variable amount of time, and suggests to be very careful when implementing RC5 on such platforms. Adding a random time to each encryption will not help as it will have little influence on the variance computations. Therefore we suggest to add the right amount of "dummy "rotations which will achieve a constant time for every encryption whatever the initial total amount of rotations.

9   Acknowledgments

We are very grateful to Henri Gilbert for the ideas and the help he gave us on this attack. We also would like to thank David Naccache for motivating this investigation, and Jacques Stern and Gérard Cohen for helpful comments.

References

1. R. L. Rivest. The RC5 Encryption Algorithm. In *Fast Software Encryption - Second International Workshop, Leuven, Belgium, LNCS 1008*, pages 86-96, Springer-Verlag, 1995.
2. B. S. Kaliski and Y. L. Yin. On Differential and Linear Cryptanalysis of the RC5 Encryption Algorithm. In *Advances in Cryptology - Crypto'95, LNCS 963*, pages 171-184. Springer-Verlag, 1995.
3. L. R. Knudsen and W. Meier. Improved Differential Attacks on RC5. In *Advances in Cryptology - Crypto'96, LNCS 1109*, pages 216-228, Springer-Verlag, 1996.
4. Biryukov and Kushilevitz. Improved Cryptanalysis of RC5. In *Advances in Cryptology - Eurocrypt'98, LNCS*, pages 85-99, Springer-Verlag, 1998.
5. A. A. Selcuk. New results in linear cryptanalysis of RC5. In *Fast Software Encryption - Fifth International Workshop, Paris, France, LNCS*, pages 1-16, Springer-Verlag, 1998.
6. H.M. Heys, Linearly Weak Keys of RC5. *IEE Electronics Letters*, vol. 33, no. 10, pp. 836-837, 1997.
7. Paul C. Kocher. Timing Attacks on Implementations of Diffie-Hellman, RSA, DSS, and Other Systems. In *Advances in Cryptology - Crypto'96, LNCS 1109*, pages 104-113, Springer-Verlag, 1996.
8. H. Handschuh, A Timing Attack on RC5. In *Workshop Record of SAC '98*, Queen's University, Kingston, Canada, pages 318-329, 1998.
9. H.M. Heys, A Timing Attack on RC5. In *Workshop Record of SAC '98*, Queen's University, Kingston, Canada, pages 330-343, 1998.

Cryptanalysis of SPEED

Chris Hall[1], John Kelsey[1], Vincent Rijmen[2],
Bruce Schneier[1], and David Wagner[3]

[1] Counterpane Systems, 101 E. Minnehaha Pkwy
Minneapolis, MN 55419
Tel.: (612) 823-1098
{hall,kelsey,schneier}@counterpane.com
[2] K.U.Leuven, Dept. ESAT, SISTA/COSIC Lab
K. Mercierlaan 94, B-3001 Heverlee, Belgium
vincent.rijmen@esat.kuleuven.ac.be
[3] U.C. at Berkeley, Soda Hall
Berkeley, CA 94720-1776
daw@cs.berkeley.edu

Abstract. The cipher family SPEED (and an associated hashing mode) was recently proposed in *Financial Cryptography '97*. This paper cryptanalyzes that proposal, in two parts: First, we discuss several troubling potential weaknesses in the cipher. Next, we show how to efficiently break the SPEED hashing mode using differential related-key techniques, and propose a differential attack on 48-round SPEED. These results raise some significant questions about the security of the SPEED design.

1 Introduction

In *Financial Cryptography '97*, Zheng proposed a new family of block ciphers, called SPEED [12]. One specifies a particular SPEED cipher by choosing parameters such as the block size and number of rounds; the variations are otherwise alike in their key schedule and round structure. Under the hood, SPEED is built out of an unbalanced Feistel network. Zheng also proposed a hash function based on running a SPEED block cipher in a slightly modified Davies-Meyer mode.

One of the main contributions of the SPEED design is its prominent use of carefully chosen Boolean functions which can be shown to have very good non-linearity, as well as other desirable theoretical properties. One might therefore hope that SPEED rests on a solid theoretical foundation in cryptographic Boolean function theory. Nonetheless, this paper describes serious weaknesses in the cipher; many lead to practical attacks on SPEED.

This paper is organized as follows. Section 2 briefly summarizes the SPEED design. In Section 3, we discuss some preliminary analysis of the SPEED design, including comments on differential characteristics in SPEED, and on the non-surjectivity of the SPEED round function. Then we shift emphasis: Section 4 discusses differential characteristics for SPEED with 48 rounds. There appears to be an obvious 1-bit characteristic with probability 2^{-50} after 40 rounds; however, this characteristic does not actually work. We discuss this and other failed

S. Tavares and H. Meijer (Eds.): SAC'98, LNCS 1556, pp. 319–338, 1999.

characteristics in Section 4. In Section 5 we describe how a differential attack
can be mounted despite these problems. Section 6 gives differential related-key
attacks on block cipher and shows how to apply them to efficiently find collisions in the SPEED hash function. Section 7 assesses the practical implications
of these attacks, proposing a rule of thumb to characterize when we can consider a parameterized cipher family "broken." Finally, we conclude this paper in
Section 8.

2   Background

SPEED is a parameterized unbalanced Feistel cipher with a variable block width
w, variable key length l, and a variable number of rounds R (R must be a multiple
of 4 and w a multiple of 8). The block is split up into eight equally-sized pieces:
$B_7, \ldots, B_0$. The round function is then characterized by

$$t(B_7, \ldots, B_1, B_0) = (B_6, \ldots, B_1, B_0, r(B_7) \boxplus K_i \boxplus v(F_i(B_6, \ldots, B_1, B_0)))$$

where $\boxplus$ denotes addition modulo $2^{w/8}$, F_i is a round-dependent function with a
$w/8$-bit output, v is a data-dependent rotation, K_i is a round-dependent subkey,
and r is a bijective function from $\{0, 1\}^{w/8}$ to $\{0, 1\}^{w/8}$ (r is always right rotate
by $w/16 - 1$ by bits). See Figure 1 for one round of SPEED.

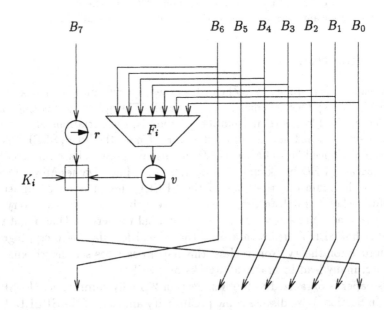

Fig. 1. One round of SPEED.

We sometimes refer to the seven input pieces, $B_6, \ldots, B_0$, as the source block and the modified piece, B_7, the target block. The number of rounds is a parameter; the paper suggests using at least 48 rounds for adequate security for the block cipher, and at least 32 rounds for the hash function. We assume that the underlying CPU operates on $w/8$-bit words.

The round dependent function F_i takes seven $w/8$-bit inputs and produces one $w/8$-bit output. The function v takes a $w/8$-bit input and produces a $w/8$-bit output by rotating the input a number of bits dependent on the input. The function that results from combining F_i and v will be denoted h_i.

$$h_i(x_6, \ldots, x_0) = v(F_i(x_6, \ldots, x_0))$$

The exact details of the rotation in v are not important to our analysis and may be found in [12] In this paper, we will write $a^{\gg b}$ to stand for the result of rotating a right by b bits.

There are four different F functions F_i for $i \in \{1, 2, 3, 4\}$, each is used for a different set of rounds. Each F_i is built out of a single (1-bit) Boolean function f_i on 7 (1-bit) variables, which is extended to a $w/8$-bit function by bitwise parallel evaluation. In other words, bit position i of the output depends only on seven input bits, each at position i in one of the input words. For example, F_1 is $F_1(x_6, \ldots, x_0) = x_6 x_3 \oplus x_5 x_1 \oplus x_4 x_2 \oplus x_1 x_0 \oplus x_0$ where $x_i x_j$ denotes bitwise AND. SPEED uses F_1 for the first $R/4$ rounds, then F_2 for the next $R/4$ rounds, and so on.

In summary, each round of SPEED uses one evaluation of F, two rotations, and an addition modulo $2^{w/8}$ to update the block.

3   Preliminary Analysis

In this section, we begin analysis of SPEED with respect to potential cryptanalytic attacks.

3.1   Non-Surjectivity of the Round Function

The round function involves generating a word to add (modulo the wordsize) to the target block. However, due to a flaw in the round function's design, we can significantly limit the possible output values. All output values are possible from F. However, the data dependent rotation in v limits the possible values actually added to the target block. In other words, the combined function $v \circ F$ is non-surjective. Rijmen et al [7] identified several attacks on ciphers with non-surjective round functions, so this property of SPEED is worrisome.

Applying the attack of [7] is not as straightforward as one might hope. There Rijmen et al depend on the fact that the range of the h-function is known and hence one can perform hypothesis testing based upon the combined output of several h-functions. However, for SPEED the output of each h-function is combined with a subkey and hence the range is unknown. Fortunately we know the shape that the distribution should take and it would appear that one can modify the analysis to perform hypothesis testing using the shape of the entire distribution rather than individual values of the distribution.

We have performed a preliminary analysis of a modified version of the cipher in which the data independent rotate is removed from each round. This allows us to write out an equation which is the sum of 6 subkeys (assuming a 48-round cipher), the outputs of 6 h-functions, part of the input block, and part of the output block. It appears that one simply performs $2^{w/8}$ hypothesis tests, one for each possible value of the sum of the 6 subkeys, selecting the value which produces the closest distribution to the output of 6 h-functions.

Analysis gets substantially more difficult when the data independent rotate is taken into account, since the carry bits that "cross the rotation boundary" spoil naive attempts to isolate the sum of the h-function outputs from the sum of the subkeys. Nonetheless, for larger word sizes the spread of the carry bits is limited. We conjecture that it may be possible to extend the technique to handle the data independent rotations in these cases, though the analysis will almost certainly be much messier.

3.2   Implications of Correlated Outputs

The outputs of successive round functions are correlated. For instance,

$$F_1(x_6, \ldots, x_0) = F_1(x_7, \ldots, x_1)$$

with probability $1/2 + 1/32$ over all choices of $x_0, \ldots, x_7$. This shows that there are non-trivial correlations between successive outputs of F_1; this could potentially lead to correlations between successive outputs of h. Similar correlations occur for F_3 and F_4, though this property does not seem to hold for F_2.

We have not been able to extend this troublesome weakness to a full attack on SPEED. Nonetheless, this gives another indication of how the SPEED design may get a great deal less strength than expected from the very strong Boolean functions chosen for it. If successive h-function outputs are correlated strongly enough, this could make linear or differential-linear attacks possible on the cipher. We leave this as an open question for future research.

3.3   Differential Characteristics

The key insight necessary for mounting a differential attack on this round function is that F is a very good nonlinear function, but it applies it to each bit position of each source block word independently. In other words, F exhibits very poor diffusion across bit positions. Therefore, we see immediately that flipping any one bit in the input of F can only affect one output bit of F. In particular, if the underlying Boolean function behaves "randomly," flipping an input bit of F should leave its output unchanged with probability $1/2$.

This would appear, at first glance, to yield a very simple eight-round iterative differential characteristic with probability approximately 2^{-8}. However, there is a problem. The straightforward attack doesn't work; we explain why below.

Complications There are two complications to our differential attack. First, the Boolean functions aren't really random, and so don't have quite the probability distribution we would have expected. Table 1 lists the probability that the output of F remains unchanged after flipping one input bit in the input at position i.

	0	1	2	3	4	5	6
F_1	.5	.5	.5	.5	.5	.5	.5
F_2	.5	.5	.5	.25	.5	.5	.75
F_3	.5	.5	.5	.5	.5	.5	.5
F_4	.5	.5	.5	.5	.5	.5	.5

Table 1. Probability that the output of F_i remains unchanged after flipping one input bit.

The second complication is much, much more problematic. SPEED is, in the terminology of [9], a source-heavy UFN. Furthermore, there is no key addition before the input of the F-function. This means that the inputs to the Feistel F-function in successive rounds can't be assumed to be independent, as they generally can be in a balanced Feistel network or in a target-heavy UFN.

If the inputs to successive rounds' F-functions aren't independent, then it's possible that the success probabilities of each round's differential characteristic are also not independent. In fact, taking this effect into account, we find that the probability for six of the eight possible eight-round (one-bit) iterative characteristics is precisely 0—the inter-round dependence makes it impossible for the characteristic to penetrate more than eight rounds. When extended to an eleven-round characteristic (across F_2), the characteristic always has a probability of 0. However, later in this paper we will show how to fix the attack by tweaking the differential characteristic. See Section 4.

The problem of inter-round dependence was mentioned as a theoretical possibility in [9]; here we give a practical example where it arises. The dependence also arises again in Section 6, where precisely this difficulty complicates a related-key attack on the cipher.

4   More on Differential Characteristics

Differential characteristics are also possible with more than one bit at the same bit position set. The flipped bits rotate through different bit positions (due to the constant rotate in the round function), but end up in the same bit position for most rounds. We will discuss the use of such characteristics later in this paper.

As we have already noted, the obvious 1-bit differential attack against the cipher does not work. The problem is that we will have a hard time ramming our differential through four rounds where F_2 is the F-function used for each round.

Suppose that we have the 3-round characteristic in Table 2, starting with round i.

r	$\Delta m_{i,7}$	$\Delta m_{i,6}$	$\Delta m_{i,5}$	$\Delta m_{i,4}$	$\Delta m_{i,3}$	$\Delta m_{i,2}$	$\Delta m_{i,1}$	$\Delta m_{i,0}$	P
-	0	0	0	0	0	0	0	A	-
i	0	0	0	0	0	0	A	0	1/2
i+1	0	0	0	0	0	A	0	0	1/2
i+2	0	0	0	0	A	0	0	0	1/2

Table 2. Failed 1-bit Differential Characteristic. $\Delta m_{i,j}$ is the value of the difference in data block j at the input of round i.

The problem is that this characteristic is impossible and we will not have the desired differences after round $i + 1$. The reason this characteristic always fails is that successive outputs of F_2 are correlated, as we saw in Section 3.2. This means that the characteristic's round probabilities are not independent, so we cannot simply multiply them to obtain the full probability. It is fairly easy to see that any 1-bit characteristic across 11 rounds of F_2 will necessarily have this 3-round characteristic, and hence the differential fails to attack the cipher with 44 or more rounds.

Unfortunately when trying to find a differential characteristic that would work we found that even slightly more complicated differentials also failed. Consider the 2-bit (8 round) differential with probability 2^{-10} given in Table 3. We

r	$\Delta m_{i,7}$	$\Delta m_{i,6}$	$\Delta m_{i,5}$	$\Delta m_{i,4}$	$\Delta m_{i,3}$	$\Delta m_{i,2}$	$\Delta m_{i,1}$	$\Delta m_{i,0}$	P
-	0	0	0	0	0	0	A	A	-
i	0	0	0	0	0	A	A	0	1/2
i+1	0	0	0	0	A	A	0	0	1/2
i+2	0	0	0	A	A	0	0	0	1/2
i+3	0	0	A	A	0	0	0	0	1/2
i+4	0	A	A	0	0	0	0	0	1/2
i+5	A	A	0	0	0	0	0	0	1/4
i+6	A	0	0	0	0	0	0	B	1/4
i+7	0	0	0	0	0	0	B	B	1/2

Table 3. Failed 2-bit Differential Characteristic.

found that repeatedly concatenating this differential causes it to fail after at most 12 rounds (assuming a 48-round cipher). In the next section we discuss the analysis we used to determine that both of the above differentials would fail.

4.1 A Failure Test for Differential Characteristics

If we make the assumption that all subkeys are independent and random, then there is a rather simple analysis technique we can apply to determine when a characteristic might fail. The technique will never fail a legitimate characteristic, but it may give a false positive. Therefore, passing the test does not imply that a characteristic will work.

The first observation is that with high probability, in characteristics such as those found in Table 2 and Table 3, bit j of $\Delta m_{i,7}$ will only be affected by bit j of the remaining $m_{i,k}$ and bit j of the subkey. Hence one can construct a state transition diagram in which each node contains the round number, bit j of the appropriate subkey, and bit j of each $m_{i,k}$ (for both halves of a pair satisfying the characteristic). We connect all nodes using directed arcs if one node is a legitimate successor of the other. This requires that the round numbers differ by 1, the $m_{i,k}$ satisfy the appropriate relations (as defined by the round function), and the output difference of the appropriate F-function applied to the $m_{i,k}$ is 0.

Once we have constructed such a graph we can view it as several different layers — one for each round of the characteristic. Clearly all edges originating from layer i in this construction will end in layer $i+1$. Therefore we can construct an adjacency matrix A_i for the transition from layer i to $i + 1$ and an overall adjacency matrix $A = \prod_i A_i$. The test we propose is to check whether A is the zero matrix. If it is, there will be no transition from a starting state to an ending state which satisfies our characteristic. Therefore we can eliminate those characteristics for which A is the zero matrix.

There is a small complication in that the last few rows of both of the characteristics we proposed above show non-zero bit differences in other bit positions. However, we observed that the remaining bits adjacent to bit j' (where $B = 2^{j'}$) are effectively random. Therefore to simplify our analysis we viewed the resulting 40-round characteristic as several superimposed n-round characteristics. We made the simplying assumption that each characteristic was independent of the others and hence were only concerned that each of the characteristics was independently possible. This assumption seems well justified, especially given that it will not eliminate legitimate characteristics (and we are only presently concerned with eliminating bogus characteristics).

Using these techniques, we examined each the characteristics given in Table 2 and Table 3. Our analysis found that each of these characteristics would not work so we were able to eliminate them. In fact, we analyzed the next most obvious 2-bit characteristic given in Table 4 but found that it also fails.

Fortunately, the differential given in Table 5 did not fail. In fact, one can construct independent round keys, a plaintext, and a ciphertext for which the differential holds. However, there appears to be one small difficulty even with this differential in that it will not work for *all* keys. Further research may reveal a differential (or family of differentials) which will work against all keys.

r	$\Delta m_{i,7}$	$\Delta m_{i,6}$	$\Delta m_{i,5}$	$\Delta m_{i,4}$	$\Delta m_{i,3}$	$\Delta m_{i,2}$	$\Delta m_{i,1}$	$\Delta m_{i,0}$	P
-	0	0	0	0	0	A	0	A	-
i	0	0	0	0	A	0	A	0	1/2
i+1	0	0	0	A	0	A	0	0	1/2
i+2	0	0	A	0	A	0	0	0	1/2
i+3	0	A	0	A	0	0	0	0	1/2
i+4	A	0	A	0	0	0	0	0	1/4
i+5	0	A	0	0	0	0	0	B	1/4
i+6	A	0	0	0	0	0	B	0	1/4
i+7	0	0	0	0	0	B	0	B	1/2

Table 4. Another Failed 2-bit Differential Characteristic.

r	$\Delta m_{i,7}$	$\Delta m_{i,6}$	$\Delta m_{i,5}$	$\Delta m_{i,4}$	$\Delta m_{i,3}$	$\Delta m_{i,2}$	$\Delta m_{i,1}$	$\Delta m_{i,0}$	P
-	0	0	0	0	0	A	A	A	-
i	0	0	0	0	A	A	A	0	1/2
i+1	0	0	0	A	A	A	0	0	1/2
i+2	0	0	A	A	A	0	0	0	1/2
i+3	0	A	A	A	0	0	0	0	1/2
i+4	A	A	A	0	0	0	0	0	1/4
i+5	A	A	0	0	0	0	0	B	1/8
i+6	A	0	0	0	0	0	B	B	1/4
i+7	0	0	0	0	0	B	B	B	1/2

Table 5. A Partially Successful 3-bit Differential Characteristic.

5   Mounting an Effective Differential Attack

The important point that we should remember from Section 4 is that in general, input differences with a small Hamming weight with large probability cause output differences with a small Hamming weight. Even if a pair does not follow one of the described characteristics, the Hamming weight of the output difference will usually be about the same as the Hamming weight of the input difference. This behavior is quite similar to that of RC2 [4] and RC5 [8]. Therefore we will use a variant of the differential attack that has been developed to attack RC5 [2] and is also used on RC2 [4].

5.1 Differentials and Characteristics

It has been observed [5] that the effectiveness of a differential attack depends on the probability of a *differential*, rather than on the probability of a characteristic. Indeed, when applying a differential attack to a block cipher, we are only concerned with the values of the differences in the first and last few rounds. The intermediate differences are not important. The analysis of RC2 has shown that in ciphers with limited diffusion, the difference in probability between characteristics and differentials may be significant.

We verified experimentally that there exist several 12-round differentials that go from a one-bit input difference to a one-bit output difference with significant probability, even for the cases where it is difficult to find a differential characteristic with nonzero probability (cf. Section 4). These 12-round differentials can be combined to produce longer differentials. For example, the 48-round differential with input difference $(0, 0, 0, 0, 0, 40, 0, 0)$ (in base-16) and output difference $(80, 0, 0, 0, 0, 0, 0, 0)$ (in base-16) has probability 2^{-60} (this holds exactly for 64-bit blocks, but the probability stays approximately the same for larger block lengths).

In fact, in our attack we will loosen the restrictions on the output difference, and consider for the last round only the Hamming weight of the difference, rather than specific values.

5.2 Key Recovery

The key recovery procedure works as follows. We choose pairs of plaintexts with a non-zero difference in one bit of B_0 only. We select the texts such that the output difference of h in the first round is zero. This is made particularly easy by the absence of a key addition at the inputs of F. Whether the output of F_1 in the second round is also zero, as required by the characteristic, depends on the plaintexts and the unknown value of the first round key only.

When we find a pair with a small output difference, we assume that it follows the characteristic in the second round. This gives us information about the value of the first round key. By judicious choice of the plaintexts, we can determine the bits of the first round key a few at a time. More requirements could be imposed on the plaintext pairs, in order to make sure that all pairs pass the first two rounds with probability one. This would, however, complicate the key recovery phase.

5.3 Filtering Issues

A basic, non-optimal approach is to use a characteristic that determines the differences until the last round of the cipher. This ensures that all wrong pairs are filtered, but the probability of the characteristic is quite low. The fact that differences spread very slowly through the rounds of SPEED can be used to relax the conditions on the pairs. Instead of requiring that B_7 of the output difference equals 80 (in base-16) and that the remaining B_i have difference zero we will just accept any pair where the Hamming weight of the output difference is below some threshold H. This improves the probability of the characteristic, because pairs are allowed to "fan out" in the last rounds. The disadvantage is that it becomes possible that wrong pairs are accepted. In order to get correct results, a value of H has to be selected such that the expected number of wrong pairs after filtering is below the expected number of good pairs. A block length of 128 or 256 allows for a larger H. Therefore, versions of SPEED with a block length of 128 or 256 bits require a higher number of rounds than the 64-bit version to be secure against the differential attack. The signal-to-noise ratio of the attack can be improved by using more sophisticated filtering techniques that are described in [2].

5.4   Experimental Results

We implemented differential attacks on versions of SPEED with a reduced number of rounds. The results are given in Table 6. For versions with more than 28 rounds, the plaintext requirements become impractical. From the obtained results, we estimate that a differential attack on SPEED with R rounds requires at least $2^{2(R-11)}$ plaintext pairs. Because of the effects described in Section 4, the plaintext requirements will probably increase even more if $R \geq 44$. This means that SPEED with 48 rounds and a block length of 64 bit is secure against our differential attack. For versions with a block length of 128 or 256 bit, more than 64 rounds are needed to obtain adequate resistance.

# rounds	success rate	# pairs
16	100%	2^{10}
20	100%	2^{18}
24	100%	2^{25}
28	80%	2^{31}

Table 6. Experimental results for reduced versions of SPEED. The numbers in the first three rows are obtained from 100 experiments each, the numbers in the last row are obtained from 10 experiments.

6   Related Key Attack

Related key attacks were first described in [1]. They are a class of attacks in which one examines the results of encrypting the same ciphertext under related keys. We perform such an analysis to produce collisions in the encryption function: two keys which encrypt the same plaintext to the same ciphertext.

In [12] the author suggests using SPEED in a variant of Davies-Meyer hashing in order to transform SPEED into a hash function. Specifically, a message is padded to a multiple of 256 bits by appending unique padding and a length (the exact details of the padding are beyond the scope of this paper). The resulting message M, is split into 256-bit chunks $M_0, M_1, \ldots, M_{n-1}$. The hash is D_n where $D_0 = 0$ and $D_i = D_{i-1} + E_{M_{i-1}}(D_{i-1})$. $E_K(X)$ denotes the encryption of X with key K. (Addition is defined slightly differently, but the exact definition does not affect our attack so we omit it.)

The cipher may be 64, 128, or 256 bits wide and hence so may the corresponding hash (although 64 bits would easily fall to a conventional birthday attack). Furthermore, the author of [12] suggests using 32 to 48 rounds for efficiency. We have successfully produced hash collisions for the 128-bit hashes with 32 rounds and also for 48 rounds. Using the reference implementation obtained from the author, we found the following collision for the 128-bit hash with 32 rounds (in base-16):

$$M = \text{21EA FE8E 1637 19F7 22D2 8CCB 3724 3437}$$
$$\text{B00F 7607 3C91 3710 2B69 C9C9 58FB 0823}$$
$$\text{AEC2 CD05 } \underline{\text{FD80}} \text{ 14E6 B11E } \underline{\text{43C0}} \text{ 5767 76F7}$$
$$\text{FF07 17EC } \underline{\text{FCBA}} \text{ 224E 9627 } \underline{\text{A16A}} \text{ 8D6E 83A9}$$

$$M' = \text{21EA FE8E 1637 19F7 22D2 8CCB 3724 3437}$$
$$\text{B00F 7607 3C91 3710 2B69 C9C9 58FB 0823}$$
$$\text{AEC2 CD05 } \underline{\text{FDC0}} \text{ 14E6 B11E } \underline{\text{4380}} \text{ 5767 76F7}$$
$$\text{FF07 17EC } \underline{\text{7CBA}} \text{ 224E 9627 } \underline{\text{216A}} \text{ 8D6E 83A9}$$

This leads to the following values when hashing (in base-16):

$$D_0 = \text{0000 0000 0000 0000 0000 0000 0000 0000}$$
$$D_1 = \text{90DA 7F34 46FA A373 B048 11F7 F8D9 BB3D}$$
$$D_2 - D_1 = \text{9781 9517 B5CC A046 D0F1 3719 ED9B A0B6}$$

$$D_0 = \text{0000 0000 0000 0000 0000 0000 0000 0000}$$
$$D_1 = \text{90DA 7F34 46FA A373 B048 11F7 F8D9 BB3D}$$
$$D_2 - D_1 = \text{9781 9517 B5CC A046 D0F1 3719 ED9B A0B6}$$

We also found the following collision for the 128-bit hash with 48 rounds (in base-16):

$$M = \text{3725 6571 48D5 CF52 DAE1 4065 7115 11A0}$$
$$\text{E3C5 9428 7BFD 18CB EF79 82BB 1D7F 2F55}$$
$$\underline{\text{36F2}}\ \underline{\text{CD58}}\ \text{9058}\ \underline{\text{FE57}}\ \underline{\text{D696}}\ \text{EA4C BD75 F7C9}$$
$$\underline{\text{1989}}\ \underline{\text{A048}}\ \text{39FB}\ \underline{\text{9B76}}\ \underline{\text{9011}}\ \text{CAC0 65F6 EBC7}$$

$$M' = \text{3725 6571 48D5 CF52 DAE1 4065 7115 11A0}$$
$$\text{E3C5 9428 7BFD 18CB EF79 82BB 1D7F 2F55}$$
$$\underline{\text{38F2}}\ \underline{\text{CB58}}\ \text{9058}\ \underline{\text{FC57}}\ \underline{\text{D896}}\ \text{EA4C BD75 F7C9}$$
$$\underline{\text{1985}}\ \underline{\text{A04C}}\ \text{39FB}\ \underline{\text{9B7A}}\ \underline{\text{900D}}\ \text{CAC0 65F6 EBC7}$$

This leads to the following values when hashing (in base-16):

$$D_0 = \text{0000 0000 0000 0000 0000 0000 0000 0000}$$
$$D_1 = \text{DA2B A119 A4F8 AA70 59ED 6FE4 188B 7969}$$
$$D_2 - D_1 = \text{CAB1 DA86 B6D3 1442 E05C A005 7B26 C432}$$

To produce collisions, we combine two different attacks:

1. A differential attack against the key schedule which produces two key schedules with a desired difference.
2. A related key attack against the cipher using the two related key schedules.

We feel that the attack is more illuminating when we address these two attacks in the opposite order. Therefore we will describe the related key attack first in order to give a motivation for the attack against the key schedule.

6.1 Related-Key Attack Against the Cipher

The fundamental observation that makes this attack possible is that a 1-bit input difference to any of the four F-functions will produce a zero output difference with probability 1/2 (actually this doesn't quite hold for F_2, but this doesn't seem to strongly affect experimental results). In our attack, we attempt to introduce 1-bit differences into the encryption state through the key schedule. We do this in such a way so that after several rounds, the introduced differences negate each other and the resulting encryption state is the same for both keys. In Table 7 of Appendix 1, we have shown all 32 rounds of our related-key attack. In summary, we encrypt a message with two different keys where the subkeys for rounds 1, 4, 9, 12, 17, and 25 have specific differences so that the encryption state under the two keys will be identical in rounds 13–16 and rounds 26–32. Initially the two encryption states are the same, and after round 25, the two encryption states will be identical with total probability 2^{-19}. Since the remaining keywords of the key schedule are identical, the probability that the same plaintext will encrypt to the same ciphertext under the two different key schedules is 2^{-19}.

Note, there are four variations on this attack in which we add 1 to 4 rounds prior to round 1 in which the two key schedules are identical. Hence the subkeys for rounds $t+1$, $t+4$, $t+9$, $t+12$, $t+17$, and $t+25$ with $t \in \{0, 1, 2, 3, 4\}$ are given the specified differences. If we define variant t to be the above differential with t additional initial rounds, then the desired key differences are:

$$\Delta K[i] = \begin{cases} 2^j & \text{if } i \in \{t+1, t+17\} \\ -2^j & \text{if } i \in \{t+4\} \\ -(2^j)^{\gg 7} & \text{if } i \in \{t+9, t+25\} \\ (2^j)^{\gg 7} & \text{if } i \in \{t+12\} \\ 0 & \text{otherwise} \end{cases} \tag{1}$$

The collisions we found at the beginning of this section were for variant 2 with $j = 64$.

variant 2

Subtle Difficulties with the Attack

It is not hard to see that our attack makes the simplifying assumption that the inputs to each round are independent and uniformly distributed. Hence any pair of inputs that will lead to the desired difference is possible. However, in many cases this is an incorrect assumption and it can strongly affect our attack.

To make the discussion easier, we will examine a smaller version of the cipher with 1-bit words. This is a fair thing to do since adjacent bits in a word affect each other within the F-function only through the data dependent rotate. As stated in Section 2, the F-function is composed of a non-linear function F_i ($i \in \{1, 2, 3, 4\}$) composed with a data-dependent rotate v. If the output difference for F_i is zero, then given two different inputs the output difference of the data-dependent rotate will also be zero. This means that a 1-bit difference in any word will not change affect any adjacent bits if the output difference of F_i is zero.

Once we have made the reduction, we can regard the sequence of rounds as the successive states of a non-linear shift register. Specifically, we have a sequence of states X_i where X_0 is the input to our encryption function, $X_{i+1} = (X_i || (k_{i+1} \oplus F_k(X_i)))_{0...6}$, where F_k can be F_1, F_2, F_3 or F_4, depending on the particular round number. The output of the cipher is then X_r where r is the number of rounds.

In our 32-round related-key attack, $k_i \neq k_i'$ for a small number of i. If one examines the sequence of states produced by the same initial state X_0 and two related keys k_i and k_i', then for variant j we want that $X_{j+16} \oplus X_{j+16}' = 0$ and $X_{j+17} \oplus X_{j+17}' = 2^0 = 1$. For $i = 0, \ldots, 6$, $k_{i+j+17} = k_{i+j+17}'$ so we can examine a simplified shift register whose feedback function is $Y_{i+1} = (Y_i || (k_{i+1} \oplus G_k(Y_i)))$, where $G_k = F_3$ if $0 < i \leq 6 - j$, and $G_k = F_4$ if $6 - j < i \leq 6$, $Y_0 = X_{j+17}$, $Y_0' = X_{j+17}'$, and k is an arbitrary key. That is, we can consider these 7 rounds (round $j + 17$ to $j + 23$) in isolation since the subkeys are the same for both values of our key.

We want to examine the sequence of states Y_i and Y_i' where $Y_0 \oplus Y_0' = 1$. In order for our related-key attack to work, we must have that $Y_i \oplus Y_i' = 2^i$ for $0 \leq i \leq 6$. Unfortunately it turns out that for certain j (e.g. $j = 0, 1$), there are no input states Y_0 and Y_0' and key k for which the resulting sequences have this property. We determined this by performing a brute-force search with a computer. There are only 64 different k (k_6 really has no influence) and 64 different Y_0 to consider. Hence we need only examine $64 \cdot 64 = 4096$ different cases.

Our analysis showed that variants 0 and 1 will not produce the desired collisions, but that variants 2, 3, and 4 will. In performing a computer search we found that we were not able to find collisions within the expected time for variants 0 and 1, providing evidence for the correctness of our analysis. The collision we provided above was for variant 2, showing that there does exist a satisfying sequence of states.

Extending the Attack to 48-rounds

Unfortunately it is not as straightforward as one might hope to extend our attack to 48 rounds. By duplicating the attack of rounds 17–32 for rounds 33–48, in Table 7, we obtain a plausible looking related-key attack for the 48-round version of the cipher. In summary, we have an additional sixteen subkeys, two of which have specific differences (the subkeys for rounds 33 and 42). However, in attempting to perform a computer search for collisions, we found that we were unable to produce collisions after the first 32 rounds, let alone the entire 48 rounds. This is what initially led us to the analysis performed in the previous section.

It turns out that for the 48-round version of the cipher, the five different variants of the attack result in a shift register with feedback function:

$$H_k(X, i) = \begin{cases} F_2(X) \oplus k_i & \text{if } 0 < i \leq \min(7 - j, 6) \\ F_3(X) \oplus k_i & \text{if } 7 - j < i \leq 6 \end{cases}$$

where j is the variant we wish to examine. For the resulting shift register, there are no initial states Y_0 and Y_0' and key k such that $Y_i \oplus Y_i' = 2^i$ for $0 \leq i < 7$. Consequently our related-key attack will fail to produce a zero output difference after round 32.

A slightly modified version of our attack does work. We present the first 32 rounds of the attack in Table 8 (in Appendix 1). The last 16 rounds are simply a copy of rounds 17–33. The associated differential attack on the key schedule is presented in Table 10 (also in Appendix 1). In essence we overlap two differential attacks so that differences are introduced for 9 rounds instead of 8 rounds as before. The total probability that a key will satisfy the differential is 2^{-18}. The total probability that two related keys will produce a collision is 2^{-32}.

6.2   Differential-Attack on the Key Schedule

In order to carry out our related key attack, we must find two different keys which will produce key schedules with the differences specified in (1). We performed a straightforward differential attack on the key schedule to produce the desired pair. The specifications for using SPEED as a hash function require a 256-bit key. Since we are using a 128-bit blockwidth, this implies that the first 16 subkeys will be our key and the remaining subkeys will be derived from the key.

The key scheduling algorithm is straightforward and we include a modified copy from [12] here:

Step 1. Let $kb[0], kb[1], \ldots, kb[31]$ be an array of double-bytes where $kb[0], \ldots,$ $kb[15]$ are the original 16 double-byte values of the key.

Step 2. This step constructs $kb[16], \ldots, kb[31]$ from the user key data $kb[0], \ldots,$ $kb[15]$. It employs three double-byte constants $Q_{l,0}, Q_{l,1},$ and $Q_{l,2}$.

    1. Let $S_0 = Q_{l,0}, S_1 = Q_{l,1},$ and $S_2 = Q_{l,2}$.
    2. For i from 16 to 31 do the following:
        (a) $T = G(S_2, S_1, S_0)$.
        (b) Rotate T to the right by 11 bits.
        (c) $T = T + S_2 + kb[j] \pmod{2^{16}}$, where $j = i \pmod{16}$.
        (d) $kb[i] = T$.
        (e) $S_2 = S_1, S_1 = S_0, S_0 = T$.

where

$$G(X_0, X_1, X_2) = (X_0 \oplus X_1) \wedge (X_0 \oplus X_2) \wedge (X_1 \oplus X_2).$$

The two crucial observations about the key scheduling algorithm are:

1. Adjacent bits in the input have minimal influence on each other. Hence we can work as if it took 1-bit inputs and produced a 1-bit output.
2. When viewed as a function on 1-bit inputs, flipping one or two inputs will flip the output with probability exactly $1/2$.

7 Practical Considerations: Performance and Security

SPEED is defined as having both a variable block length and a variable number of rounds. Since almost any Feistel network is secure after enough rounds, what does it mean to say that SPEED is "broken"? For a cryptographic engineer, this means that the still-secure variants of SPEED are significantly slower than other secure alternatives.

Table 2 compares the throughput of SPEED with the throughput of other block ciphers. Benchmarks for the other algorithms were taken from [10,6][1]. We only include the SPEED block-length and round-number pairings that we believe are secure.

---

[1] In [12], the author compares different variants of SPEED with IDEA. Since [10,6] both benchmark IDEA twice as fast as [12], we performed our own efficiency analysis on SPEED. We estimate that a fully optimized version of SPEED on a Pentium will take 20 clock cycles per round, for word sizes of 64 bits, 128 bits, and 256 bits.

Algorithm	Block width	# rounds	Clocks/byte of output
Blowfish	64	16	19.8
Square	128	8	20.3
RC5-32/16	64	16	24.8
CAST-128	64	16	29.5
DES	64	16	43
SAFER (S)K-128	64	8	52
IDEA	64	8	74
Triple-DES	64	48	116
SPEED	64	64	160
SPEED	64	80	200
SPEED	64	96	240
SPEED	128	64	80
SPEED	128	80	100
SPEED	128	96	120
SPEED	256	64	40
SPEED	256	80	50
SPEED	256	96	60

Fig. 2. Comparison of Different Block Ciphers.

8   Conclusions

In this paper, we have discussed the SPEED proposed block cipher in terms of cryptanalytic attack. We have pointed out a few potential weaknesses, demonstrated an attack on the Davies-Meyer hashing mode of SPEED with 32 and 48 rounds, and explored other attacks on the 48-round SPEED block cipher.

It is interesting to note that SPEED, though built using very strong component functions, doesn't appear to be terribly secure. The SPEED design apparently relied upon the high quality of the binary functions used, the fact that different functions were used at different points in the cipher, and the data-dependent rotations to provide resistance to cryptanalysis. Unfortunately, the most effective attacks aren't made much less powerful by any of these defenses.

It is also interesting to note the new difficulties that occur in attacking this kind of cipher. The source-heavy UFN construction of SPEED forced us to reconsider our assumptions about carrying out differential attacks, and added somewhat to the difficulty of thinking about these attacks. Surprisingly many of the obvious differential attacks did not work against the cipher, although it's not obvious that a different choice of F-functions would present the same problems.

References

1. E. Biham, "New Types of Cryptanalytic Attacks Using Related Keys," *Journal of Cryptology*, v. 7, n. 4 (1994), pp. 229–246.
2. A. Biryukov and E Kushilevitz, "Improved cryptanalysis of RC5," *Advances in Cryptology, Proc. Eurocrypt'97, LNCS*, to appear.
3. J. Kelsey, B. Schneier, and D. Wagner, "Key-Schedule Cryptanalysis of IDEA, G-DES, GOST, SAFER, and Triple-DES," *Advances in Cryptology— CRYPTO '96*, Springer-Verlag, 1996, pp. 237–251.
4. L.R. Knudsen, V. Rijmen, R.L. Rivest and M.J.B. Robshaw, "On the design and security of RC2," *Fast Software Encryption, LNCS 1372*, S. Vaudenay, Ed., Springer-Verlag, 1998, pp. 206–221.
5. X. Lai, J.L. Massey, and S. Murphy, "Markov ciphers and differential cryptanalysis," *Advances in Cryptology, Proc. Eurocrypt'91, LNCS 547*, D.W. Davies, Ed., Springer-Verlag, 1992, pp. 17–38.
6. B. Preneel, V. Rijmen, and A. Bosselaers, "Recent developments in the design of conventional cryptographic algorithms," *Computer Security and Industrial Cryptography - State of the Art and Evolution, LNCS*, Springer-Verlag, to appear.
7. V. Rijmen, B. Preneel, E. De Win, "On weaknesses of non-surjective round functions," *Designs, Codes, and Cryptography*, Vol. 12, No. 3, November 1997, pp. 253-266.
8. R.L. Rivest, "The RC5 encryption algorithm," *Fast Software Encryption, LNCS 1008*, B. Preneel, Ed., Springer-Verlag, 1995, pp. 86–96.
9. B. Schneier, J. Kelsey, "Unbalanced Feistel Networks and Block Cipher Design," *Fast Software Encryption–Third International Workshop*, Springer-Verlag, 1996.
10. B. Schneier, D. Whiting, "Fast Software Encryption: Designing Encryption Algorithms for Optimal Software Speed on the Intel Pentium Processor," *Fast Software Encryption–Fourth International Workshop*, Springer-Verlag, 1997.
11. R. Winternitz and M. Hellman, "Chosen-key Attacks on a Block Cipher," *Cryptologia*, v. 11, n. 1, Jan 1987, pp. 16–20.
12. Y. Zheng, "The SPEED Cipher," in Proceedings of *Financial Cryptography '97*, Springer-Verlag.

Appendix 1

This section contains the details for the two related key and differential attacks we performed against SPEED.

Note, the first column of Table 7 (and also Table 8) shows the additive differences between the respective words of the two key schedules. For example, $K'[0] - K[0] = 64$. Note that in the 128-bit version of the cipher, each word is 16 bits wide. Hence $m_{i,7}$ (the target block) will be rotated right $16/2 - 1 = 7$ bits and $64^{\gg 1} = 2^{15}$. With probability 2^{-12}, two different key schedules with the specified differences will have the exact same encryption state after round 12.

R	$\Delta K[i]$	$\Delta m_{i,7}$	$\Delta m_{i,6}$	$\Delta m_{i,5}$	$\Delta m_{i,4}$	$\Delta m_{i,3}$	$\Delta m_{i,2}$	$\Delta m_{i,1}$	$\Delta m_{i,0}$	P
-	-	0	0	0	0	0	0	0	0	1^A
1	64	0	0	0	0	0	0	0	64	$1/4^B$
2	0	0	0	0	0	0	0	64	0	$1/2$
3	0	0	0	0	0	0	64	0	0	$1/2$
4	-64	0	0	0	0	64	0	0	-64	$1/4^C$
5	0	0	0	0	64	0	0	-64	0	$1/2$
6	0	0	0	64	0	0	-64	0	0	$1/2$
7	0	0	64	0	0	-64	0	0	0	$1/2$
8	0	64	0	0	-64	0	0	0	0	$1/2$
9	2^{15}	0	0	-64	0	0	0	0	0	$1/2^D$
10	0	0	-64	0	0	0	0	0	0	$1/2$
11	0	-64	0	0	0	0	0	0	0	1
12	2^{15}	0	0	0	0	0	0	0	0	1
13	0	0	0	0	0	0	0	0	0	1
14	0	0	0	0	0	0	0	0	0	1
15	0	0	0	0	0	0	0	0	0	1
16	0	0	0	0	0	0	0	0	0	1
17	64	0	0	0	0	0	0	0	64	$1/4$
18	0	0	0	0	0	0	0	64	0	$1/2$
19	0	0	0	0	0	0	64	0	0	$1/2$
20	0	0	0	0	0	64	0	0	0	$1/2$
21	0	0	0	0	64	0	0	0	0	$1/2$
22	0	0	0	64	0	0	0	0	0	$1/2$
23	0	0	64	0	0	0	0	0	0	$1/2$
24	0	64	0	0	0	0	0	0	0	1
25	2^{15}	0	0	0	0	0	0	0	0	1
26	0	0	0	0	0	0	0	0	0	1
27	0	0	0	0	0	0	0	0	0	1
28	0	0	0	0	0	0	0	0	0	1
29	0	0	0	0	0	0	0	0	0	1
30	0	0	0	0	0	0	0	0	0	1
31	0	0	0	0	0	0	0	0	0	1
32	0	0	0	0	0	0	0	0	0	1

A The initial input to the cipher is the same for both keys.

B The probability that an additive difference of 64 produces an XOR difference of 64 is 1/2. 1/4 comes from multiplying by the probability that a 1-bit input difference will produce a zero output difference for the F-function.

C The probability that an additive difference of -64 produces an XOR difference of 64 is 1/2. The XOR difference of 64 and -64 occur in the same bit-position so the probability that the output difference of the F-function is zero is 1/2.

D $(64^{\gg 7}) + 2^{15} \equiv 0 \pmod{2}^{15}$. Hence $\Delta m_{9,0} = 0$ with probability 1 (assuming $\Delta f = 0$, which occurs with probability 1/2).

Table 7. Related Key Attack Against 32-round Cipher.

R	$\Delta K[i]$	$\Delta m_{i,7}$	$\Delta m_{i,6}$	$\Delta m_{i,5}$	$\Delta m_{i,4}$	$\Delta m_{i,3}$	$\Delta m_{i,2}$	$\Delta m_{i,1}$	$\Delta m_{i,0}$	P
-	-	0	0	0	0	0	0	0	0	1
0	64	0	0	0	0	0	0	0	64	1/2
1	64	0	0	0	0	0	0	64	64	1/4
2	0	0	0	0	0	0	64	64	0	1/2
3	-64	0	0	0	0	64	64	0	-64	1/4
4	-64	0	0	0	64	64	0	-64	-64	1/4
5	0	0	0	64	64	0	-64	-64	0	1/2
6	0	0	64	64	0	-64	-64	0	0	1/2
7	0	64	64	0	-64	-64	0	0	0	1/2
8	2^{15}	64	0	-64	-64	0	0	0	0	1/2
9	2^{15}	0	-64	-64	0	0	0	0	0	1/2
10	0	-64	-64	0	0	0	0	0	0	1/2
11	2^{15}	-64	0	0	0	0	0	0	0	1
12	2^{15}	0	0	0	0	0	0	0	0	1
13	0	0	0	0	0	0	0	0	0	1
14	0	0	0	0	0	0	0	0	0	1
15	0	0	0	0	0	0	0	0	0	1
16	64	0	0	0	0	0	0	0	64	1/2
17	64	0	0	0	0	0	0	64	64	1/4
18	0	0	0	0	0	0	64	64	0	1/2
19	0	0	0	0	0	64	64	0	0	1/2
20	0	0	0	0	64	64	0	0	0	1/2
21	0	0	0	64	64	0	0	0	0	1/2
22	0	0	64	64	0	0	0	0	0	1/2
23	0	64	64	0	0	0	0	0	0	1/2
24	2^{15}	64	0	0	0	0	0	0	0	1
25	2^{15}	0	0	0	0	0	0	0	0	1
26	0	0	0	0	0	0	0	0	0	1
27	0	0	0	0	0	0	0	0	0	1
28	0	0	0	0	0	0	0	0	0	1
29	0	0	0	0	0	0	0	0	0	1
30	0	0	0	0	0	0	0	0	0	1
31	0	0	0	0	0	0	0	0	0	1

Table 8. First 32 Rounds of 48-round Related-Key Attack.

j	$\Delta K[j]$	ΔT_a^a	ΔT_b	ΔT_c	$\Delta S2$	$\Delta S1$	$\Delta S0$	ΔG	$\Delta K[j+16]$	P
-	-	-	-	-	0	0	0	0	-	1
1	2^l	0	0	2^l	0	0	2^l	0	2^l	1/2
2	0	0	0	0	0	2^l	0	0	0	1/2
3	0	0	0	0	2^l	0	0	0	0	1/2
4	-2^l	0	0	0	0	0	0	0	0	1/2
5	0	0	0	0	0	0	0	0	0	1
6	0	0	0	0	0	0	0	0	0	1
7	0	0	0	0	0	0	0	0	0	1
8	0	0	0	0	0	0	0	0	0	1
9	$A = -(2^l)^{\ggg 7}$	0	0	A	0	0	A	0	A	1/2
10	0	0	0	0	0	A	0	0	0	1/2
11	0	0	0	0	A	0	0	0	0	1/2
12	$(2^l)^{\ggg 7}$	0	0	0	0	0	0	0	0	1
13	0	0	0	0	0	0	0	0	0	1
14	0	0	0	0	0	0	0	0	0	1
15	0	0	0	0	0	0	0	0	0	1
16	0	0	0	0	0	0	0	0	0	1

a. Denotes the value of T after step 2a. Similarly, T_b and T_c denote the value of T after steps 2b and 2c respectively.

Table 9. Differential Attack Against 32-round Key Schedule.

j	$\Delta K[j]$	ΔT_a^a	ΔT_b	ΔT_c	$\Delta S2$	$\Delta S1$	$\Delta S0$	ΔG	$\Delta K[j+16]$	P
-	-	-	-	-	0	0	0	0	-	1
0	64	0	0	64	0	0	64	0	64	1/2
1	64	0	0	64	0	64	64	0	64	1/2
2	0	0	0	0	64	64	0	0	0	1/2
3	-64	0	0	0	64	0	0	0	0	1/2
4	-64	0	0	0	0	0	0	0	0	1/2
5	0	0	0	0	0	0	0	0	0	1
6	0	0	0	0	0	0	0	0	0	1
7	0	0	0	0	0	0	0	0	0	1
8	2^{15}	0	0	2^{15}	0	0	2^{15}	0	2^{15}	1/2
9	2^{15}	0	0	2^{15}	0	2^{15}	2^{15}	0	2^{15}	1/2
10	0	0	0	0	2^{15}	2^{15}	0	0	0	1/2
11	2^{15}	0	0	0	2^{15}	0	0	0	0	1/2
12	2^{15}	0	0	0	0	0	0	0	0	1
13	0	0	0	0	0	0	0	0	0	1
14	0	0	0	0	0	0	0	0	0	1
15	0	0	0	0	0	0	0	0	0	1

Table 10. First 16 Rounds of Differential Attack Against 48-round Key Schedule.

Authenticated Diffie–Hellman Key Agreement Protocols

Simon Blake-Wilson[1] and Alfred Menezes[2]

[1] Certicom Research, 200 Matheson Blvd. W., Suite 103
Mississauga, Ontario, Canada L5R 3L7
sblakewi@certicom.com
[2] Department of Combinatorics & Optimization
University of Waterloo, Waterloo, Ontario, Canada N2L 3G1
ajmeneze@cacr.math.uwaterloo.ca

Abstract. This paper surveys recent work on the design and analysis of key agreement protocols that are based on the intractability of the Diffie–Hellman problem. The focus is on protocols that have been standardized, or are in the process of being standardized, by organizations such as ANSI, IEEE, ISO/IEC, and NIST. The practical and provable security aspects of these protocols are discussed.

1 Introduction

Authenticated key establishment protocols are designed to provide two or more specified entities communicating over an open network with a shared secret key which may subsequently be used to achieve some cryptographic goal such as confidentiality or data integrity. Secure authenticated key establishment protocols are important as effective replacements for traditional key establishment achieved using expensive and inefficient couriers.

Key establishment protocols come in various flavors. In *key transport* protocols, a key is created by one entity and securely transmitted to the second entity, while in *key agreement* protocols both entities contribute information which is used to derive the shared secret key. In *symmetric* protocols the two entities a priori possess common secret information, while in *asymmetric* protocols the two entities share only public information that has been authenticated. This paper is concerned with two-party authenticated key agreement protocols in the asymmetric setting.

The design of asymmetric authenticated key agreement protocols has a checkered history. Over the years, numerous protocols have been proposed to meet a variety of desirable security and performance requirements. Many of these protocols were subsequently found to be flawed, and then either were modified to resist the new attacks, or were totally abandoned. After a series of attacks and modifications, only those surviving protocols which had received substantial public scrutiny and were believed to resist all known attacks were deemed secure for practical usage. Protocols that evolve from this 'attack-response' methodology are said to provide *heuristic* security.

S. Tavares and H. Meijer (Eds.): SAC'98, LNCS 1556, pp. 339–361, 1999.
© Springer-Verlag Berlin Heidelberg 1999

There are two primary drawbacks of protocols which provide heuristic security. First, their security attributes are typically unclear or not completely specified. Second, they offer no assurances that new attacks will not be discovered in the future. These drawbacks make a notion of *provable* security desirable. This would entail specification of a formal model of computing which accurately captures the characteristics of the participating entities and a real-life powerful adversary, a formal definition of the security goals within this model, a clear statement of any assumptions made, and, finally, a rigorous proof that the protocol meets these goals within the model.

While provable security may appear to be the highest possible level of security for a key agreement protocol, the approach does have some limitations. Most significantly, it is difficult to judge whether or not real-life threats and security goals are adequately reflected in a given model. That is, the provable security of a protocol is meaningful only if one finds the model, definitions, and underlying assumptions to be appropriate for one's purposes. Nevertheless, significant progress has been made in recent years, and authenticated key agreement protocols which are both provably secure and efficient have been devised and are being used in practice.

This paper focuses on asymmetric authenticated key agreement protocols whose security is based on the intractability of the Diffie–Hellman problem. We discuss the practical and provable security aspects of some protocols which are being standardized by accredited standards organizations such as ANSI (American National Standards Institute), IEEE (Institute of Electrical and Electronics Engineers), ISO/IEC (International Standards Organization/International Electrotechnical Commission), and the U.S. government's NIST (National Institute of Standards and Technology). Such cryptographic standards have significant practical impact because they facilitate the widespread use of sound techniques, and promote interoperability between different implementations.

The remainder of this paper is organized as follows. §2 summarizes the desirable security and performance attributes of a key agreement protocol. In §3, we review the basic ephemeral (short-term) and static (long-term) Diffie–Hellman key agreement protocols and point out their limitations. §4 presents the KEA, Unified Model, and MQV authenticated key agreement protocols, while in §5 we discuss protocols for authenticated key agreement with key confirmation. The protocols are compared in §6. Recent progress in defining and proving the security of key agreement protocols is reviewed in §7. §8 concludes with some directions for future research.

2   Goals of key agreement

This section discusses in more detail the goals of asymmetric authenticated key establishment protocols. The complexity and variety of these goals explains in part the difficulties involved in designing secure protocols.

The fundamental goal of any authenticated key establishment protocol is to distribute keying data. Ideally, the established key should have precisely the same attributes as a key established face-to-face — for example, it should be shared by the (two) specified entities, it should be distributed uniformly at random from the key space, and no unauthorized (and computationally bounded) entity should learn anything about the key. A protocol achieving this idealistic goal could then be used as a drop-in replacement for face-to-face key establishment without the need to review system security in much the same way as pseudorandom bit generators can replace random bit generators.

Unfortunately, such an abstract goal is not easily attained and it is not an easy task to identify and enunciate the precise security requirements of authenticated key establishment. Nonetheless over the years several concrete security and performance attributes have been identified as desirable. These are informally described in the remainder of this section. Recent more formal attempts at capturing concrete security definitions are discussed in §7.

The first step is to identify what types of attacks it is vital for a protocol to withstand. Since protocols are used over open networks like the Internet, a secure protocol should be able to withstand both *passive* attacks (where an adversary attempts to prevent a protocol from achieving its goals by merely observing honest entities carrying out the protocol) and *active* attacks (where an adversary additionally subverts the communications by injecting, deleting, altering or replaying messages).

The second step is to identify what concrete security goals it is vital for a protocol to provide. The fundamental security goals described below are considered to be vital in any application. The other security and performance attributes are important in some environments, but less important in others.

Fundamental security goals. Let A and B be two honest entities, i.e., legitimate entities who execute the steps of a protocol correctly.

1. *implicit key authentication.* A key agreement protocol is said to provide implicit key authentication (of B to A) if entity A is assured that no other entity aside from a specifically identified second entity B can possibly learn the value of a particular secret key. Note that the property of implicit key authentication does not necessarily mean that A is assured of B actually possessing the key.

2. *explicit key authentication.* A key agreement protocol is said to provide *explicit key confirmation* (of B to A) if entity A is assured that the second entity B has actually computed the agreed key. The protocol provides *implicit key confirmation* if A is assured that B can compute the agreed key. While explicit key confirmation appears to provide stronger assurances to A than implicit key confirmation (in particular, the former implies the latter), it appears that, for all practical purposes, the assurances are in fact the same. That is, the assurance that A requires in practice is merely that B can compute the key rather than that B has actually computed the key. Indeed in practice, even if a protocol does provide explicit key confirmation, it cannot guarantee to A that B will not lose the key between key establishment and key use. Thus it would indeed seem that implicit key confirmation and

explicit key confirmation are in practice very similar, and the remainder of this paper will not distinguish between the two.

Key confirmation by itself is not a useful service — it is only desirable when accompanied with implicit key authentication. A key agreement protocol is said to provide explicit key authentication (of B to A) if both implicit key authentication and key confirmation (of B to A) are provided.

A key agreement protocol which provides implicit key authentication to both participating entities is called an *authenticated key agreement (AK)* protocol, while one providing explicit key authentication to both participating entities is called an *authenticated key agreement with key confirmation (AKC)* protocol.

Key agreement protocols in which the services of implicit key authentication or explicit key authentication are provided to only one (*unilateral*) rather than both (*mutual*) participating entities are also useful in practice, for example in encryption applications where only authentication of the intended recipient is required. Such unilateral key agreement protocols (e.g., ElGamal key agreement [33, Protocol 12.52]) are not considered in this paper.

Other desirable security attributes. A number of other desirable security attributes have also been identified. Typically the importance of supplying these attributes will depend on the application. In the following, A and B are two honest entities.

1. *known-key security*. Each run of a key agreement protocol between A and B should produce a unique secret key; such keys are called *session* keys. Session keys are desirable in order to limit the amount of data available for cryptanalytic attack (e.g., ciphertext generated using a fixed session key in an encryption application), and to limit exposure in the event of (session) key compromise. A protocol should still achieve its goal in the face of an adversary who has learned some other session keys.

2. *forward secrecy*. If long-term private keys of one or more entities are compromised, the secrecy of previous session keys established by honest entities is not affected. A distinction is sometimes made between the scenario in which a single entity's private key entity is compromised (*half* forward secrecy) and the scenario in which the private keys of both participating entities are compromised (*full* forward secrecy).

3. *key-compromise impersonation*. Suppose A's long-term private key is disclosed. Clearly an adversary that knows this value can now impersonate A, since it is precisely this value that identifies A. However, it may be desirable in some circumstances that this loss does not enable the adversary to impersonate other entities to A.

4. *unknown key-share*. Entity B cannot be coerced into sharing a key with entity A without B's knowledge, i.e., when B believes the key is shared with some entity $C \neq A$, and A (correctly) believes the key is shared with B.

   A hypothetical scenario where an unknown key-share attack can have damaging consequences is the following; this scenario was first described by Diffie, van Oorschot and Wiener [21]. Suppose that B is a bank branch and A is an account holder. Certificates are issued by the bank headquarters and within each certificate is the account information of the holder. Suppose that the protocol for electronic deposit of funds is to exchange a key with a bank branch via a mutually authenticated key agreement. Once B has authenticated the transmitting entity, encrypted funds are deposited to the account number in the certificate. Suppose that no further authentication is done in the encrypted deposit message (which might be the case to save bandwidth). If the attack mentioned above is successfully launched then the deposit will be made to C's account instead of A's account.

Desirable performance attributes. These include:

1. Minimal number of *passes* (the number of messages exchanged).
2. Low *communication overhead* (total number of bits transmitted).
3. Low *computation overhead* (total number of arithmetical operations required).
4. Possibility of *precomputation* (to minimize on-line computational overhead).

Other desirable attributes. These include:

1. *Anonymity* of the entities participating in a run of the protocol.
2. *Role symmetry* (the messages transmitted have the same structure).
3. *Non-interactiveness* (the messages transmitted between the two entities are independent of each other).
4. Non-reliance on encryption in order to meet export restrictions.
5. Non-reliance on hash functions since these are notoriously hard to design.
6. Non-reliance on timestamping since it is difficult to implement securely in practice.

3   Diffie–Hellman key agreement

This section describes the basis of Diffie–Hellman based key agreement protocols and motivates the modern protocols we describe in §4 and §5 by illustrating some of the deficiencies of early protocols.

The mathematical tool commonly used for devising key agreement protocols is the *Diffie–Hellman problem*: given a cyclic group G of prime order n, a generator g of G, and elements g^x, $g^y \in G$ (where $x, y \in_R [1, n-1]$), find g^{xy}. (We use $x \in_R S$ to denote that x is chosen uniformly at random from the set S.) This problem is closely related to the widely-studied *discrete logarithm problem* (given G, n, g, and g^x where $x \in_R [0, n-1]$, find x), and there is strong evidence that the two problems are computationally equivalent (e.g., see [16] and [32]).

For concreteness, this paper deals with the case where G is a prime order subgroup of $\mathbb{Z}_p^*$, the multiplicative group of the integers modulo a prime p. However, the discussion applies equally well to any group of prime order in which the discrete logarithm problem is computationally intractable, for example prime order subgroups of the group of points on an elliptic curve over a finite field. The following notation is used throughout the paper.

A, B Honest entities.

p 1024-bit prime.

q 160-bit prime divisor of $p - 1$.

g An element of order q in $\mathbb{Z}_p^*$.

a, b Static private keys of A and B; $a, b \in_R [1, q - 1]$.

Y_A, Y_B Static public keys of A and B; $Y_A = g^a \bmod p$, $Y_B = g^b \bmod p$.

x, y Ephemeral private keys of A and B; $x, y \in_R [1, q - 1]$.

R_A, R_B Ephemeral public keys of A and B; $R_A = g^x \bmod p$, $R_B = g^y \bmod p$.

H A cryptographic hash function (e.g., SHA-1 [35]).

MAC   A message authentication code algorithm (e.g., [4, 6, 7]).

The operator mod p will henceforth be omitted.

The *domain* parameters (p, q, g) are common to all entities. For the remainder of this paper, we will assume that static public keys are exchanged via certificates. Cert$_A$ denotes A's public-key certificate, containing a string of information that uniquely identifies A (such as A's name and address), her static public key Y_A, and a certifying authority CA's signature over this information. Other information may be included in the data portion of the certificate, including the domain parameters if these are not known from context. Any other entity B can use his authentic copy of the CA's public key to verify A's certificate, thereby obtaining an authentic copy of A's static public key.

We assume that the CA has verified that A possess the private key a corresponding to her static public key Y_A. This is done in order to prevent potential unknown key-share attacks whereby an adversary E registers A's public key Y_A as its own and subsequently deceives B into believing that A's messages originated from E (see [15] for more details). Checking knowledge of private keys is in general a sensible precaution and is often vital for theoretical analysis. We also assume that the CA has verified the validity of A's static public key Y_A, i.e., the CA has verified that $1 < Y_A < p$ and that $(Y_A)^q \equiv 1 \pmod{p}$; this process is called *public key validation* [26]. Rationale for performing public key validation is provided in §4.1.

The first asymmetric key agreement protocol was proposed by Diffie and Hellman in their seminal 1976 paper [20]. We present two versions of the basic protocol, one where the entities exchange *ephemeral* (short-term) public keys, and the other where the entities exchange *static* (long-term) public keys.

Protocol 1 (Ephemeral Diffie–Hellman)

1. A selects $x \in_R [1, q - 1]$ and sends $R_A = g^x$ to B.
2. B selects $y \in_R [1, q - 1]$ and sends $R_B = g^y$ to A.
3. A computes $K = (R_B)^x = g^{xy}$.
4. B computes $K = (R_A)^y = g^{xy}$.

While the ephemeral Diffie–Hellman protocol provides implicit key authentication in the presence of passive adversaries, it does not on its own provide any useful services in the presence of active adversaries since neither entity is provided with any assurances regarding the identity of the entity it is communicating with. (See also Table 1 in §6.) This drawback can be overcome by using public keys that have been certified by a trusted CA.

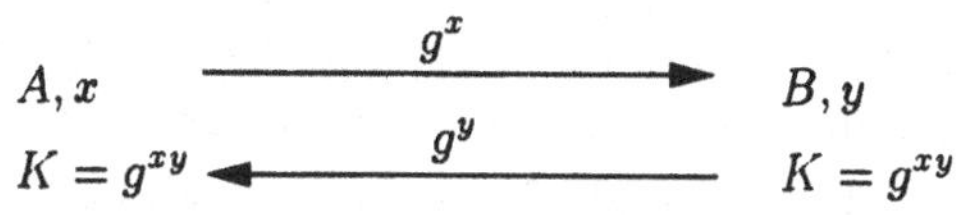

$$A, x \xrightarrow{\quad g^x \quad} B, y$$
$$K = g^{xy} \xleftarrow{\quad g^y \quad} K = g^{xy}$$

Fig. 1. Protocol 1 (Ephemeral Diffie–Hellman).

Protocol 2 (Static Diffie–Hellman)

1. A sends Cert_A to B.
2. B sends Cert_B to A.
3. A computes $K = (Y_B)^a = g^{ab}$.
4. B computes $K = (Y_A)^b = g^{ab}$.

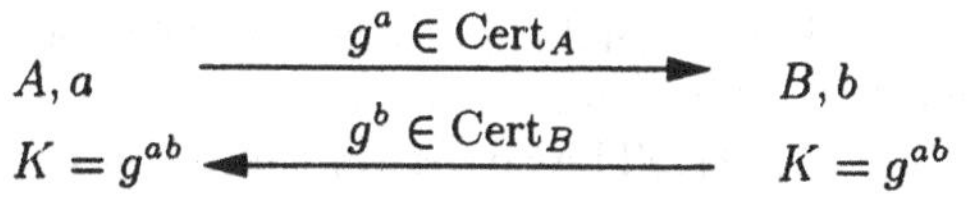

$$A, a \xrightarrow{\quad g^a \in \text{Cert}_A \quad} B, b$$
$$K = g^{ab} \xleftarrow{\quad g^b \in \text{Cert}_B \quad} K = g^{ab}$$

Fig. 2. Protocol 2 (Static Diffie–Hellman).

Since each entity is assured that it possesses an authentic copy of the other entity's public key, the static Diffie–Hellman protocol offers implicit key authentication. A major drawback, however, is that A and B compute the same shared secret $K = g^{ab}$ for each run of the protocol.

The drawbacks of the ephemeral and static Diffie–Hellman protocols can be alleviated by using both static and ephemeral keying material in the formation of shared secrets. An example of an early protocol designed in this manner is the MTI/C0 protocol [31].

Protocol 3 (MTI/C0)

1. A selects $x \in_R [1, q-1]$ and sends $T_A = (Y_B)^x$ to B.
2. B selects $y \in_R [1, q-1]$ and sends $T_B = (Y_A)^y$ to A.
3. A computes $K = (T_B)^{a^{-1}x} = g^{xy}$.
4. B computes $K = (T_A)^{b^{-1}y} = g^{xy}$.

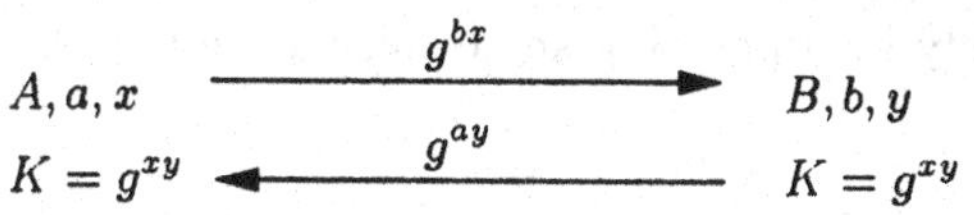

$$A, a, x \xrightarrow{\quad g^{bx} \quad} B, b, y$$
$$K = g^{xy} \xleftarrow{\quad g^{ay} \quad} K = g^{xy}$$

Fig. 3. Protocol 3 (MTI/C0).

This protocol appears secure at first glance. Unfortunately, it turns out that this attempt to combine static and ephemeral Diffie–Hellman protocols has introduced some subtle problems. As an example, consider the following instance of the *small subgroup* attack [29] on the MTI/C0 protocol. An adversary E replaces T_A and T_B with the identity element 1. Both A and B now form $K = 1$, which is also known to E. This attack demonstrates that the MTI/C0 protocol (as described above) does not offer implicit key authentication.

The 3 protocols described in this section demonstrate some of the subtleties involved in designing secure authenticated key agreement protocols. Other kinds of attacks that have been identified besides small subgroup attacks include:

1. *intruder-in-the-middle attack* [37]. In this classic attack on ephemeral Diffie–Hellman, the adversary replaces A's and B's ephemeral keys g^x and g^y with keys $g^{\bar{x}}$ and $g^{\bar{y}}$ of its choice. E can then compute the session keys formed by A and B ($g^{x\bar{y}}$ and $g^{\bar{x}y}$, respectively), and use these to translate messages exchanged between A and B that are encrypted under the session keys.
2. *reflection attack* [34]. A's challenges are replayed back to A as messages purportedly from B.
3. *interleaving attack* [12,21]. The adversary reuses messages transmitted during a run of the protocol in other runs of the protocol.

Such attacks are typically very subtle and require little computational overhead. They highlight the necessity of some kind of formal analysis to avoid the use of flawed protocols.

4 AK protocols

This section discusses some AK protocols currently proposed in standards. We present the two-pass KEA, Unified Model, and MQV protocols, and their one-pass variants.

Before we present the AK protocols it is worth reminding the reader that, as discussed in §2, it is highly desirable for key establishment protocols to provide explicit key authentication. Thus, when AK protocols are used in practice, key confirmation should usually be added to the protocols. Nonetheless it is worth presenting the raw AK protocols since key confirmation can be achieved in a variety of ways and it is sometimes desirable to separate key confirmation from implicit key authentication and move the burden of key confirmation from the key establishment mechanism to the application. For example, if the key is to be subsequently used to achieve confidentiality, then encryption with the key can begin on some (carefully chosen) known data. Other systems may provide key confirmation during a 'real-time' telephone conversation. We present a generic method for securely incorporating key confirmation into AK protocols in §5.

4.1 KEA

The Key Exchange Algorithm (KEA) was designed by the National Security Agency (NSA) and declassified in May 1998 [36]. It is the key agreement protocol in the FORTEZZA suite of cryptographic algorithms designed by NSA in 1994. KEA is very similar to the Goss [23] and MTI/A0 [31] protocols.

Protocol 4 (KEA)

1. A and B obtain authentic copies of each other's public keys Y_A and Y_B.
2. A selects $x \in_R [1, q - 1]$ and sends $R_A = g^x$ to B.
3. B selects $y \in_R [1, q - 1]$ and sends $R_B = g^y$ to A.
4. A verifies that $1 < R_B < p$ and $(R_B)^q \equiv 1 \pmod{p}$. If any check fails, then A terminates the protocol run with failure. Otherwise, A computes the shared secret $K = (Y_B)^x + (R_B)^a \bmod p$. If $K = 0$, then A terminates the protocol run with failure.
5. B verifies that $1 < R_A < p$ and $(R_A)^q \equiv 1 \pmod{p}$. If any check fails, then B terminates the protocol run with failure. Otherwise, B computes the shared secret $K = (Y_A)^y + (R_A)^b \bmod p$. If $K = 0$, then B terminates the protocol run with failure.
6. Both A and B compute the 80-bit session key $k = \mathrm{kdf}(K)$, where kdf is a key derivation function derived from the symmetric-key encryption scheme SKIPJACK (see [36] for further details).

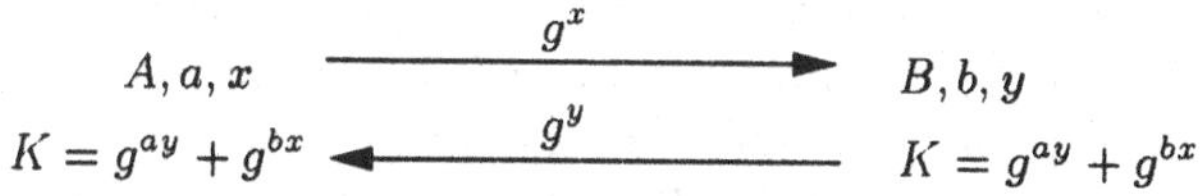

$$A, a, x \quad \xrightarrow{\quad g^x \quad} \quad B, b, y$$
$$K = g^{ay} + g^{bx} \quad \xleftarrow{\quad g^y \quad} \quad K = g^{ay} + g^{bx}$$

Fig. 4. Protocol 4 (KEA).

To illustrate the need for the features of KEA, we demonstrate how the protocol is weakened when certain modifications are made. This serves to further illustrate that designing secure key agreement protocols is a delicate and difficult task, and that subtle changes to a protocol can render it insecure.

VALIDATION OF PUBLIC KEYS – VERIFYING THAT THEY LIE IN THE SUBGROUP OF ORDER q. Suppose that A does not verify that $(R_B)^q \equiv 1 \pmod{p}$. Then, as observed by Lim and Lee [30], it may be possible for a malicious B to learn information about A's static private key a as follows using a variant of the small subgroup attack. Suppose that $p - 1$ has a prime factor l of small bitlength (e.g., 40 bits). Let $\beta \in \mathbb{Z}_p^*$ be of order l. If B sends $R_B = \beta$ to A, then A computes $K = g^{bx} + \beta^a \bmod p$ and $k = \mathrm{kdf}(K)$. Suppose now that A sends B an encrypted message $c = E_k(m)$, where E is a symmetric-key encryption scheme and the plaintext m has some recognizable structure. For each d, $0 \le d \le l - 1$, B computes $K' = g^{bx} + \beta^d \bmod p$, $k' = \mathrm{kdf}(K')$, and $m' = E_{k'}^{-1}(c)$. If m' possesses the requisite structure, then B concludes that $d = a \bmod l$, thus learning some partial information about a. This can be repeated for different small prime factors l of $p - 1$.

VALIDATION OF PUBLIC KEYS – VERIFYING THAT THEY LIE IN THE INTERVAL $[2, p-1]$. Suppose that A does not verify that $1 < Y_B < p$ and $1 < R_B < p$. Then an adversary E can launch the following unknown key-share attack. E gets $Y_E = 1$ certified as its static public key. E then forwards A's ephemeral public key R_A to B alleging it came from E. After B replies to E with R_B, E sends $R'_B = 1$ to A alleging it came from B. A computes $K_{AB} = g^{bx} + 1$ and B computes $K_{BE} = g^{bx} + 1$. Thus B is coerced into sharing a key with A without B's knowledge.

USE OF A KEY DERIVATION FUNCTION. The key derivation function kdf is used to derive a session key from the shared secret key K. One reason for doing this is to mix together *strong* bits and potential *weak* bits of K — weak bits are certain bits of information about K that can be correctly predicted with non-negligible advantage.

Another reason is to destroy the algebraic relationships between the shared secret K and the static and ephemeral public keys. This can help prevent against some kinds of known-key attacks, such as Burmester's triangle attack [17] which we describe next. An adversary E, whose static key pair is (c, g^c), observes a run of protocol between A and B in which ephemeral public keys g^x and g^y are exchanged; the resulting shared secret is $K_{AB} = g^{ay} + g^{bx}$. E then initiates a run of the protocol with A, replaying g^y as its ephemeral public key; the resulting secret which only A can compute is $K_{AE} = g^{ay} + g^{c\overline{x}}$, where $g^{\overline{x}}$ is A's ephemeral public key. Similarly, E initiates a run of the protocol with B, replaying g^x as its ephemeral public key; the resulting secret which only B can compute is $K_{BE} = g^{bx} + g^{c\overline{y}}$, where $g^{\overline{y}}$ is B's ephemeral public key. If E can somehow learn K_{AE} and K_{BE} (this is the known-key portion of the attack), then E can compute $K_{AB} = K_{AE} + K_{BE} - g^{c\overline{x}} - g^{c\overline{y}}$.

THE CHECK THAT $K \neq 0$. This check is actually unnecessary as the following argument shows. Since $(g^b)^q \equiv (g^y)^q \equiv 1 \pmod{p}$, we have that $(g^{bx})^q \equiv (g^{ay})^q \equiv 1 \pmod{p}$. Now, $K = 0$ if and only if $g^{bx} \equiv -g^{ay} \pmod{p}$. But this is impossible since otherwise $(g^{bx})^q \equiv (-g^{ay})^q \equiv (-1)^q \equiv -1 \pmod{p}$.

SECURITY NOTES. KEA does not provide (full) forward secrecy since an adversary who learns a and b can compute all session keys established by A and B. See also Table 1 in §6.

4.2   The Unified Model

The Unified Model, proposed by Ankney, Johnson and Matyas [1], is an AK protocol that is in the draft standards ANSI X9.42 [2], ANSI X9.63 [3], and IEEE P1363 [24]. One of its advantages is that it is conceptually simple and consequently easier to analyze (see §7.1).

Protocol 5 (Unified Model)

1. A selects $x \in_R [1, q-1]$ and sends $R_A = g^x$ and Cert_A to B.
2. B selects $y \in_R [1, q-1]$ and sends $R_B = g^y$ and Cert_B to A.

3. A verifies that $1 < R_B < p$ and $(R_B)^q \equiv 1 \pmod{p}$. If any check fails, then A terminates the protocol run with failure. Otherwise, A computes the session key $k = H((Y_B)^a \| (R_B)^x)$.

4. B verifies that $1 < R_A < p$ and $(R_A)^q \equiv 1 \pmod{p}$. If any check fails, then B terminates the protocol run with failure. Otherwise, B computes the session key $k = H((Y_A)^b \| (R_A)^y)$.

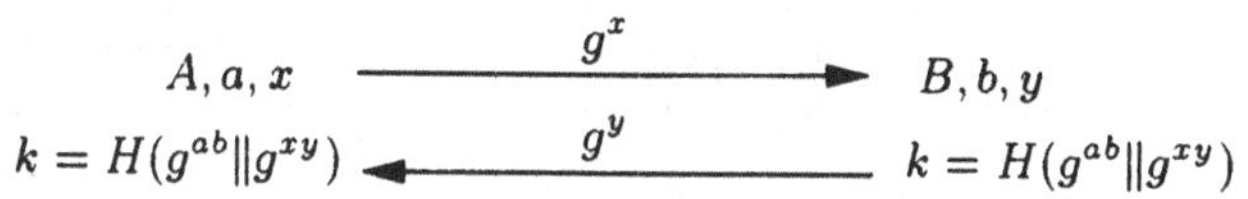

$$A, a, x \quad \xrightarrow{\quad g^x \quad} \quad B, b, y$$
$$k = H(g^{ab} \| g^{xy}) \quad \xleftarrow{\quad g^y \quad} \quad k = H(g^{ab} \| g^{xy})$$

Fig. 5. Protocol 5 (Unified model).

SECURITY NOTES. The Unified Model does not provide the service of key compromise impersonation, since an adversary who learns a can impersonate any other entity B to A. See also Table 1 in §6.

4.3   MQV

The so-called MQV protocol [29] is an AK protocol that is in the draft standards ANSI X9.42 [2], ANSI X9.63 [3], and IEEE P1363 [24]. The following notation is used. If $X \in [1, p-1]$, then $\overline{X} = (X \bmod 2^{80}) + 2^{80}$; more generally, $\overline{X} = (X \bmod 2^{\lceil f/2 \rceil}) + 2^{\lceil f/2 \rceil}$, where f is the bitlength of q. Note that $(\overline{X} \bmod q) \neq 0$.

Protocol 6 (MQV)

1. A selects $x \in_R [1, q-1]$ and sends $R_A = g^x$ and Cert_A to B.
2. B selects $y \in_R [1, q-1]$ and sends $R_B = g^y$ and Cert_B to A.
3. A verifies that $1 < R_B < p$ and $(R_B)^q \equiv 1 \pmod{p}$. If any check fails, then A terminates the protocol run with failure. Otherwise, A computes $s_A = (x + a\overline{R_A}) \bmod q$ and the shared secret $K = (R_B(Y_B)^{\overline{R_B}})^{s_A}$. If $K = 1$, then A terminates the protocol run with failure.
4. B verifies that $1 < R_A < p$ and $(R_A)^q \equiv 1 \pmod{p}$. If any check fails, then B terminates the protocol run with failure. Otherwise, B computes $s_B = (y + b\overline{R_B}) \bmod q$ and the shared secret $K = (R_A(Y_A)^{\overline{R_A}})^{s_B}$. If $K = 1$, then B terminates the protocol run with failure.
5. The session key is $k = H(K)$.

SECURITY NOTES. The expression for $\overline{R_A}$ uses only half the bits of R_A. This was done in order to increase the efficiency of computing K because the modular exponentiation $(Y_A)^{\overline{R_A}}$ can be done in half the time of a full exponentiation. The modification does not appear to affect the security of the protocol. The definition of $\overline{R_A}$ implies that $\overline{R_A} \neq 0$; this ensures that the contribution of the static private key a is not being cancelled in the formation of s_A.

The check $K = 1$ ensures that K has order q.

$$A, a, x \xrightarrow{\qquad g^x \qquad} B, b, y$$

$s_A = (x + a\overline{g^x}) \bmod q \qquad\qquad \xleftarrow{\quad g^y \quad} \qquad\qquad s_B = (y + b\overline{g^y}) \bmod q$

$K = g^{s_A s_B} \qquad\qquad\qquad\qquad\qquad\qquad\qquad\qquad K = g^{s_A s_B}$

Fig. 6. Protocol 6 (MQV).

Kaliski [27] has recently observed that Protocol 6 does not possess the unknown key-share attribute. This is demonstrated by the following on-line attack. An adversary E intercepts A's ephemeral public key R_A intended for B, and computes $R_E = R_A(Y_A)^{\overline{R}_A} g^{-1}$, $e = (\overline{R}_E)^{-1} \bmod q$, and $Y_E = g^e$. E then gets Y_E certified as her static public key (note that E knows the corresponding private key e), and transmits R_E to B. B responds by sending R_B to E, which E forwards to A. Both A and B compute the same session key k, however B mistakenly believes that he shares k with E. We emphasize that lack of the unknown key-share attribute does not contradict the fundamental goal of mutual implicit key authentication — by definition the provision of implicit key authentication is only considered in the case where B engages in the protocol with an honest entity (which E isn't). If an application using Protocol 6 is concerned with the lack of the unknown key-share attribute under such on-line attacks, then appropriate key confirmation should be added, for example as specified in Protocol 8 in §5.

4.4   One-pass variants

The purpose of a one-pass AK protocol is for entities A and B to agree upon a session key by only having to transmit one message from A to B — this assumes that A a priori has an authentic copy of B's static public key. One-pass protocols can be useful in applications where only one entity is on-line, such as secure email. Their main security drawbacks are that they do not offer known-key security (since an adversary can replay A's ephemeral public key to B) and forward secrecy (since entity B does not contribute a random per-message component).

The 3 two-pass AK protocols (KEA, Unified Model, MQV) presented in this section can be converted to one-pass AK protocols by simply setting B's ephemeral public key equal to his static public key. We illustrate this next for the one-pass variant of the MQV protocol. A summary of the security services of the 3 one-pass variants is provided in Table 1 in §6.

Protocol 7 (One-pass MQV)

1. A selects $x \in_R [1, q-1]$ and sends $R_A = g^x$ and Cert$_A$ to B.
2. A computes $s_A = (x + a\overline{R}_A) \bmod q$ and the shared secret $K = (Y_B(Y_B)^{\overline{Y}_B})^{s_A}$. If $K = 1$, then A terminates the protocol run with failure.

3. B verifies that $1 < R_A < p$ and $(R_A)^q \equiv 1 \pmod{p}$. If any check fails, then B terminates the protocol run with failure. Otherwise, B computes $s_B = (b + b\overline{Y}_B) \bmod q$ and the shared secret $K = (R_A(Y_A)^{\overline{R}_A})^{s_B}$. If $K = 1$, then B terminates the protocol run with failure.
4. The session key is $k = H(K)$.

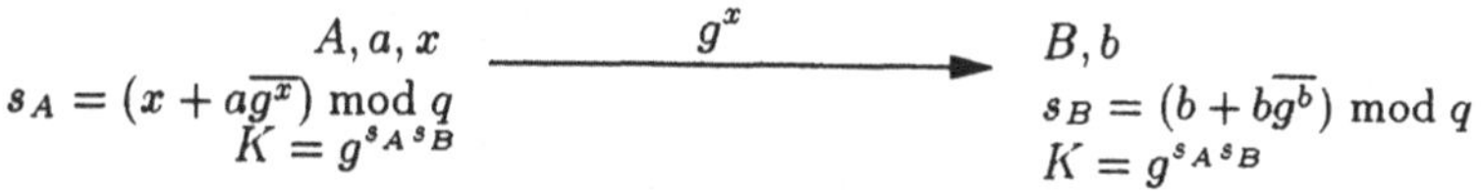

$$A, a, x \qquad\qquad\qquad\qquad g^x \qquad\qquad\qquad\qquad B, b$$
$$s_A = (x + a\overline{g^x}) \bmod q \qquad\longrightarrow\qquad s_B = (b + b\overline{g^b}) \bmod q$$
$$K = g^{s_A s_B} \qquad\qquad\qquad\qquad\qquad\qquad K = g^{s_A s_B}$$

Fig. 7. Protocol 7 (One-pass MQV).

5   AKC protocols

This section discusses AKC protocols and describes a method to derive AKC protocols from AK protocols.

The following three-pass AKC protocol [13] is derived from the Unified Model AK protocol (Protocol 5) by adding the MACs of the flow number, identities, and the ephemeral public keys. Here, H_1 and H_2 are 'independent' hash functions. In practice, one may choose $H_1(m) = H(10, m)$ and $H_2(m) = H(01, m)$, where H is a cryptographic hash function.

The MACs are computed under the shared key k', which is different from the session key k; Protocol 8 thus offers *implicit* key confirmation. If *explicit* key confirmation were to be provided by using the session key k as the MAC key, then a passive adversary would learn some information about k — the MAC of a known message under k. The adversary can use this to distinguish k from a key selected uniformly at random from the key space. This variant therefore sacrifices the desired goal that a protocol establish a computationally indistinguishable key. The maxim that a key establishment protocol can be used as a drop-in replacement for face-to-face key establishment therefore no longer applies and in theory security must be analyzed on a case-by-case basis. We therefore prefer Protocol 8.

Protocol 8 (Unified Model with key confirmation)

1. A selects $x \in_R [1, q-1]$ and sends $R_A = g^x$ and Cert_A to B.
2. (a) B verifies that $1 < R_A < p$ and $(R_A)^q \equiv 1 \pmod{p}$. If any check fails, then B terminates the protocol run with failure.
   (b) B selects $y \in_R [1, q-1]$, and computes $R_B = g^y$, $k' = H_1((Y_A)^b \| (R_A)^y)$, $k = H_2((Y_A)^b \| (R_A)^y)$, and $m_B = \text{MAC}_{k'}(2, B, A, R_B, R_A)$.
   (c) B sends R_B, Cert_B, and m_B to A.
3. (a) A verifies that $1 < R_B < p$ and $(R_B)^q \equiv 1 \pmod{p}$. If any check fails, then A terminates the protocol run with failure.

(b) A computes $k' = H_1((Y_B)^a\|(R_B)^x)$ and $m'_B = \mathrm{MAC}_{k'}(2, B, A, R_B, R_A)$, and verifies $m'_B = m_B$.

(c) A computes $m_A = \mathrm{MAC}_{k'}(3, A, B, R_A, R_B)$ and $k = H_2((Y_B)^a\|(R_B)^x)$, and sends R_A and m_A to B.

4. B computes $m'_A = \mathrm{MAC}_{k'}(3, A, B, R_A, R_B)$ and verifies that $m'_A = m_A$.
5. The session key is k.

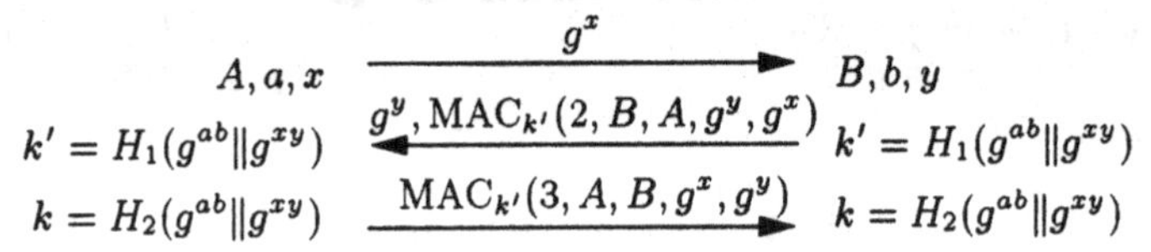

$$
\begin{array}{ccc}
A, a, x & \xrightarrow{\quad g^x \quad} & B, b, y \\[4pt]
k' = H_1(g^{ab}\|g^{xy}) & \xleftarrow{\;g^y,\,\mathrm{MAC}_{k'}(2, B, A, g^y, g^x)\;} & k' = H_1(g^{ab}\|g^{xy}) \\[4pt]
k = H_2(g^{ab}\|g^{xy}) & \xrightarrow{\;\mathrm{MAC}_{k'}(3, A, B, g^x, g^y)\;} & k = H_2(g^{ab}\|g^{xy})
\end{array}
$$

Fig. 8. Protocol 8 (Unified model with key confirmation).

In a similar manner, one can derive three-pass AKC protocols from the KEA (*KEA with key confirmation*) and MQV (*MQV with key confirmation*) AK protocols. The AKC variants of the Unified Model and MQV protocols are being considered for inclusion in ANSI X9.63 [3].

A summary of the security services provided by the 3 AKC variants is given in Table 1 in §6. This table illustrates why AKC protocols may be preferred over AK protocols in practice. First, the incorporation of key confirmation may provide additional security attributes which are not present in the AK protocol. For example, addition of key confirmation in the manner described above makes the MQV protocol resistant to unknown key-share attacks. Second, the security properties of AKC protocols appear to be better understood; see also the discussion in §7.1. Note that since the MACs can be computed efficiently, this method of adding key confirmation to an AK protocol does not place a significant computational burden on the key establishment mechanism.

6   Comparison

This section compares the security and efficiency of the protocols presented in §4 and §5.

Security services. Table 1 contains a summary of the services that are believed to be provided by the AK and AKC protocols discussed in §4 and §5. Although only implicit and explicit key authentication are considered vital properties of key establishment, any new results related to other information in this table would be interesting.

The services are discussed in the context of an entity A who has successfully executed the key agreement protocol over an open network wishing to establish keying data with entity B. In the table:

Scheme	IKA	EKA	K-KS	FS	K-CI	UK-S
Ephemeral Diffie–Hellman	×	×	$\sqrt{?^a}$	n/a	n/a	×
Ephemeral Diffie–Hellman (against passive attack)	√√	×	√√	n/a	n/a	√√
Static Diffie–Hellman	√√	×	×	×	×	√√
One-pass KEA	√√	×	×	×	$\sqrt{I}$	√√
KEA	√√	×	√√	×	√√	√√
KEA with Key Confirmation	√√	√√	√√	×	√√	√√
One-pass Unified Model	√√	×	×	×	$\sqrt{I}$	√√
Unified Model	√√	×	$\sqrt{?^a}$	$\sqrt{?^b}$	×	√√
Unified Model with Key Confirmation	√√	√√	√√	√√	×	√√
One-pass MQV	√√	×	×	×	$\sqrt{I}$	×
MQV	√√	×	√√	$\sqrt{?^b}$	√√	×
MQV with Key Confirmation	√√	√√	√√	√√	√√	√√

[a] Here the technicality hinges on the definition of what contributes 'another session key'. The service of known-key security is certainly provided if the protocol is extended so that explicit authentication of all session keys is supplied.

[b] Again the technicality concerns key confirmation. Both protocols provide forward secrecy if explicit authentication is supplied for all session keys. If not supplied, then the service of forward secrecy cannot be guaranteed.

Table 1. Security services offered by authenticated key agreement protocols.

- √√ indicates that the assurance is provided to A no matter whether A initiated the protocol or not.
- √? indicates that the assurance is provided modulo a theoretical technicality.
- $\sqrt{I}$ indicates that the assurance is provided to A only if A is the protocol's initiator.
- × indicates that the assurance is not provided to A by the protocol.

The names of the services have been abbreviated to save space: IKA denotes implicit key authentication, EKA explicit key authentication, K-KS known-key security, FS forward secrecy, K-CI key-compromise impersonation, and UK-S unknown key-share.

The provision of these assurances is considered in the case that both A and B are honest and have always executed the protocol correctly. The requirement that A and B are honest is certainly necessary for the provision of any service by a key establishment protocol: no key establishment protocol can protect against a dishonest entity who chooses to reveal the session key... just as no encryption scheme can guard against an entity who chooses to reveal confidential data.

Efficiency. The work done by each entity is dominated by the time to perform the modular exponentiations. The total number of modular exponentiations per entity for the KEA, Unified Model, and MQV AK protocols is 4, 4, and 3.5, respectively. If precomputations (of quantities involving the entity's static and ephemeral keys and the other entity's static keys) are discounted, then the total number of on-line modular exponentiations per entity reduces to 2, 2, and 2.5, respectively.

As noted in §5, MACs can be computed efficiently and hence the AKC variants have essentially the same computational overhead as their AK counterparts. They do, however, require an extra flow.

7   Provable security

This section discusses methods that have been used to formally analyze key agreement protocols. The goal of these methods is to facilitate the design of secure protocols that avoid subtle flaws like those described in §3. We examine two approaches, provable security and formal methods, focusing on the former.

Provable security was invented in the 1980's and applied to encryption schemes and signature schemes. The process of proving security of a protocol comes in five stages:

1. Specification of model.
2. Definition of goals within this model.
3. Statement of assumptions.
4. Description of protocols.
5. Proof that the protocol meets its goals within the model.

As discussed in §1, the emphasis of work in provable security of a protocol should be how appropriate the model, definitions, and underlying assumptions are, rather than the mere statement that a protocol attains provable security — after all, all protocols are provably secure in some model, under some definitions, or under some assumptions.

History of provable security. Building on earlier informal work of Bird et al. [12] for the symmetric setting and Diffie, van Oorschot and Wiener [21] for the asymmetric setting, Bellare and Rogaway [9] provided a model of distributed computing and rigorous security definitions, proposed concrete two-party authenticated key transport protocols in the symmetric setting, and proved them secure under the assumption that a pseudorandom function family exists. They then extended the model to handle the three-party (*Kerberos*) case [10]; see also Shoup and Rubin [39] for an extension of this work to the smart card world. Blake-Wilson and Menezes [14] and Blake-Wilson, Johnson and Menezes [13] extended the Bellare-Rogaway model to the asymmetric setting, and proposed and proved the security of some authenticated key transport, AK, and AKC protocols (see §7.1). More recently, Bellare, Canetti and Krawczyk [5] provided a systematic method for transforming authentication protocols that are secure in a model of idealized authenticated communications into protocols that are secure against active attacks; their work is discussed further in §7.2.

Formal methods. These are methods for analyzing cryptographic protocols in which the communications system is described using a formal specification language which has some mathematical basis, from which security properties of the protocol can be inferred. (See [38] and [28] for surveys on formal methods.) The most widely used of these methods are those related to the *BAN logic* of Burrows, Abadi and Needham [18], which was extended by van Oorschot [40] to enable the formal analysis of authenticated key agreement protocols in the asymmetric setting. Such methods begin with a set of beliefs for the participants and use logical inference rules to derive a belief that the protocol goals have been obtained.

Such formal methods have been useful in uncovering flaws and redundancies in protocols. However, they suffer from a number of shortcomings when considered as tools for designing high-assurance protocols. First, a proof that a protocol is logically correct does not imply that it is secure. This is especially the case because the process of converting a protocol into a formal specification may itself be subject to subtle flaws. Second, there is no clear security model associated with the formal systems used and thus it is hard to assess whether the implied threat model corresponds with the requirements of an application. Therefore, we believe that provable security techniques offer greater assurance than formal methods and we focus on provable security for the remainder of this section.

7.1   Bellare-Rogaway model of distributed computing

Work on the design of provably secure authenticated key agreement has largely focused on the Bellare-Rogaway model of distributed computing [9, 10].

The Bellare-Rogaway model, depicted in Figure 9, is a formal model of communication over an open network in which the adversary E is afforded enormous power. She controls all communication between entities, and can at any time ask an entity to reveal its static private key. Furthermore, she may at any time initiate sessions between any two entities, engage in multiple sessions with the same entity at the same time, and ask an entity to enter a session with itself. We provide an informal description of the Bellare-Rogaway model, and informal definitions of the goals of secure AK and AKC protocols. For complete descriptions, see [9, 10, 13].

In the model, E is equipped with a collection of $\Pi^s_{A,B}$ oracles. $\Pi^s_{A,B}$ models entity A who believe she is communicating with entity B for the sth time. E is allowed to make three types of queries of its oracles:

Send($\Pi_{A,B}, x$): E gives a particular oracle x as input and learns the oracle's response.

Reveal($\Pi_{A,B}$): E learns the session key (if any) the oracle currently holds.

Corrupt(A): E learns A's static private key.

When E asks an oracle a query, the oracle computes its response using the description of the protocol. Security goals are defined in the context of running E in the presence of these oracles.

Secure key agreement is now captured by a test involving an additional **Test** query. At the end of its experiment, E selects a *fresh* oracle $\Pi^s_{A,B}$ — this is an oracle which has accepted a session key k, and where the adversary has not learned k by trivial means (either by corrupting A or B, or by issuing a **Reveal** query to $\Pi^s_{A,B}$ or to any $\Pi^t_{B,A}$ oracle which has had a matching conversation with $\Pi^s_{A,B}$) — and asks it **Test** query. The oracle replies with either its session key k or a random key, and the adversary's job is to decide which key it has been given.

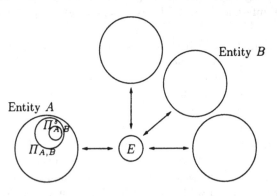

Fig. 9. The Bellare-Rogaway model of distributed computing.

Definition 1 ([13]). *(Informal) An AK protocol is* secure *if:*

(i) The protocol successfully distributes keys in the absence of an adversary.

(ii) No adversary E can distinguish a session key held by a fresh $\Pi^s_{A,B}$ oracle from a key selected uniformly at random.

A secure AKC protocol is defined by amalgamating the notion of entity authentication with the notion of a secure AK protocol.

Definition 2 ([13]). *(Informal) An AKC protocol is* secure *if in addition to conditions (i) and (ii) of Definition 1:*

(iii) The only way an adversary E can induce a $\Pi^s_{A,B}$ oracle to accept a session key is by honestly transmitting messages between $\Pi^s_{A,B}$ and some $\Pi^t_{B,A}$.

The security of the Unified Model (Protocol 5) and the Unified Model with key confirmation (Protocol 8) in the Bellare-Rogaway model was proven under certain assumptions in [13].

Theorem 1 ([13]). *Protocol 5 is a secure AK protocol in the Bellare-Rogaway model provided that:*

(i) the adversary makes no Reveal queries;

(ii) the Diffie–Hellman problem is hard; and
(iii) H is a random oracle.

Theorem 2 ([13]). *Protocol 8 is a secure AKC protocol in the Bellare-Rogaway model provided that:*

(i) the Diffie–Hellman problem is hard;
(ii) the MAC is secure; and
(iii) H_1 and H_2 are independent random oracles.

A random oracle is a 'black-box' random function which is supplied to all entities, including the adversary. The assumption that H, H_1 and H_2 are random oracles is a very powerful one and facilitates security analysis. This so-called *random oracle model* was introduced and popularized by Bellare and Rogaway [8]. In practice, the random oracles can be instantiated with hash functions — therefore the security proofs in the random model are no longer valid in the practical implementation. Nonetheless, and despite recent results demonstrating the limitations of the random oracle model [19], it is a thesis that protocols proven secure in the random oracle provide higher security assurances than protocols deemed secure by ad-hoc means.

To see that Protocol 5 is not a secure AK protocol in the Bellare-Rogaway model if the adversary is allowed to make **Reveal** queries, consider the following interleaving/reflection attack. Suppose that A initiates 2 runs of the protocol; let A's ephemeral public keys be g^x and $g^{\overline{x}}$ in the first and second runs, respectively. The adversary E then replays $g^{\overline{x}}$ and g^x to A in the first and second rounds respectively, purportedly as B's ephemeral public keys. A computes both session keys as $k = H(g^{ab}\|g^{x\overline{x}})$. E can now **Reveal** one session key, and thus also learn the other.

It is conjectured in [13] that the modification of Protocol 5 in which the session key is formed as $k = H(g^{ay}\|g^{bx})$ is a secure AK protocol assuming only that the Diffie–Hellman problem is hard and that H is a random oracle.

7.2   A modular approach

Recently, Bellare, Canetti and Krawczyk [5] have suggested an approach to the design of provably secure key agreement protocols that differs from the Bellare-Rogaway model. Their approach is a modular approach and starts with protocols that are secure in a model of idealized authenticated communication and then systematically transforms them into protocols which are secure in the realistic unauthenticated setting. This approach has the advantage that a new proof of security is not required for each protocol — instead once the approach is justified it can be applied to any protocol that works in the ideal model. On the other hand, it is less clear what practical guarantees are provided so the evaluation of whether the guarantees are appropriate in an application is perhaps less understood. The following is an informal overview of their approach.

AUTHENTICATORS. Authenticators are key to the systematic transformations at the heart of the modular approach. They are compilers that take as input a protocol designed for authenticated networks, and transforms it into an 'equivalent' protocol for unauthenticated networks. The notion of *equivalence* or *emulation* is formalized as follows. A protocol P' designed for unauthenticated networks is said to *emulate* a protocol P designed for authenticated networks, if for each adversary E' of P' there exists an adversary E of P such that for all inputs x, the views $V_{P,E}(x)$ and $V_{P',E'}(x)$ are computationally indistinguishable. (The *view* $V_{P,E}(x)$ of a protocol P which is run on input x in the presence of an adversary E is the random variable describing the cumulative outputs of E and all the legitimate entities.)

MT-AUTHENTICATORS. In [5], authenticators are realized using the simpler idea of an *MT-authenticator* which emulates the most straightforward *message transmission (MT)* protocol in which a single message is passed from A to B as depicted in Figure 10. Figure 11 illustrates the protocol λ_{sig} which is proven in [5] to be an MT-authenticator. In the figure, $\mathrm{sign}_A()$ denotes A's signature using a signature scheme that is secure against chosen message attacks (e.g., [11, 22]). Now an MT-authenticator λ can be used to construct a *compiler* C_λ as follows: given a protocol P, $P' = C_\lambda(P)$ is the protocol obtained by applying λ to each message transmitted by P. It is proven in [5] that C_λ is indeed an authenticator.

$$A \xrightarrow{\quad A, B, m \quad} B$$

Fig. 10. Message transmission protocol (MT).

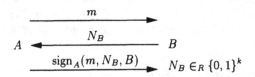

Fig. 11. MT-authenticator λ_{sig}.

KEY ESTABLISHMENT. Finally, this MT-authenticator is used to build a secure authenticated key agreement protocol. It is first shown in [5] that ephemeral Diffie–Hellman EDH (Protocol 1) is a secure key establishment protocol for authenticated networks by showing that it emulates traditional face-to-face key establishment as described in §2. Then, EDH is emulated using $C_{\lambda_{\mathrm{sig}}}$. The result $C_{\lambda_{\mathrm{sig}}}(EDH)$ is a secure six-pass authenticated key agreement protocol. Combining messages from different flows, and replacing the challenges N_A and N_B with the ephemeral public keys g^x and g^y, respectively, yields the three-pass *BCK protocol*, depicted in Figure 12.

The BCK protocol is similar to Key Agreement Mechanism 7 in ISO/IEC 11770-3 [25]. In the latter, the MACs of the signatures under the shared secret $K = g^{xy}$ are also included in flows 2 and 3, thus providing *explicit* key confirmation, instead of just *implicit* key confirmation as provided by the BCK protocol.

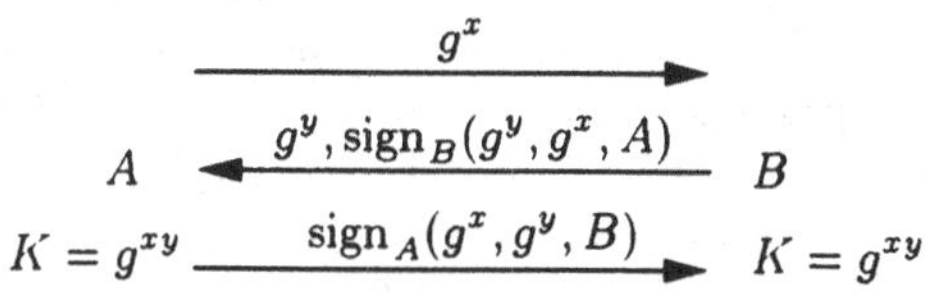

$$g^x \longrightarrow$$

$$A \quad \xleftarrow{\quad g^y, \mathrm{sign}_B(g^y, g^x, A) \quad} \quad B$$

$$K = g^{xy} \quad \xrightarrow{\quad \mathrm{sign}_A(g^x, g^y, B) \quad} \quad K = g^{xy}$$

Fig. 12. BCK protocol.

8   Conclusions and future work

This paper surveyed practical and provable security aspects of some authenticated Diffie–Hellman key agreement protocols that are being considered for standardization.

A number of questions can be asked. Can the MQV protocol be proven secure in a reasonable model of computing? Are the definitions of secure AK and AKC protocols in §7.1 the right ones? How do the models and security definitions presented in §7.1 and §7.2 compare? Are the security proofs meaningful in practice? That is, can the reductions used in the proofs be untilized to obtain meaningful measures of exact security [11]? (*Exact security* is a concrete quantification of the security guaranteed by a protocol in terms of the perceived security of the underlying cryptographic primitives, e.g., the Diffie–Hellman problem or a secure MAC algorithm.)

Two important tasks that remain are to devise a provably secure two-pass AK protocol, and to provide formal definitions for secure one-pass key agreement protocols.

References

1. R. Ankney, D. Johnson and M. Matyas, "The Unified Model", contribution to X9F1, October 1995.
2. ANSI X9.42, *Agreement of Symmetric Algorithm Keys Using Diffie–Hellman*, working draft, May 1998.
3. ANSI X9.63, *Elliptic Curve Key Agreement and Key Transport Protocols*, working draft, July 1998.
4. M. Bellare, R. Canetti and H. Krawczyk, "Keying hash functions for message authentication", *Advances in Cryptology – Crypto '96*, LNCS **1109**, 1996, 1-15.

5. M. Bellare, R. Canetti and H. Krawczyk, "A modular approach to the design and analysis of authentication and key exchange protocols", *Proceedings of the 30th Annual ACM Symposium on the Theory of Computing*, 1998. A full version of this paper is available at http://www-cse.ucsd.edu/users/mihir

6. M. Bellare, R. Guerin and P. Rogaway, "XOR MACs: New methods for message authentication using finite pseudorandom functions", *Advances in Cryptology – Crypto '95*, LNCS **963**, 1995, 15-28.

7. M. Bellare, J. Kilian and P. Rogaway, "The security of cipher block chaining", *Advances in Cryptology – Crypto '94*, LNCS **839**, Springer-Verlag, 1994, 341-358.

8. M. Bellare and P. Rogaway, "Random oracles are practical: a paradigm for designing efficient protocols", *1st ACM Conference on Computer and Communications Security*, 1993, 62-73. A full version of this paper is available at http://www-cse.ucsd.edu/users/mihir

9. M. Bellare and P. Rogaway, "Entity authentication and key distribution", *Advances in Cryptology – Crypto '93*, LNCS **773**, 1994, 232-249. A full version of this paper is available at http://www-cse.ucsd.edu/users/mihir

10. M. Bellare and P. Rogaway, "Provably secure session key distribution — the three party case", *Proceedings of the 27th Annual ACM Symposium on the Theory of Computing*, 1995, 57-66.

11. M. Bellare and P. Rogaway, "The exact security of digital signatures — how to sign with RSA and Rabin", *Advances in Cryptology – Eurocrypt '96*, LNCS **1070**, 1996, 399-416. A full version of this paper is available at http://www-cse.ucsd.edu/users/mihir

12. R. Bird, I. Gopal, A. Herzberg, P. Janson, S. Kutten, R. Molva, and M. Yung, "Systematic design of two-party authentication protocols", *Advances in Cryptology – Crypto '91*, LNCS **576**, 1992, 44-61.

13. S. Blake-Wilson, D. Johnson and A. Menezes, "Key agreement protocols and their security analysis", *Proceedings of the sixth IMA International Conference on Cryptography and Coding*, LNCS **1355**, 1997, 30-45. A full version of this paper is available at http://www.cacr.math.uwaterloo.ca

14. S. Blake-Wilson and A. Menezes, "Entity authentication and authenticated key transport protocols employing asymmetric techniques", *Proceedings of the 5th International Workshop on Security Protocols*, LNCS **1361**, 1997, 137-158.

15. S. Blake-Wilson and A. Menezes, "Unknown key-share attacks on the station-to-station (STS) protocol", Technical report CORR 98-42, University of Waterloo, 1998. Also available at http://www.cacr.math.uwaterloo.ca/

16. D. Boneh and R. Lipton, "Algorithms for black-box fields and their application to cryptography", *Advances in Cryptology – Crypto '96*, LNCS **1109**, 1996, 283-297.

17. M. Burmester, "On the risk of opening distributed keys", *Advances in Cryptology – Crypto '94*, LNCS **839**, 1994, 308-317.

18. M. Burrows, M. Abadi and R. Needham, "A logic of authentication", *ACM Transactions on Computer Systems*, **8** (1990), 18-36.

19. R. Canetti, O. Goldreich and S. Halevi, "The random oracle methodology, revisited", *Proceedings of the 30th Annual ACM Symposium on the Theory of Computing*, 1998.

20. W. Diffie and M. Hellman, "New directions in cryptography", *IEEE Transactions on Information Theory*, **22** (1976), 644-654.

21. W. Diffie, P. van Oorschot and M. Wiener, "Authentication and authenticated key exchanges", *Designs, Codes and Cryptography*, **2** (1992), 107-125.

22. C. Dwork and M. Naor, "An efficient existentially unforgeable signature scheme and its applications", *Journal of Cryptology*, **11** (1998), 187-208.

23. K.C. Goss, "Cryptographic method and apparatus for public key exchange with authentication", U.S. patent 4,956,865, September 11 1990.

24. IEEE P1363, *Standard Specifications for Public-Key Cryptography*, working draft, July 1998.

25. ISO/IEC 11770-3, *Information Technology – Security Techniques – Key Management – Part 3: Mechanisms Using Asymmetric Techniques*, draft, (DIS), 1996.

26. D. Johnson, Contribution to ANSI X9F1 working group, 1997.

27. B. Kaliski, Contribution to ANSI X9F1 and IEEE P1363 working groups, June 1998.

28. R. Kemmerer, C. Meadows and J. Millen, "Three systems for cryptographic protocol analysis", *Journal of Cryptology*, **7** (1994), 79-130.

29. L. Law, A. Menezes, M. Qu, J. Solinas and S. Vanstone, "An efficient protocol for authenticated key agreement", Technical report CORR 98-05, University of Waterloo, 1998. Also available at http://www.cacr.math.uwaterloo.ca/

30. C. Lim and P. Lee, "A key recovery attack on discrete log-based schemes using a prime order subgroup", *Advances in Cryptology – Crypto '97*, LNCS **1294**, 1997, 249-263.

31. T. Matsumoto, Y. Takashima and H. Imai, "On seeking smart public-key distribution systems", *The Transactions of the IECE of Japan*, **E69** (1986), 99-106.

32. U. Maurer and S. Wolf, "Diffie–Hellman oracles", *Advances in Cryptology – Crypto '96*, LNCS **1109**, 1996, 283-297.

33. A. Menezes, P. van Oorschot and S. Vanstone, *Handbook of Applied Cryptography*, CRC Press, 1997.

34. C. Mitchell, "Limitations of challenge-response entity authentication", *Electronics Letters*, **25** (August 17, 1989), 1195-1196.

35. National Institute of Standards and Technology, "Secure Hash Standard (SHS)", FIPS Publication 180-1, April 1995.

36. National Security Agency, "SKIPJACK and KEA algorithm specification", Version 2.0, May 29 1998. Also available at http://csrc.nist.gov/encryption/skipjack-kea.htm

37. R. Rivest and A. Shamir, "How to expose an eavesdropper", *Communications of the ACM*, **27** (1984), 393-395.

38. A. Rubin and P. Honeyman, "Formal methods for the analysis of authentication protocols", CITI Technical Report 93-7, Information Technology Division, University of Michigan, 1993. Also available at http://cs.nyu.edu/~rubin/

39. V. Shoup and A. Rubin, "Session key distribution using smart cards", *Advances in Cryptology – Eurocrypt '96*, LNCS **1070**, 1996, 321-331.

40. P. van Oorschot, "Extending cryptographic logics of belief to key agreement protocols", *1st ACM Conference on Computer and Communications Security*, 1993, 232-243.

Initial Observations on Skipjack: Cryptanalysis of Skipjack-3XOR

Eli Biham[1], Alex Biryukov[2], Orr Dunkelman[1],
Eran Richardson[3], and Adi Shamir[4]

[1] Computer Science Department
Technion – Israel Institute of Technology
Haifa 32000, Israel
biham@cs.technion.ac.il
http://www.cs.technion.ac.il/~biham/
[2] Applied Mathematics Department
Technion – Israel Institute of Technology
Haifa 32000, Israel
[3] Electrical Engineering Department
Technion – Israel Institute of Technology
Haifa 32000, Israel
[4] Department of Applied Mathematics and Computer Science
Weizmann Institute of Science
Rehovot 76100, Israel
shamir@wisdom.weizmann.ac.il

Abstract. Skipjack is the secret key encryption algorithm developed by the NSA for the Clipper chip and Fortezza PC card. It uses an 80-bit key, 128 table lookup operations, and 320 XOR operations to map a 64-bit plaintext into a 64-bit ciphertext in 32 rounds. This paper describes an efficient attack on a variant, which we call *Skipjack-3XOR* (Skipjack minus 3 XORs). The only difference between Skipjack and Skipjack-3XOR is the removal of 3 out of the 320 XOR operations. The attack uses the ciphertexts derived from about 500 plaintexts and its total running time is equivalent to about one million Skipjack encryptions, which can be carried out in seconds on a personal computer. We also present a new cryptographic tool, which we call the *Yoyo game*, and efficient attacks on Skipjack reduced to 16 rounds. We conclude that Skipjack does not have a conservative design with a large margin of safety.
Key words: Cryptanalysis, Skipjack, Yoyo Game, Clipper chip, Fortezza PC card.

1 Introduction

Skipjack is the secret key encryption algorithm developed by the NSA for the Clipper chip and Fortezza PC card. It was implemented in tamper-resistant hardware and its structure was kept secret since its introduction in 1993.

S. Tavares and H. Meijer (Eds.): SAC'98, LNCS 1556, pp. 362–375, 1999.
© Springer-Verlag Berlin Heidelberg 1999

To increase confidence in the strength of Skipjack and the Clipper chip initiative, five well known cryptographers were assigned in 1993 to analyze Skipjack and report their findings[4]. They investigated the strength of Skipjack using differential cryptanalysis[3] and other methods, and concentrated on reviewing NSA's design and evaluation process. They reported that Skipjack is a "representative of a family of encryption algorithms developed in 1980 as part of the NSA suite of "Type I" algorithms, suitable for protecting all levels of classified data. The specific algorithm, SKIPJACK, is intended to be used with sensitive but unclassified information." They concluded that "Skipjack is based on some of NSA's best technology" and quoted the head of the NSA evaluation team who confidently concluded "I believe that Skipjack can only be broken by brute force - there is no better way."

On June 24th, 1998, Skipjack was declassified, and its description was made public in the web site of NIST [7]. It uses an 80-bit key, $32 \cdot 4 = 128$ table lookup operations, and $32 \cdot 10 = 320$ XOR operations to map a 64-bit plaintext into a 64-bit ciphertext in 32 rounds.

This paper summarizes our initial analysis. We study the differential[3] and linear[6] properties of Skipjack, together with other observations on the design of Skipjack. Then, we use these observations to present a differential attack on Skipjack reduced to 16 rounds, using about 2^{22} chosen plaintexts and steps of analysis. Some of these results are based on important observations communicated to us by David Wagner [8].

We present a new cryptographic tool, which we call the *Yoyo game*, applied to Skipjack reduced to 16 rounds. This tool can be used to identify pairs satisfying a certain property, and be used as a tool for attacking Skipjack reduced to 16 rounds using only 2^{14} adaptive chosen plaintexts and ciphertexts and 2^{14} steps of analysis. This tool can also be used as a distinguisher to decide whether a given black box contains this variant of Skipjack, or a random permutation.

We then present the main result of this paper, which is an exceptionally simple attack on a 32-round variant, which we call *Skipjack-3XOR* (Skipjack minus 3 XORs). The only difference between the actual Skipjack and Skipjack-3XOR is the removal of 3 out of the 320 XOR operations. The attack uses the ciphertexts derived from about 500 plaintexts which are identical except for the second 16 bit word. Its total running time is equivalent to about one million Skipjack encryptions, which can be carried out in seconds on a personal computer. We thus believe that Skipjack does not have a conservative design with a large margin of safety.

This paper is organized as follows: In Section 2 we describe the structure of Skipjack, and the main variants that we analyze in this paper. In Section 3 we present useful observations on the design, which we later use in our analysis. In Section 4 we describe a differential attack on a 16-round variant of Skipjack. The

Yoyo game and its applications are described in Section 5. Finally, in Section 6 we present our main attack on Skipjack-3XOR.

2   Description of Skipjack

The published description of Skipjack characterizes the rounds as either Rule A or Rule B. Each round is described in the form of a linear feedback shift register with an additional non linear keyed G permutation. Rule B is basically the inverse of Rule A with minor positioning differences. Skipjack applies eight rounds of Rule A, followed by eight rounds of Rule B, followed by another eight rounds of Rule A, followed by another eight rounds of Rule B. The original definitions of Rule A and Rule B are given in Figure 1, where *counter* is the

Rule A	Rule B
$w_1^{k+1} = G^k(w_1^k) \oplus w_4^k \oplus counter^k$	$w_1^{k+1} = w_4^k$
$w_2^{k+1} = G^k(w_1^k)$	$w_2^{k+1} = G^k(w_1^k)$
$w_3^{k+1} = w_2^k$	$w_3^{k+1} = w_1^k \oplus w_2^k \oplus counter^k$
$w_4^{k+1} = w_3^k$	$w_4^{k+1} = w_3^k$

Fig. 1 Rule A and Rule B.

round number (in the range 1 to 32), G is a four-round Feistel permutation whose F function is defined as an 8x8-bit S box, called *F Table*, and each round of G is keyed by eight bits of the key. The key scheduling of Skipjack takes a 10-byte key, and uses four of them at a time to key each G permutation. The first four bytes are used to key the first G permutation, and each additional G permutation is keyed by the next four bytes cyclically.

The description becomes simpler (and the software implementation becomes more efficient) if we unroll the rounds, and keep the four elements in the shift register stationary. In this form the code is simply a sequence of alternate G operations and XOR operations of cyclically adjacent elements. In this representation the main difference between Rule A and Rule B is the direction in which the adjacent elements are XORed (left to right or right to left).

The XOR operations of Rule A and Rule B after round 8 and after round 24 (on the borders between Rule A and Rule B) are consecutive without application of the G permutation in between. In the unrolled description these XORs are of the form

$$W2 = G(W2, subkey) \qquad\qquad \text{-- Rule A}$$

$$W1 = W1 \oplus W2 \oplus 8$$
$$W2 = W2 \oplus W1 \oplus 9 \qquad \text{-- Rule B}$$
$$W1 = G(W1, subkey)$$

which is equivalent to exchanging the words $W1$ and $W2$, and leaving $W2$ as the original $W1 \oplus 1$:

$$W2 = G(W2, subkey)$$

exchange W1 and W2

$$W1 = W1 \oplus W2 \oplus 8$$
$$W2 = W2 \oplus 1$$
$$W1 = G(W1, subkey)$$

(the same situation occurs after round 24 with the round numbers 8 and 9 replaced by 24 and 25). Figure 2 describes this representation of Skipjack (only the first 16 rounds out of the 32 are listed; the next 16 rounds are identical except for the counter values).

Also, on the border between Rule B and Rule A (after round 16), there are two parallel applications of the G permutation on two different words, with no other linear mixing in between.

Note that Rule A mixes the output of the G permutation into the input of the next G permutation, while Rule B mixes the input of a G permutation into the output of the previous G permutation (similarly in decryption of Rule A), and thus during encryption Rule B rounds add little to the avalanche effect, and during decryption Rule A rounds add little to the avalanche effect.

In this paper we consider variants of Skipjack which are identical to the original version except for the removal of a few XOR operations. We use the name *Skipjack-$(i_1, \ldots, i_k)$* to denote the variant in which the XOR operations mixing two data words at rounds $i_1, \ldots, i_k$ are removed, and the name *Skipjack-3XOR* as a more mnemonic name for Skipjack-(4,16,17), which is the main variant we attack. Note that the removal of these XOR operations does not remove the effect of any other operation (as could happen if we removed the XORs of the Feistel structure of G, which would eliminate the effect of the corresponding F tables).

3   Useful Observations

3.1   Observations Regarding the Key Schedule

The key schedule is cyclic in the sense that the same set of four bytes of the subkeys (entering a single G permutation) are repeated every five rounds, and

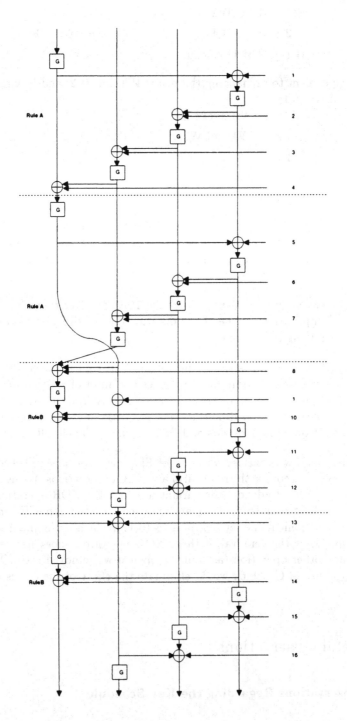

Fig. 2 Skipjack.

there are only five such sets. In addition, the key bytes are divided into two sets: the even bytes and the odd bytes. The even bytes always enter the even rounds of the G permutation, while the odd bytes always enter the odd rounds of the G permutation.

3.2   Decryption

As in most symmetric ciphers, decryption can be done using encryption with minor modifications. These modifications are (1) reordering the key bytes to $K^* = (cv_7, cv_6, ..., cv_0, cv_9, cv_8)$, (2) reversing the order of the round counters, and then (3) encrypting the reordered ciphertext $C^* = (cb_3, cb_2, cb_1, cb_0, cb_7, cb_6, cb_5, cb_4)$ gives the reordered plaintext $P^* = (pb_3, pb_2, pb_1, pb_0, pb_7, pb_6, pb_5, pb_4)$.

The mixings with the round numbers (counters) are often used to protect against related key attacks. In Skipjack, if these mixings are removed, the following stronger property would hold: Given a plaintext $P = (pb_0, pb_1, ..., pb_7)$, a key $K = (cv_0, ..., cv_9)$ and a ciphertext $C = (cb_0, ..., cb_7)$ such that $C = $ Skipjack$_K(P)$, then decryption can be performed using encryption by $P^* = $ Skipjack$_{K^*}(C^*)$, where $K^* = (cv_7, cv_6, ..., cv_0, cv_9, cv_8)$, $P^* = (pb_3, pb_2, pb_1, pb_0, pb_7, pb_6, pb_5, pb_4)$, and $C^* = (cb_3, cb_2, cb_1, cb_0, cb_7, cb_6, cb_5, cb_4)$.

This property could be used to reduce the complexity of exhaustive search of this Skipjack variant by a factor of almost 2 (26% of the key space rather than 50% in average) in a similar way to the complementation property of DES: Given the encrypted ciphertext $C1$ of some plaintext P, and the decrypted plaintext $C2$ of the related P^* under the same unknown key, perform trial encryptions with 60% of the keys K (three keys of each cycle of 5 keys of the rotation by two key bytes operations; efficient implementations first try two keys of each cycle, and only if all of them fail, they try the third keys of the cycles). For each of these keys compare the ciphertext to $C1$, and to $C2^*$ (i.e., $C2$ in which the bytes are reordered as above). If the comparison fails, the unknown key is neither K nor K^*. If it succeeds, we make two or three trial encryptions, and in case they succeed we found the key.

3.3   Complementation Properties of the G Permutation

The G permutation has $2^{16} - 1$ complementation properties: Let $G_{K0,K1,K2,K3}(x1, x2) = (y1, y2)$, where $K0, K1, K2, K3, x1, x2, y1, y2$ are all byte values, and let $d1, d2$ be two byte values. Then,

$$G_{K0 \oplus d1, K1 \oplus d2, K2 \oplus d1, K3 \oplus d2}(x1 \oplus d2, x2 \oplus d1) = (y1 \oplus d2, y2 \oplus d1).$$

G has exactly one fixpoint for every subkey (this was identified by Frank Gifford, and described in sci.crypt). Moreover, we observed that for every key and every value v of the form $(0, b)$ or $(b, 0)$ where 0 is a zero byte and b is an arbitrary byte value, G has exactly one value x for which $G(x) = x \oplus v$. It is unknown whether this property can aid in the analysis of Skipjack.

3.4   Differential and Linear Properties of the F Table

We generated the differential and linear distribution tables of the F table, and found that in the difference distribution table:

1. The maximal entry is 12 (while the average is 1).
2. 39.9% of the entries have non-zero values.
3. The value 0 appears in 39360 entries, 2 in 20559, 4 in 4855, 6 in 686, 8 in 69, 10 in 5, and 12 in 2 entries.
4. One-bit to one-bit differences are possible, such as $01_x \rightarrow 01_x$ (where the subscript x denotes a hexadecimal representation) with probability 2/256.

In the linear approximation table:

1. The maximal biases are 28 and -28 (i.e., probabilities of $1/2 + 28/256$ and $1/2 - 28/256$).
2. 89.3% of the entries have non-zero values.
3. The absolute value of the bias is 0 in 7005 entries, 2 in 12456, 4 in 11244, 6 in 9799, 8 in 7882, 10 in 6032, 12 in 4354, 14 in 2813, 16 in 1814, 18 in 1041, 20 in 567, 22 in 317, 24 in 154, 26 in 54, and 28 in 3 entries.
4. Unbalanced one-bit to one-bit linear approximations exist, such as $80_x \rightarrow 80_x$ with probability $1/2 + 20/256$.

3.5   Differential and Linear Properties of the G Permutation

Consider the F table, and let a and b be two byte values such that both $a \rightarrow b$ and $b \rightarrow a$ occur with a non-zero probability. We can prove that the best possible characteristic of G must be of the form: input difference: $(a, 0)$, output differences: $(0, b)$, with the intermediate differences $(a, 0) \rightarrow (a, 0) \rightarrow (a, b) \rightarrow (0, b) \rightarrow (0, b)$. There are 10778 pairs of such a and b, of which four have probability $48/2^{16} = 2^{-10.42}$. They are

1. $a = 52_x, b = f5_x,$
2. $a = f5_x, b = 52_x,$
3. $a = 77_x, b = 92_x,$ and
4. $a = 92_x, b = 77_x.$

Most other characteristics of this form have probability 2^{-14} (6672 pairs) and 2^{-13} (3088 pairs). The remaining characteristics of this form have probabilities between 2^{-13} and $2^{-10.67}$.

Given a and b, there are additional characteristics with three active F tables (rather than only two), and for the above values of a and b the probabilities are between $2^{-15.4}$ and $2^{-15.83}$. These characteristics of G are $(0, b) \rightarrow (a, b)$ and $(a, b) \rightarrow (a, 0)$. We can combine these characteristics with the characteristics of the previous form and get cycles of three characteristics which have the form $(a, 0) \rightarrow (0, b) \rightarrow (a, b) \rightarrow (a, 0)$.

We studied the differential corresponding to these characteristics, and computed their exact probabilities by summing up the probabilities of all the characteristics with the same external differences. We found that the characteristic $(a, 0) \rightarrow (0, b)$ has the same probability as a differential and as a characteristic, as there are no other characteristics with the same external differences. $(0, b) \rightarrow (a, b)$ and $(a, b) \rightarrow (a, 0)$ with the same a and b as above have over a thousand small-probability counterparts with the same external differences, whose total probability is slightly smaller than the probabilities of the original characteristics. Thus, the probability of the differentials are almost twice that of the original characteristics (e.g., $137088/2^{32} = 2^{-14.94}$ instead of $73728/2^{32} = 2^{-15.83}$ in one of the cases).

We had also investigated other differentials of G. The characteristics we described with probability of around $2^{-10.42}$ (and other lower probability characteristics with zero differences in the first and fourth rounds of the G permutation) do not have any counterparts, and thus the corresponding differentials have the same probabilities as the characteristics. The best other differential we are aware of is $002A_x \rightarrow 0095_x$ with probability $2^{-14.715}$, and the best possible differential with the same input and output differences is $7F7F_x \rightarrow 7F7F_x$ with probability $2^{-15.84}$.

We next consider the case of linear cryptanalysis. As the characteristics are built in a similar way where XORs are replaced by duplications and duplications are replaced by XORs of the subsets of parity bits[1], we can apply the same technique for linear cryptanalysis. In this case we have 52736 possible pairs of a and b. The best linear characteristic of G is based on $a = b = 60_x$ and its probability is $1/2 + 2 \cdot 676/2^{16} = 1/2 + 2^{-5.6}$.

It is interesting to note that (due to its design) many criteria used in other ciphers are neither relevant to nor used in Skipjack. For example, a difference of one input bit in a DES S box cannot cause a difference of only one bit in its output, but there are many such instances in the F table of Skipjack.

Another observation is that due to the switchover from Rule A to Rule B iterations, the data between rounds 5 and 12 is very badly mixed. As mentioned earlier, on the border between the two rules (after rounds 8 and 24), the leftmost word is exchanged with word 2, and the new word 1 is XORed with the new word 2. We observed that the output of the G permutation in round 5 becomes the input to the G permutation in round 12, unaffected by other words (but XORed with the fixed value $8 \oplus 9 = 1$). Thus, this word is not affected by any other word during 8 consecutive rounds. A similar property occurs in word 3 from round 7 to round 11, and in word 4 from round 6 to round 10. On the other hand, from round 5 to round 12 word 2 (renamed later to word 1) is affected several times by the other words, and the G permutation is applied to it several times, but it does not affect other words. Moreover, from round 13 to round 16, this word affects directly or indirectly only two of the three other words, and therefore, the input of the second word in round 5 never affects the fourth data word twelve rounds later.[1]

4   Cryptanalysis of Skipjack Reduced to 16 Rounds

4.1   Differential Cryptanalysis of Skipjack with Reduced Number of Rounds

The differential attack we describe here for 16-round Skipjack is considerably faster than exhaustive search. This attack is based on our original attack [2] with additional improvements based on Wagner's observations [8].

The best characteristics of 16-round Skipjack that we are aware of use the characteristics of the G permutation described above. The plaintext difference is $(a, 0, a, 0, 0, 0, 0, b)$ (where a, b and 0 are eight-bit values, and a, b are the values described in Section 3.5) and only six active G permutations (in which there are a total of 14 active F tables) are required to achieve the ciphertext difference $(0, b, 0, b, a, 0, 0, 0)$. There are four such characteristics with probabilities about $2^{-72.9}$. When we replace the characteristics by the corresponding differentials of G, the probability grows to about 2^{-71}. However, when we view the two G permutations in rounds 8 and 9 (unaffected by differences from other words) as one new permutation, its probability is about 2^{-16}, and thus the probability of the differential grows to about 2^{-58}.

Given the ciphertexts of many plaintext pairs with the difference $(a, 0, a, 0, 0, 0, 0, b)$, it is easy to identify and discard most of the wrong pairs in a 0R-attack. Such an attack requires about 2^{60} pairs. We observe that only a four-round characteristic of the first four rounds is required, with probability about 2^{-21}, and that when the characteristic holds, the truncated (word-wise) differences in rounds 5–16 are fixed. In this case we choose about 2^{22} chosen plaintext pairs,

---
[1] This property was found by Wagner[8].

and can discard most of the wrong pairs, except for a fraction of 2^{-16} of them. Thus, about $2^6 = 64$ pairs remain.

Now we use a second observation that the same set of subkeys is used in the first and the 16th rounds. We try all the 2^{32} possible sets of subkeys and for each remaining pair we encrypt the first round and verify that the characteristic of G holds, and decrypt the last round and verify whether the expected difference (i.e., the difference of the third ciphertext word) holds in the input of the last G permutation. The probability that a wrong set of subkeys does not discard a pair is $2^{-16} \cdot 2^{-10.4} = 2^{-26.4}$, and thus only the correct 32-bit subkey is expected to be proposed twice, by two different remaining pairs, and thus can be identified. This attack can be applied efficiently in 2^{16} steps for each analyzed pair, i.e., a total complexity of 2^{22} steps. Similar techniques (or even exhaustive search of the remaining 48 bits of the key) can complete the cryptanalysis.

4.2 Linear Cryptanalysis of Skipjack with Reduced Number of Rounds

Linear characteristics are built in a similar way where XORs are replaced by duplications and duplications are replaced by XORs of the subsets of parity bits[1]. As Rule A and Rule B differ essentially in this way, we can have similar analysis for linear cryptanalysis (except that we use linear characteristics rather than differentials). The probability of the best linear characteristic we found is about $1/2 + 2^{-35.5}$, and thus the attack seems to require more known plaintexts than the total number of possible plaintexts. However, this number can be reduced below 2^{64} by using shorter characteristics.

4.3 Modified Variants of Skipjack

Skipjack uses alternately eight rounds of Rule A and eight rounds of Rule B. In this section we investigate whether other mixing orders strengthen or weaken the cipher. A simple example of a modified design uses alternately four 'Rule A' rounds and four 'Rule B' rounds. We found an attack on this 16-round cipher which requires only about 2^{10} chosen plaintexts and about 2^{32} steps of analysis to find the subkey of round 3.

When Rule A rounds and Rule B rounds appear in reverse order (i.e., Rule B is applied first), and four rounds of each are applied consecutively, then only two pairs are required to find the last subkey.

These few examples indicate that the order of Rule A and Rule B rounds can have a major impact on the security of modified variants of Skipjack. Further study of modified variants will shed more light on Skipjack's design principles.

5   A New Cryptographic Tool: The Yoyo Game

Consider the first 16 rounds of Skipjack, and consider pairs of plaintexts $P = (w_1, w_2, w_3, w_4)$ and $P^* = (w_1^*, w_2^*, w_3^*, w_4^*)$ whose partial encryptions differ only in the second word in the input of round 5 (we will refer to it as the *property* from now on). As this word does not affect any other word until it becomes word 1 in round 12, the other three words have difference zero between rounds 5 and 12.

We next observe that given a pair with such a property, we can exchange the second words of the plaintexts (which cannot be equal if the property holds), and the new pair of plaintexts (w_1, w_2^*, w_3, w_4) and $(w_1^*, w_2, w_3^*, w_4^*)$ still satisfies the property, i.e., differs only in the second word in the input of round 5. Given the ciphertexts we can carry out a similar operation of exchanging words 1.

The Yoyo game starts by choosing an arbitrary pair of distinct plaintexts P_0 and P_0^*. The plaintexts are encrypted to C_0 and C_0^*. We exchange the first words of the two ciphertexts as described above, receiving C_1 and C_1^*, and decrypt them to get P_1, P_1^*. Now we exchange the second words of the plaintexts, receiving P_2 and P_2^*, and encrypt them to get C_2 and C_2^*. The Yoyo game repeats this forever.

In this game, whenever we start with a pair of plaintexts which satisfies the property, all the resultant pairs of encryptions must also satisfy the property, and if we start with a pair of plaintexts which does not satisfy the property, all the resultant encryptions cannot satisfy it.

It is easy to identify whether the pairs in a Yoyo game satisfy the above property, by verifying whether some of the pairs achieved in the game have a non-zero difference in the third word of the plaintexts or in the fourth word of the ciphertexts. If one of these differences is non-zero, the pair cannot satisfy the property. On the other hand, if the pair does not satisfy the property, there is only a probability of 2^{-16} that the next pair in the game has difference zero, and thus it is possible to stops games in which the property is not satisfied after only a few steps. If the game is not stopped within a few steps, we conclude with overwhelming probability that the property is satisfied.

This game can be used for several purposes. The first is to identify whether a given pair satisfies the above property, and to generate many additional pairs satisfying the property.

This can be used to attack Skipjack reduced to 16 rounds in just 2^{14} steps. For the sake of simplicity, we describe a suboptimal implementation with complexity 2^{17}. In this version we choose 2^{17} plaintexts whose third word is fixed. This set of plaintexts defines about 2^{33} possible pairs, of which about 2^{17} candidate pairs have difference zero in the fourth word of the ciphertexts, and of which about one or two pairs are expected to satisfy the property. Up to this point, this attack is similar to Wagner's attack on 16-round Skipjack [8]. We then use the Yoyo game to reduce the complexity of analysis considerably. We play the game for each of the 2^{17} candidate pairs, and within a few steps of the game discard all the pairs which do not satisfy the property. We are left with one pair which satisfies the property, and with several additional pairs generated during the Yoyo game which also satisfy the property. Using two or three of these pairs, we can analyze the last round of the cipher and find the unique subkey of the last round that satisfies all the requirements with complexity about 2^{16}. The rest of the key bytes can be found by similar techniques.

This game can also be used as a distinguisher which can decide whether an unknown encryption algorithm (given as an oracle) is Skipjack reduced to 16 rounds or a random permutation.

The above Yoyo game keeps three words with difference zero in each pair. We note that there is another (less useful) Yoyo game for Skipjack reduced to 14 rounds (specifically, rounds 2 to 15), which keeps only one word with difference zero. Consider pairs of encryptions $P = (w_1, w_2, w_3, w_4)$ and $P^* = (w_1^*, w_2^*, w_3^*, w_4^*)$ which have the same data at the leftmost word in the input of round 5. As this word is not affected by any other word until it becomes word 2 in round 12, we can conclude that both encryptions have the same data in word 2 after round 12. Given a pair with such an equality in the data, we can exchange the first word of the plaintexts, and the new pair of plaintexts (w_1^*, w_2, w_3, w_4) and $(w_1, w_2^*, w_3^*, w_4^*)$ still has the same property of equality at the input of round 5. Moreover, if the first words of the plaintexts are equal (i.e., $w_1 = w_1^*$ and thus exchanging them does nothing) we can exchange the second words (w_2 with w_2^*) and get the same property. If they are also equal, we can exchange w_3 with w_3^* and get the same property. If they are also equal, we exchange w_4 with w_4^*. However, if the property holds, this last case is impossible, as at least two words of the two plaintexts must be different. Given the ciphertexts we can carry out a similar operation of exchanging words 2. If words 2 are equal, exchange words 1, then words 4, and then words 3. Also in this case a difference of only one word ensures that the property is not satisfied. This Yoyo game is similar to the previous game, except for its modified exchange process, and it behaves similarly with respect to the new difference property.

6   Cryptanalysis of Skipjack-3XOR

In this section we analyze Skipjack-3XOR, which is identical to the original 32-round Skipjack except for the removal of the three XOR operations which mix 16-bit data words with their neighbors at rounds 4, 16 and 17. We show that this version is completely insecure, since it can be broken in one million steps using only about 500 chosen plaintexts.

The starting point of the attack is Wagner's observation[8] that the differential characteristic we used in the previous section can use truncated (i.e., wordwise) differences [5]. The attack uses the following characteristic of Skipjack-3XOR: For any 16-bit non-zero value a, the plaintext difference $(0, a, 0, 0)$ leads to the difference $(b, c, 0, 0)$ after round 16 with probability 1, which in turn leads to a difference $(d, 0, 0, 0)$ after round 28 with probability 2^{-16}, for some unspecified non-zero values b, c, and d. This difference leads to some difference $(e, f, g, 0)$ in the ciphertexts, for some e, f, and g.

The attack requires two pairs of plaintexts with such a differential behavior. To get them, encrypt $2^9 = 512$ distinct plaintexts which are identical except at their second word. They give rise to about $2^{18}/2 = 2^{17}$ pairs, and each pair has the required property with probability 2^{-16}. The two right pairs can be easily recognized since the two ciphertexts in each pair must be equal in their last 16 bits.

The basic steps of the attack are:

1. We know the input differences and the actual outputs of the 32nd G permutation. Each right pair yields a subset of about 2^{16} possible key bytes $cv_4, \ldots, cv_7$, and the intersection of the two subsets is likely to define these 32 key bits (almost) uniquely. This part can be implemented in about 2^{16} evaluations of G.
2. The 29th G permutation shares two key bytes cv_4, cv_5 with the 32nd G permutation, which are already known. 2^{16} possible combinations of the two key bytes cv_2, cv_3 and the inputs to the 30th G permutation in both pairs can be found. A careful implementation of this step requires a time complexity which is equivalent to 2^{17} evaluations of G.
3. For each of the 2^{16} combinations we still miss the key bytes cv_8, cv_9 entering the last two F tables in round 30, and the key bytes cv_0 and cv_1 entering the first two F tables in round 31. Together they are equivalent to a single G, which we call G'. In each right pair, the two encryptions have the same values in G'. We view both right pairs as a super pair of two G' evaluations, whose actual inputs and outputs are known. The analysis of G' takes about the equivalent of 2^9 G evaluations, and thus the total complexity is equivalent to about 2^{25} G evaluations.

Since each Skipjack encryption contains $2^5 = 32$ G evaluations, the total time complexity of this cryptanalytic attack is equivalent to about one million Skipjack encryptions, and can be carried out in seconds on a personal computer.

Acknowledgments

We are grateful to David Wagner, Lars Knudsen, and Matt Robshaw for sharing various beautiful observations and results with us. We are also grateful to Rivka Zur, the Technion CS secretary, for preparing Figure 2.

References

1. Eli Biham, *On Matsui's Linear Cryptanalysis*, Lecture Notes in Computer Science, Advances in Cryptology, proceedings of EUROCRYPT'94, pp. 341–355, 1994.
2. Eli Biham, Alex Biryukov, Orr Dunkelman, Eran Richardson, Adi Shamir, *Initial Observations on the Skipjack Encryption Algorithm*, June 25, 1998, available at http://www.cs.technion.ac.il/~biham/Reports/SkipJack/.
3. Eli Biham, Adi Shamir, *Differential Cryptanalysis of the Data Encryption Standard*, Springer-Verlag, 1993.
4. Ernest F. Brickell, Dorothy E. Denning, Stephen T. Kent, David P. Maher, Walter Tuchman, *SKIPJACK Review, Interim Report, The SKIPJACK Algorithm*, July 28, 1993. Available at http://www.austinlinks.com/Crypto/skipjack-review.html.
5. Lars R. Knudsen, Thomas A. Berson, *Truncated differentials of SAFER-K64*, proceedings of Fast Software Encryption, Cambridge, Lecture Notes in Computer Science, pp. 15–26, 1996.
6. Mitsuru Matsui, *Linear Cryptanalysis Method for DES Cipher*, Lecture Notes in Computer Science, Advances in Cryptology, proceedings of EUROCRYPT'93, pp. 386–397, 1993.
7. *Skipjack and KEA Algorithm Specifications*, Version 2.0, 29 May 1998. Available at the National Institute of Standards and Technology's web page, http://csrc.nist.gov/encryption/skipjack-kea.htm.
8. David Wagner, *Further Attacks on 16 Rounds of SkipJack*, private communication, July 1998.

Author Index

Springer
and the
environment

At Springer we firmly believe that an international science publisher has a special obligation to the environment, and our corporate policies consistently reflect this conviction.
We also expect our business partners – paper mills, printers, packaging manufacturers, etc. – to commit themselves to using materials and production processes that do not harm the environment. The paper in this book is made from low- or no-chlorine pulp and is acid free, in conformance with international standards for paper permanency.

 Springer

Lecture Notes in Computer Science

For information about Vols. 1–1506
please contact your bookseller or Springer-Verlag

Vol. 1544: C. Zhang, D. Lukose (Eds.), Multi-Agent Systems. Proceedings, 1998. VII, 195 pages. 1998. (Subseries LNAI).

Vol. 1545: A. Birk, J. Demiris (Eds.), Learning Robots. Proceedings, 1996. IX, 188 pages. 1998. (Subseries LNAI).

Vol. 1546: B. Möller, J.V. Tucker (Eds.), Prospects for Hardware Foundations. Survey Chapters, 1998. X, 468 pages. 1998.

Vol. 1547: S.H. Whitesides (Ed.), Graph Drawing. Proceedings 1998. XII, 468 pages. 1998.

Vol. 1548: A.M. Haeberer (Ed.), Algebraic Methodology and Software Technology. Proceedings, 1999. XI, 531 pages. 1999.

Vol. 1550: B. Christianson, B. Crispo, W.S. Harbison, M. Roe (Eds.), Security Protocols. Proceedings, 1998. VIII, 241 pages. 1999.

Vol. 1551: G. Gupta (Ed.), Practical Aspects of Declarative Languages. Proceedings, 1999. VIII, 367 pgages. 1999.

Vol. 1552: Y. Kambayashi, D.L. Lee, E.-P. Lim, M.K. Mohania, Y. Masunaga (Eds.), Advances in Database Technologies. Proceedings, 1998. XIX, 592 pages. 1999.

Vol. 1553: S.F. Andler, J. Hansson (Eds.), Active, Real-Time, and Temporal Database Systems. Proceedings, 1997. VIII, 245 pages. 1998.

Vol. 1554: S. Nishio, F. Kishino (Eds.), Advanced Multimedia Content Processing. Proceedings, 1998. XIV, 454 pages. 1999.

Vol. 1555: J.P. Müller, M.P. Singh, A.S. Rao (Eds.), Intelligent Agents V. Proceedings, 1998. XXIV, 455 pages. 1999. (Subseries LNAI).

Vol. 1556: S. Tavares, H. Meijer (Eds.), Selected Areas in Cryptography. Proceedings, 1998. IX, 377 pages. 1999.

Vol. 1557: P. Zinterhof, M. Vajteršic, A. Uhl (Eds.), Parallel Computation. Proceedings, 1999. XV, 604 pages. 1999.

Vol. 1558: H. J.v.d. Herik, H. Iida (Eds.), Computers and Games. Proceedings, 1998. XVIII, 337 pages. 1999.

Vol. 1559: P. Flener (Ed.), Logic-Based Program Synthesis and Transformation. Proceedings, 1998. X, 331 pages. 1999.

Vol. 1560: K. Imai, Y. Zheng (Eds.), Public Key Cryptography. Proceedings, 1999. IX, 327 pages. 1999.

Vol. 1561: I. Damgård (Ed.), Lectures on Data Security. VII, 250 pages. 1999.

Vol. 1563: Ch. Meinel, S. Tison (Eds.), STACS 99. Proceedings, 1999. XIV, 582 pages. 1999.

Vol. 1567: P. Antsaklis, W. Kohn, M. Lemmon, A. Nerode, S. Sastry (Eds.), Hybrid Systems V. X, 445 pages. 1999.

Vol. 1568: G. Bertrand, M. Couprie, L. Perroton (Eds.), Discrete Geometry for Computer Imagery. Proceedings, 1999. XI, 459 pages. 1999.

Vol. 1569: F.W. Vaandrager, J.H. van Schuppen (Eds.), Hybrid Systems: Computation and Control. Proceedings, 1999. X, 271 pages. 1999.

Vol. 1570: F. Puppe (Ed.), XPS-99: Knowledge-Based Systems. VIII, 227 pages. 1999. (Subseries LNAI).

Vol. 1572: P. Fischer, H.U. Simon (Eds.), Computational Learning Theory. Proceedings, 1999. X, 301 pages. 1999. (Subseries LNAI).

Vol. 1574: N. Zhong, L. Zhou (Eds.), Methodologies for Knowledge Discovery and Data Mining. Proceedings, 1999. XV, 533 pages. 1999. (Subseries LNAI).

Vol. 1575: S. Jähnichen (Ed.), Compiler Construction. Proceedings, 1999. X, 301 pages. 1999.

Vol. 1576: S.D. Swierstra (Ed.), Programming Languages and Systems. Proceedings, 1999. X, 307 pages. 1999.

Vol. 1577: J.-P. Finance (Ed.), Fundamental Approaches to Software Engineering. Proceedings, 1999. X, 245 pages. 1999.

Vol. 1578: W. Thomas (Ed.), Foundations of Software Science and Computation Structures. Proceedings, 1999. X, 323 pages. 1999.

Vol. 1579: W.R. Cleaveland (Ed.), Tools and Algorithms for the Construction and Analysis of Systems. Proceedings, 1999. XI, 445 pages. 1999.

Vol. 1580: A. Včkovski, K.E. Brassel, H.-J. Schek (Eds.), Interoperating Geographic Information Systems. Proceedings, 1999. XI, 329 pages. 1999.

Vol. 1581: J.-Y. Girard (Ed.), Typed Lambda Calculi and Applications. Proceedings, 1999. VIII, 397 pages. 1999.

Vol. 1582: A. Lecomte, F. Lamarche, G. Perrier (Eds.), Logical Aspects of Computational Linguistics. Proceedings, 1997. XI, 251 pages. 1999. (Subseries LNAI).

Vol. 1584: G. Gottlob, E. Grandjean, K. Seyr (Eds.), Computer Science Logic. Proceedings, 1998. X, 431 pages. 1999.

Vol. 1586: J. Rolim et al. (Eds.), Parallel and Distributed Processing. Proceedings, 1999. XVII, 1443 pages. 1999.

Vol. 1587: J. Pieprzyk, R. Safavi-Naini, J. Seberry (Eds.), Information Security and Privacy. Proceedings, 1999. XI, 327 pages. 1999.

Vol. 1590: P. Atzeni, A. Mendelzon, G. Mecca (Eds.), The World Wide Web and Databases. Proceedings, 1998. VIII, 213 pages. 1999.

Vol. 1592: J. Stern (Ed.), Advances in Cryptology – EUROCRYPT '99. Proceedings, 1999. XII, 475 pages. 1999.

Vol. 1593: P. Sloot, M. Bubak, A. Hoekstra, B. Hertzberger (Eds.), High-Performance Computing and Networking. Proceedings, 1999. XXIII, 1318 pages. 1999.

Vol. 1594: P. Ciancarini, A.L. Wolf (Eds.), Coordination Languages and Models. Proceedings, 1999. IX, 420 pages. 1999.

Vol. 1596: R. Poli, H.-M. Voigt, S. Cagnoni, D. Corne, G.D. Smith, T.C. Fogarty (Eds.), Evolutionary Image Analysis, Signal Processing and Telecommunications. Proceedings, 1999. X, 225 pages. 1999.

Vol. 1597: H. Zuidweg, M. Campolargo, J. Delgado, A. Mullery (Eds.), Intelligence in Services and Networks. Proceedings, 1999. XII, 552 pages. 1999.

Vol. 1605: J. Billington, M. Diaz, G. Rozenberg (Eds.), Application of Petri Nets to Communication Networks. IX, 303 pages. 1999.